Doing Social Research and Publishing Results

Candauda Arachchige Saliya

Doing Social Research and Publishing Results

A Guide to Non-native English Speakers

 Springer

Candauda Arachchige Saliya
Sri Lanka Institute of Information
Technology
Malabe, Colombo, Sri Lanka

ISBN 978-981-19-3782-8   ISBN 978-981-19-3780-4 (eBook)
https://doi.org/10.1007/978-981-19-3780-4

This Springer imprint is published by the registered company Springer Nature Singapore Pte Ltd.
The registered company address is: 152 Beach Road, #21-01/04 Gateway East, Singapore 189721,
Singapore

This book is dedicated to the researchers who promote equality, justice and diversity.

Foreword

The experience that I have gathered over four decades of work in research including teaching, guiding, and supervising the research work of postgraduate/doctoral students in the fields of management, business, health, and education has repeatedly showed that successful researchers are (a) sharp in thinking, (b) integrative in design, (c) pragmatic in approach, (d) skilled in techniques, and (e) professional in character. Being sharp thinkers, they recognize researchable problems and conceptualize study frameworks; being integrative designers, they relate problems to previous theory, develop concepts, and formulate study frameworks showing interconnections among the dependent and independent variables; being pragmatic, they operationalize their concepts to reach reality; being skilled they apply methodological tools with ease and write their reports to communicate effectively; and being professional, they adopt international standards within ethical guidelines. This book by Dr. Saliya exemplify the above and enable the beginner to become masters of research.

Having experience in the banking sector, Dr. Saliya ventured into academic work and I believe his interactions with university students had prompted him to design and launch this book. Because of his exposure to both practical and academic work, I believe, he recognizes that successful explanation to phenomena requires the researcher to approach problems with both quantitative as well as qualitative methodologies. Therefore, the author presents both the deductive and the inductive approaches which demand the researcher to be flexible in his/her views while dealing with objective and subjective interpretations.

In many ways, Dr. Saliya's work is unique. First, I observe the simplicity of language that carefully avoids jargon in order to communicate effectively. Rather than hiding behind the jargon, the writer describes and defines the concepts for the benefit of the beginner and illustrates with real-life examples. Secondly, he covers the full range of issue areas in the subject so that this work is a compendium on research methods including problem recognition, conceptualization, operationalization, data gathering techniques, data analysis, drawing inferences, discussion in conjunction with previous knowledge, and communicating the research in conformity with international standards. Thirdly, I believe that this is the first time that a comprehensive book is made available with practical examples on how to select a suitable academic journal and address editors' and reviewers' queries.

The last, not the least, point that makes this work unique in my view is the series of case studies the writer has used wisely to illustrate difficult and interdisciplinary issues in systematic investigation of phenomena. After all, Dr. Saliya demonstrates how instructors and supervisors in research alike must apply their roles professionally. I wish that this book would become a landmark in the evolution of new research traditions in our world.

I take this opportunity to give an advice to research students: become the researcher who brings about new, reliable, and usable knowledge in society. Knowledge being the most powerful driver of change in this world, you as a researcher has a great responsibility to be objective, practical, and ethical in your contributions. A good research skill is an asset that will make you strong, and a good research finding is an asset that makes the society strong.

January 2022   Prof. Gunapala Nanayakkara, Ph.D.
                                                                                    (Carleton)
                                                                          Founder Director
                                                  Postgraduate Institute of Management
                                                    University of Sri Jayewardenepura
                                                               Sri Lanka and Chairman
                                                  Graduate School of Management and
                                                                        Entrepreneurship

Preface

There are plenty of English books available on academic research methods and philosophy. Many complain that they are harder and slower for budding researchers to use and understand. The purpose of this book is to provide a guide for these researchers, to assist them in understanding research methodology and methods. This book also provides many practical examples of how to conduct research; present research results; integrate these with existing literature to draw conclusions; and publish their research results as journal articles, book chapters, or conference papers. For this purpose, in this book, I use relatively more graphics and links to YouTube tutorials in simple English than long texts (written in jargonized language with complex English sentences). I explain how to use contemporary, complex, statistical techniques, and various qualitative research methods—from defining and articulating research problems to dealing with journal editors and peer reviewers.

A 2014 survey of over 16,000 full-time faculty in the United States revealed that almost a fifth had not read or written any scholarship in the past 2 years. In addition, almost a third had not published any piece of writing in that time. "Over half of them spent less than an hour a day reading and writing scholarship…two-thirds of the humanities faculty and one third of the social science faculty had not published even one article in the past eight years" (Larivière et al. 2010, p. 48).

Many up-and-coming academics are struggling to conduct proper research and publish their findings in international journals. These are known as International Peer-Reviewed Academic Journals (IPRAJs). Journal rankings are used to assess the impact, quality, reputation, and the prestige associated with an academic journal. These journals are considered the main official research evaluation tools in universities. They are indexed in organizations such as Web of Science, Scimago, ABDC, etc. Publishing of research findings in an IPRAJ is an essential condition of academic advancement for faculty members. Otherwise, the findings within research will not be exposed to the world, but will go to waste—"publish or perish", they say (see Fig. 1). Part VI of this book is dedicated to discussion of the process of research results publishing in an IPRAJ, including examples.

The objectives of this book are fivefold. (1) The primary purpose is to explain, as simply as possible, the methods commonly in use in the field of academic research. (2)

Fig. 1 Publish or perish

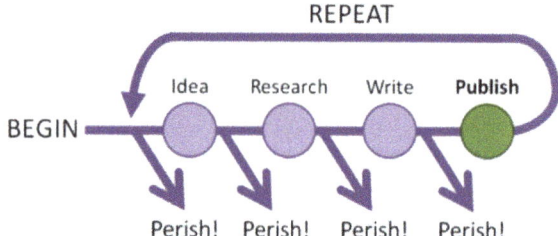

Another objective is to provide some stimulus to budding academics for successful implementation of academic research. (3) This book will also largely meet the need for an appropriate course framework and materials for teaching basics in research methodology. Advancement of information technology has now given us the opportunity to study serious complex research methods without using intimidating mathematical formulas. The most effective method used today is to reliably retrieve results instantly by mastering the feeding of data into a computer and analyze data using various software such as SPSS, AMOS, Eviews, Mplus, etc. This book is designed to encourage budding researchers to explore those opportunities and advantages. (4) The fourth objective is to discuss strategies and the process of publishing research papers, especially for beginners. Lastly (5), the case studies discussed in this book will, specifically, provide some insight into credit evaluating systems for credit officers/managers and credit customers/clients. It will be useful for everyone in society to be aware of this two-edged sword called the *credit weapon* and its impact on society.

This book combines the various methods that have been extracted from many different books and they are presented in a way that will help to put new approaches into practice. Figure 2 is one such attempt to outline the research process. Not presenting the group of components of the research process in a boxed framework indicates that the researcher should not be confined to a straitjacket of sequential activities but should be as open as possible to paradigms, perspectives, worldviews, and theories.

This book presents a brief overview of the most used statistical techniques needed to select the most appropriate strategy for your research. It is not within the scope of this book to elaborate on how to use those techniques after selecting a particular research design. To gain the necessary skills and expertise on how to use those tools and techniques effectively, you should master them using the sources/YouTube tutorials provided at the ends of chapters as further readings. I provide as many relevant real-world examples as possible in suitable places. These examples are labeled as **Example 1, 2, 3, 4,** etc. References to such items as **Tables, Figures**, and **Boxes** are shown in **bold**. The examples discussed in this book are some of the most used research methods and publishing strategies.

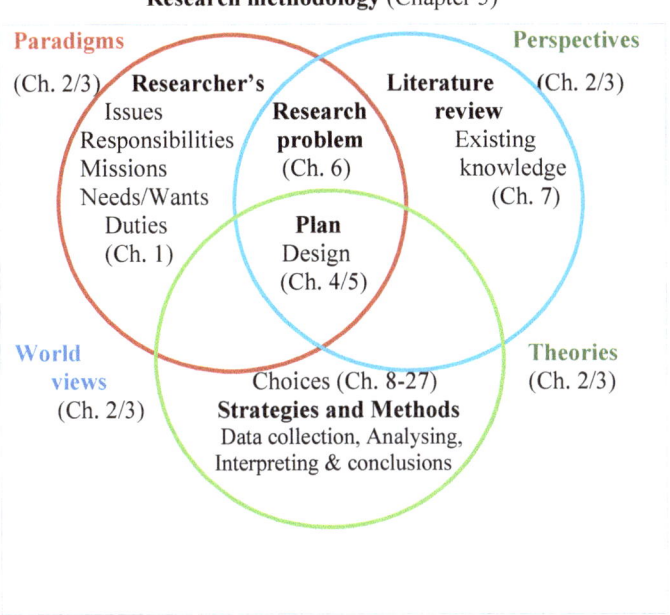

Fig. 2 Research process

I would like to mention that the main challenge I faced in compiling this book was to summarize the long process from designing research to publishing results into a single book. There are several commonly used strategies (for example, quantitative, qualitative, and mixed method) and methods (data collection, analysis, interpretation, and drawing conclusions) needed to select the methodology that is the best fit for your research. To specialize the selected methodology properly, the sources provided here will be helpful; however, many reputable scholars believe that it is important to seek experienced guidance/instruction, even though you can learn most used research methods by using study materials and YouTube tutorials on your own.

This book consists of six parts. The first part (Part I) spans five chapters and presents an overview and the background of academic research. Part II covers articulating research problem, literature review, and data-collection processes, in three chapters. The chapters in Part III discuss basic quantitative research strategies or traditions, including econometrics, with real-world examples. The chapters in Part IV discuss qualitative and mixed method research strategies/traditions with real-world examples. Part V discusses the practical implementation of various research methods (data analysing, interpreting, and drawing conclusions). Methods are presented with several research papers published in international journals as examples explaining important critical aspects of managing certain communication issues. Part VI presents several strategies for manuscript preparation and the process of publishing in international journals. There are 30 chapters altogether with introductory chapters in each part.

The contents may be useful, not only to undergraduates, postgraduate, and Ph.D. students, to budding academics who wish to publish research papers but also to employees in lending institutions as well as everyone in society who are interested in the role of research in shaping our everyday lives.

Auckland, New Zealand
January 2022

Candauda Arachchige Saliya, Ph.D., FCA, CPA

Acknowledgements

Professor KAS Wickrama of Georgia University, USA for his valuable advices.

Professor Gunapala Nanayakkara for providing the Foreword.

Professors Kelum Jayasinghe, Keith Hooper, Gamini Jayasuriya, W. A. Wijewardene for their encouraging reviews.

My parents C. A. Piyapala, Dona Kusumawathie sahabandu and brothers and sisters.

My wife Mali, son Savi, daughters Sneha and Nimna, daughter-in-law Geil, son-in-law Kalindu.

SpringerNature publishers for accepting to publish this book.

Reviews

…The world of business has become a highly competitive and dynamic place to live and work! To be able to make sound business decisions requires a comprehensive understanding of the various inter-relationships, both domestic and international. Saliya's book is the ideal companion guide for those students intending to make their career in the business world, providing a user-friendly approach to the vast array of scientific analyses. This book strikes a balance between academic rigour that instructors demand, and the simplicity that students require. This book is a 'must read' for a person engaged in undergraduate as well as postgraduate research in this field…

Prof. Gamini Jayasuriya, Professor in Economics University of Auckland, New Zealand

This book provides a complete guide to conduct research and publish the findings as an academic article in ranked international journals. It provides numerous real-world examples to demonstrate this process using wide array of scientific methods, both quantitative and qualitative.

Prof. W. A. Wijewardena, Deputy Governor (Rtd.), Central Bank of Sri Lanka

…This book provides the foundation necessary for the astute student engaged in research. It is my expectation and hope that emerging academics will enjoy research, and gain an adequate understanding of the research methodology and methods from this excellent book and apply them in a productive manner in both their work and everyday life…

Prof. Keith Hooper, Professor in Accounting, Auckland University of Technology, New Zealand

I highly recommend this book for the beginners of social science research. In particular the case studies discussed are well-presented making them very interesting to read and a very important guidance for emerging researchers in accounting and management. This book provides clickable links to numerous number of Youtube tutorials on various research methods.

Prof. Kelum Jayasinghe, Department of Accounting Essex University, UK

Chapter Summary

PART 1: Overview	PART II: Methods 1	PART III: Quantitative strategies	PART IV: Qualitative strategies	PART V: Methods 2	PART VI: Publishing
Chapter 1	Chapter 6	Chapter 9	Chapter 15	Chapter 24	Chapter 28
Preparation	Research problem	Introduction	Qualitative Strategies	Data analysis	Publishing process
Chapter 2		Chapter 10	Chapter 16		Chapter 29
Introduction	Chapter 7	Data and Variables	Narrative	Chapter 25	Cover letters
Chapter 3	Literature review	Chapter 11	Chapter 17	Interpretations and discussions	Chapter 30
Evolution		Statistic fundamental	Case studies		Addressing issues raised by reviewers
Chapter 4	Chapter 8	Chapter 12	Chapter 18	Chapter 26	
Methodology Types	Data collection,	Statistical Techniques	Phenomenology	Conclusions	
Chapter 5		Chapter 13	Chapter 19	Chapter 27	
The 10Ps model		Advanced techniques	Ethnography	Abstracts	
		Chapter 14	Chapter 20		
		Econometric	Grounded Theory		
			Chapter 21		
			Poetic Inquiry		
			Chapter 22		
			Action Research		
			Chapter 23		
			Mixed-methods Research		

Contents

LIST OF EXAMPLES (PAGE #) AND REFERENCES

Relevant discussion	Reference (Page)
A. EXAMPLE 6.1 (Page 58): Saliya, C A. (2021a). Conducting Case Study Research: A Concise Practical Guidance for Management Students. *International Journal of KIU*.	
Articulating problem statement	**Box 6.1**: Articulating a research problem (p. 58) **Box 6.2**: Articulating a Case study research problem (p .59) **Box 6.3**: Research questions (p. 60) **Box 6.4**: Discriminatory credit and critical paradigm (p. 61) **Fig. 6.1**: Illustrated Integrated Research Questions and Proposition (p. 61)
Graphical presentation	**Fig. 25.8**: Discriminatory credit decision-making model (p. 351) **Box 25.5**: Discriminatory credit decision-making explanation (p. 351)
Abstract	**Abstract 27.3:** Conducting Case-study Research (p. 385)
Letters	**Email 29.1**: Message to an editor of a journal on the instruction of another editor (p. 415)
B. EXAMPLE 6.3 (Page 65), 8.1 (Page 104): Saliya, C A. (2022). Role of Enterprising personality in motivating stock investors: Case of Fiji, *forthcoming*.	
Developing quantitative hypotheses	**Box 6.5**: Hypothesis 1 (p. 65) **Box 6.6**: Hypothesis 2 (p. 65) **Box 6.7**: Hypothesis 6 (p. 66) **Box 6.8**: Summery of the Hypotheses (p. 67) **Box 10.5**: Correlation matrix interpretation (p. 167)
Descriptive analysis	**Table 10.10**: Descriptive statistics and correlation matrix (p. 166)
Pathway analysis: SEM (structural equation modeling)	**Fig. 6.2**: The theoretical framework and hypothesized pathway analysis 1 (p. 68) **Fig. 12.7:** (p. 231)

(continued)

(continued)

Relevant discussion	Reference (Page)
Survey: Likert scale	**Fig. 8.1**: Responses to survey questionnaires Likert scale (p. 106)
Letters	**Editor's letter 28.6**: Rejection letter from a journal editor (p. 403) **Editor's letter 28.7:** A letter from an editor for a resubmission (p. 404) **Editor's letter 28.9**: Letter referring to a similar paper (same author though) and advised to avoid plagiarism (p. 405) **Editor's letter 28.8 and 28.10**: Positive responses when edited and resubmitted (p. 405)

C. EXAMPLE 6.4 (Page 69), 10.1 (Page 156), 26.5 (Page 366): Saliya, C. A. (2020a). Stock Market Development and Nexus of Market Liquidity, *International Journal of Finance and Economics.*

Articulating problem statement	**Box 6.9**: Articulating mixed method problem statement 1 (p. 69) **Box 6.10**: Articulating mixed method problem statement 2 (p. 70)
Literature Review	**Table 7.1:** Summary of macroeconomic determinants of stock market development, the studies and their impact (p. 80–81)
Descriptive statistics **Normal distribution** **Bell shape: Skewness** **Peakedness: Kurtosis**	**Table 26.1**: 15 variables: Descriptive statistics. (p. 361) **Discussion 26.1:** Normal distribution. Bell shape: Skewness Peakedness: Kurtosis) (p. 360)
F test	**Table 14.2:** ARDL bounds test F-statistics for cointegration (p. 249)
Sstationarity or stable process:	**Box 14.1:** Stationary test interpretation (p. 247) **Table 14.1:** Stationarity tests (p. 246)
Cointegration	**Box 14.2:** Cointegration interpretation (p. 249)
Descriptive statistics	**Table 10.3:** Descriptive and normality statistics (p. 157) **Box 10.1:** Normality tests interpretations (p. 157)
Autoregressive Distributed Lag (ARDL) Lagged regression Modeling lag procedure	**Box 14.3:** General regression method explanation (p. 250) **Box 14.4:** Lagged regression method explanation (p. 251) **Table 14.3:** Lagged regression statistics (p. 252) **Box 14.5:** Lagged regression analysis (p. 252)

(continued)

(continued)

Relevant discussion	Reference (Page)
Discussion	**Table 26.2**: Regression analysis results (p. 364) **Discussion 26.3**: Regression analysis interpretation (p. 363) **Discussion 26.7**: Mixed methods discussion (p. 371) **Fig. 26.5**: Per capita GDP and MCAP (Two axes graph) (p. 374) **Fig. 26.7**: FDI of Fiji compared with Mauritius (p. 375) **Fig.26.6**: Annual Returns: Changes in indices (p. 375)
Graphical presentation	**Figure 25.3**: The nexus of stock market liquidity (p. 345)
Conclusions Mixed method	**Conclusions 26.3:** Stock Market Development and Nexus of Liquidity (p. 375)
Future research	**Box 26.1:** Future research example (p. 377)
Abstract	**Abstract 27.3**: Stock Market Development and Nexus of Liquidity (p. 385)
Letters	**Editor's letter 28.1**: Informing that the manuscript is sending for reviewers (p. 392) **Editor's letter 28.2**: A rejection letter (p. 394) **Editor's letter 28.3**: A letter requesting to resubmit with minor amendments (p. 395) **Author's letter 29.1**: A cover letter: International Journal of Finance and Economics (p. 413)
Reviewers' comments and authors' responses	**Reviewers' comments and authors' responses 30.5:** Heliyon (p. 433)

D. EXAMPLE 7.2 (Page 84), 17.2 (Page 276), 24.2 (327): Saliya, C. A. (2010). *The Role of Bank Lending in Sustaining Income/Wealth Inequality in Sri Lanka.* PhD Thesis. Auckland University of Technology, New Zealand.

Literature review	**EXAMPLE 7.2:** The Role of Bank Lending in Sustaining Income/Wealth Inequality (p. 84)
Representation of cases – Typical cases	**Box 7.1**: Representation of Sri Lankan credit applicants (p. 93) **Table 7.4**: An assessment of Sri Lankan credit applicants (p. 94) **Box 7.2**: Representation of private banks in Sri Lankan (p. 94) **Table 7.5**: An assessment of representation of private banks (p. 95)
Interviews: Data collection	**Table 8.4**: Primary data collection techniques & timing, nature & amount of data (p. 118)

(continued)

(continued)

Relevant discussion	Reference (Page)
Auto-Ethnography: Participant observation	**PO.8.1:** Approach (p. 123) **PO.8.2:** The proposal (p. 123) **PO.8.3:** The negotiation (p. 124) **PO.8.4:** The decision (p. 126)
Narrative research:	**Narrative 8.1:** Background (p. 127) **Narrative 8.2:** Action against Mr. Silva (p. 128) **Narrative 8.3:** Negotiations (p. 128) **Narrative 8.4:** Decisions (p. 129)
Discussions	**Discussion 8.1:** p. 129) **Discussion 8.2:** (p. 130)
Thematic analysis	**Analysis 24.2, 24.3, 24.5:** Tony's case (p. 327–331) **Analysis 24.6:** Tony's case (p. 332) **Analysis 24.4: Tony's case (p. 327)** **Analysis 24.5: Tony's case (p. 327)**
Case study research: Case stories	**Case 17.1:** Tony's garment manufacturing business (p. 276) **Case 17.2**: Silva's janitorial business (p. 277)
Graphical presentation	Fig. 25.9: Lenses of the research and application (p. 354)

E. EXAMPLE 8.2 (Page 109): Saliya, C. A. and Hooper, K. (2023b). Pink Unicorns and religion: Reflecting Gender Diversity in Fijian Public Companies. *Critical Research on Religion.* Pending

Interviews	**Interview 8.1** Email interviews (p. 109) **Table 8.3**: Summary of the interview responses (p. 110–117)

F. EXAMPLE 18.2 (Page 284), 20.1 (Page 293), 25.1 (Page 346): Saliya, C. A. (2023b), Enterprising mothers in Fiji need comprehensive support. The Qualitative Report. *Forthcoming.*

Grounded theory: **Coding and generating concept** Emerging theories	**Analysis 24.7:** Data analysis – Initial and focus coding (p. 336) **Analysis 24.8**: Categorizing and theorizing (p. 338) **Fig. 25.4:** Data analysis using the Grounded Theory coding strategies: Emergent theories (p. 346) **Table 18.2**: Theorization of women's expectations and destiny in Fiji (p. 284)

(continued)

(continued)

Relevant discussion	Reference (Page)
Graphical presentation	**Fig. 25.10**: The Process Path analysis of Grounded Theory (p. 355) **Fig. 20.1**: Grounded theory (p. 295) **Fig. 25.4**: Coding and classification of data (p. 347) **Fig. 25.2**: Taxonomy of typology: Theorization of women's expectations and destiny in Fiji (Reproduced) (p. 344) **Box 25.2**: Theorization of women's expectations and destiny in Fiji (p. 345)
G. EXAMPLE 10.2 (Page 160): Saliya, C. A. (2021b). Driving Forces of Individual Investors in Stock Market Participation, *Review of Economics and Finance*	
Reliability test of consistency (Chronbach's Alpha - α coefficient)	**Table 10.5**: Reliability Statistics (p. 159) **Box 10.2**: Method and procedures; Chronbach's Alpha interpretation (p. 160)
Component Extraction) **Communalities table**	**Table 10.8**: Total Variance Explained (p. 162) **Box 10.3**: Total variance explained interpretation (p. 164) **Fig. 10.5**: Scree Plot (p. 164)
Data reduction **Component loadings)** **Component matrix**	**Table 10.9:** Component Matrix: : Principal component analysis (PCA). (p. 163) **Box 10.4:** PCA interpretation (p. 165)
Regression interpretation **Model coverage** R^2	**Table 11.1**: Regression Results: The Four Dimensions on IB (p. 174). **Table 12.4** (p. 226) **Box 10.2**: Method and procedures; Regression results interpretation (p. 160) **Box 12.2**: (p. 227)
Goodness of fit	**Table 11.8**: Chi-square test (p. 200) **Box 11.1**: Chi-square test results interpretation (p. 200)
SEM Model fit **RMSEA**; Root Mean Square Error of Approximation. CFI; Comparative Fit Index	**Table 11.9**: SEM model fit (p. 201) **Box 11.1**: Interpretation (p. 200)
SEM Model pathway analysis: Results	**Fig. 12.6**: Model pathway analysis; RMSEA and CFI with results (p. 230)
SEM Interpretation	**Fig. 13.1**: Model pathway analysis; Results (p. 234) **Box 13.1**: SEM results Interpretation (p. 233)
Principle Component analysis	**Box 10.3**: Total variance explained interpretation APA (p. 164)
Letters	**Editor's letter 28.9:** A letter with editor's instructions (p. 405) **Author's letter 29.2:** The cover letter: Driving Forces of Individual Investors in Stock Market Participation (p. 414)

(continued)

(continued)

Relevant discussion	Reference (Page)
H. EXAMPLE 25.4 (Page 352): Saliya, C. A. and Pandey, S. K. (2021), Financial battle against climate change – assessing effectiveness using a scorecard. *Qualitative Research in Financial Markets.*	
Data analysis **Constructing matrix**	**Table 24.1**: Data entry-Excel worksheet (p. 321) **Table 24.2:** Data matrix (p. 322)
Scoring system	**Box 12.1**: Scoring system explained (p. 208) **Fig. 12.4:** Scoring system (p. 208) **Fig. 26.1**: Scoring system, % Contribution of perspectives, % Contributions of Critical Factors (Criteria) (p. 362)
Discussion	**Discussion 26.2**: Survey results of FBACC (p. 362)
Balanced Scorecard	**Fig. 25.4:** Financial Battle Against Climate Change–FBACC (p. 347) **Box 25.6**: Balanced scorecard explained (p. 353)
Reviewers' comments and authors' responses	**Reviewers' comments and authors' responses 30.1** (RBF) (p. 420, 430) **Reviewers' comments and authors' responses 30.6** (AAAJ) (p. 440)
J. EXAMPLE 24.1 (Page 323), 26.4 (Page 364): Saliya, C. A. & Wickrama KAS (2021). Determinants of Financial-risk Preparedness for Climate Change. *Advances in Climate Change Research*	
Pathway analysis (SEM)	**Fig. 13.2** A research strategic model that combines several …(p. 235)
SEM (Model fit): RMSEA; root mean square error of approximation CFI; comparative fit index.	**Fig. 24.2**: RMSEA; root mean square error of approximation CFI; Comparative fit index (p. 324)
SEM Model pathway analysis: Results **Interpretation**	**Analysis 24.1**: Interpretation SEM results (p. 323)
Correlations between dimensions:	**Fig. 13.3**: Correlations between variables (p. 236) **Box 13.2**: Correlations between dimensions (p. 235)
Conclusions	**Conclusions 26.1:** Determinants of Financial-risk preparedness for Climate Change (p. 364)
Reviewers' comments and authors' responses	**Reviewers' comments and authors' responses 30.2**: ACCR (p. 426)
K. EXAMPLES 18.1 (Page 283), 26.5 (Page 366) : Saliya, C. A. (2019b). Dynamics of credit decision-making: a taxonomy and a typological matrix, *Review of Behavioral Finance.*	

(continued)

(continued)

Relevant discussion	Reference (Page)
Phenomenology **Typologies, Taxonomies**	**Fig. 25.2 Table 25.2:** Taxonomy of typology (p. 344) **Box 18.1**: A taxonomy and a typological matrix (p. 283) **Table 25.1:** Six Dynamics of Credit Decision-making (p. 344) **Box 25.1:** Procedures of developing a typology (p. 343) **Fig. 25.1:** Replicated Barry's Taxonomy (replication) (p. 343)
Discussion	**Discussion 26.4:** Dynamics of Credit Decision-making (p. 366) **Fig. 26.2**: Taxonomy of Influence Tactics and Personality Traits in Credit Decision-making (p. 370) **Table 26.3:** Matrix of Credit Decision-making (p. 371)
Conclusions	**Conclusions 26.2:** Dynamics of credit decision-making (p. 369)
Letters:	**Editor's letter 28.5**: An acceptance letter "as it is" (p. 398)

L. EXAMPLE 21.1 (Page 301): Saliya, C. A. (1999). *Balanced Scorecard for Commercial Banks. Association of Professional Bankers of Sri Lanka. Colombo.*

Action research	**Fig. 21.2:** Association of Professional Bankers (p. 302)
Balanced Scorecard	**Fig. 21.3**: The Five perspectives (p. 303) **Fig. 21.4**: Progress review through reflection (p. 304)

M. EXAMPLE 22.1 (Page 309): Saliya, C. A. (2012). The Oldest Profession. *Accounting, Auditing and Accountability Journal*

Poetic inquiry	**Box 22.1:** The Oldest Profession: Abstract (p. 309)
Abstract	**Abstract 27.1:** The Oldest Profession (p. 383)

N. EXAMPLE 25.2 (Page 347): Saliya, C. A. (2019a). Credit capital, employment and poverty. *International Journal of Money, Banking and Finance*

(continued)

(continued)

Relevant discussion	Reference (Page)
Graphical presentation **Diagrammatic presentations**	**Fig. 25.5:** Illustrated Integrated Research Questions and Proposition (p. 348) **Box 25.3:** Research problem with two variables (p. 347) **Fig. 25.6**: Arbitrary Decision-Making Model (p. 349) **Box 25.4:** Arbitrary Decision-Making description (p. 349) **Fig. 25.7**: Integrated Research Problem and Proposition with Research Conclusions (p. 350) **Box. 25.5**: Discriminatory credit Decision-Making (p. 351) **Fig. 25.8**: Discriminatory credit Decision-Making Model (p. 351)
Abstract	**Abstract 27.1:** Credit Capital and Employment (p. 383)

P. EXAMPLE 27.2 (Page 384): Saliya and Jayasinghe, K. (2016).Creating and Reinforcing Discrimination: The
Controversial Role of Accounting in Bank Lending, *Accounting Forum.*

Abstract	**Abstract 27.2**: Creating and Reenforcing (p. 384)

Q. EXAMPLE 28.4 (Page 396): Saliya, C. A., Hooper, K. (2020). The Role of Credit Weapon and Income/Wealth Inequality: A Sri Lankan Case Study. *Journal of Applied Economic Sciences.*

Letters	**Editor's letter 28.4**: A rejection letter after several rounds of revisions (p. 396) **Author's letter 29.2:** A letter to a Marxist journal(p. 141)
Reviewers' comments and authors' responses	**Reviewers' comments and authors' responses 28.1**:

R. EXAMPLE 30.4 (Page 432): Saliya, C. A. and Jayasinghe, K. (2016) Cultural Politics of Bank Lending for Development Financing in Sri Lanka, *Journal of Accounting in Emerging Economies*

Reviewers' comments and authors' responses	**Reviewers' comments and authors' responses 30.4**: (p. 432)

About the Author

Dr. Candauda Arachchige Saliya holds a Ph.D. in Finance/Banking from the Auckland University of Technology (AUT), New Zealand. He is a Fellow Chartered Accountant and a member of the Association of Professional Bankers (Sri Lanka), CPA (Australia), MICM (UK), and a founder member of the Society of Certified Management Accountants of Sri Lanka. He also holds a Diploma of Central Banking from the Central Bank of Sri Lanka and a TESOL qualified teacher.

He leads the AUT postgraduate team, who won the Boston Consulting Group (BCG) business strategy competition Regional Award in New Zealand, and participated in the grand final in Sydney, Australia in 2006.

Before being appointed as the Professor in Banking and Finance at the Fiji National University, he worked for the AUT University in New Zealand, the University of Southern Queensland-Australia (in Auckland), the University of the South Pacific in Fiji, and for several other tertiary polytechnics in New Zealand. He serves as an external examiner for several MBA theses, Undergraduate, Graduate, and Postgraduate research assignments for many institutions.

He has traveled widely and enjoys music, movies, and all type of sports. He lives in New Zealand with his family, he is a father of a son and two daughters.
https://saliyaca.wixsite.com/website
https://casaliya.academia.edu/

Part I
Overview

Research methodology							
Selection and justification					Research design (Plan-P5)		
Paradigms (P1)	Perspectives (P2)	Strategies, approaches, traditions (Plot-P7)			Purpose and Knowledge	Methods Procedures (P9)	Programming (P6)
		Quantitative statistical techniques based	Quantitative congnitive process based	Mixed methods			
Assumptions Epistemology Ontology Axiology	World views Theoretical frameworks				Articulation of problem (Prob. P3) Literature review (Past-P4)	Data collection. Analysis. Presentation	Approvals Collaborations Timeframes Budgets

Drawing conclusions (Persuasion-P8)		
Production of research papers	Reflections and future research	Publishing (P10)

Chapter 1
Introduction

1.1 Preparation

"An ounce of prevention is worth a pound of cure. By failing to prepare, you are preparing to fail" (Benjamin Franklin, https://www.forbes.com/quotes/1107/).

Accidents very often occur while carrying out a research project. The only way to overcome this challenge is to take appropriate precautions and be vigilant. It is important to take all precautionary steps to protect your work from accidental damage to your data. Having a strong focus and determination with regard to positive and flexible attitudes is also a very important precaution. Below is a casual conversation between two postgraduate students who are engaged in a research project which is to constitute part fulfilment of their degrees.

"I lost weeks' worth of my work."

"Why, what happened?"

"Laptop doesn't start."

"Why is that?"

"During a system update, the battery had gone low and the machine is stuck, now."

"When you do things like updates, you need to connect to the main supply ... because those updating tasks take lots of power, battery alone cannot bear it sometimes."

C. A. Saliya, *Doing Social Research and Publishing Results*,
https://doi.org/10.1007/978-981-19-3780-4_1

"Really?"

"Always keep the updates set to be downloaded automatically at night."

"What are we going to do now?"

"The laptop can be repaired but there is a chance of losing the data."

One of the biggest pitfalls you can face during a research project is that of the sudden loss of the work you have done so far (such as the data you had collected and drafts, etc.) which can be damaged or completely disappear due to accidental computer crashes, fire, or rain, etc. Here are some suggestions on how to tackle such problems.

You have to back up computer data constantly. There are several tools and techniques available for this. The best way to do this is to set your computer to save your drafts in real-time (online-real time) automatically. It can be done using technologies such as Time-machine, OneDrive, Google Drive and iCloud, etc. Data can also be updated manually using an external device such as external hard disk or a USB. Or you can send such vital information frequently or at the end of the day as an attachment from one of your e-mail accounts to another e-mail account (for example, from your Gmail to Hotmail) or to a friend's e-mail account. The information is then stored securely in one of your email accounts in the 'Sent' box of the sending account as well as in the inbox of the receiving account thus making it virtually accessible anywhere in the world through the internet.

Important documents and information can also be protected by scanning them or taking photos on your mobile phone and sending those photos back to your Messenger or Instagram account, so they will be available even if your phone is lost.

About understanding texts, it should be noted that, although there are many translating tools/websites available on the internet, online translations generally cannot take the context into account and, therefore, may not be credible. Therefore, must learn adaptation rather than translation.

1.2   Annotated Bibliography

Keeping track of references and citations in every important source that you refer can save you time and effort when you need to refer the same at a later stage. It may also be important to keep hard copies of some frequently used documents.

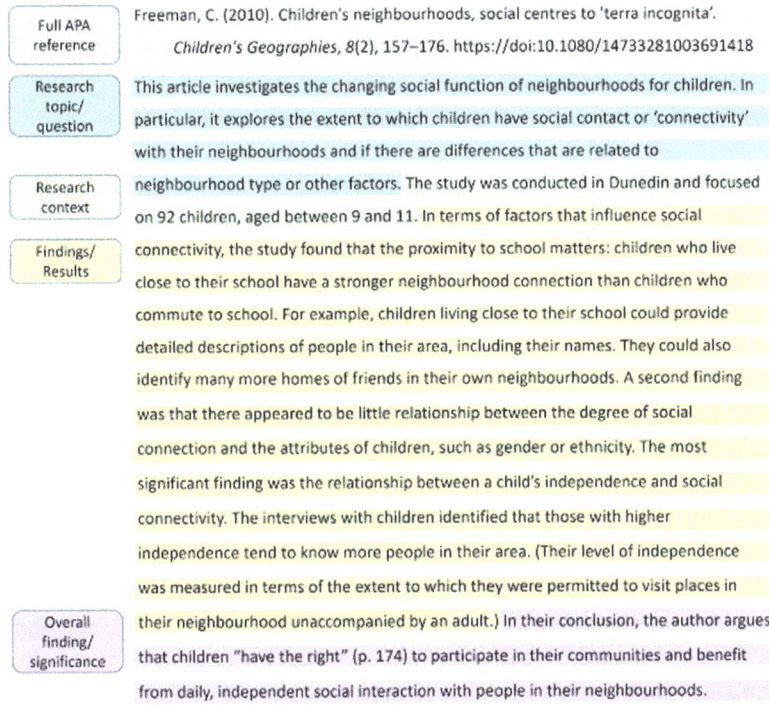

Full APA reference	Freeman, C. (2010). Children's neighbourhoods, social centres to 'terra incognita'. *Children's Geographies, 8*(2), 157–176. https://doi:10.1080/14733281003691418
Research topic/ question	This article investigates the changing social function of neighbourhoods for children. In particular, it explores the extent to which children have social contact or 'connectivity' with their neighbourhoods and if there are differences that are related to
Research context	neighbourhood type or other factors. The study was conducted in Dunedin and focused on 92 children, aged between 9 and 11. In terms of factors that influence social
Findings/ Results	connectivity, the study found that the proximity to school matters: children who live close to their school have a stronger neighbourhood connection than children who commute to school. For example, children living close to their school could provide detailed descriptions of people in their area, including their names. They could also identify many more homes of friends in their own neighbourhoods. A second finding was that there appeared to be little relationship between the degree of social connection and the attributes of children, such as gender or ethnicity. The most significant finding was the relationship between a child's independence and social connectivity. The interviews with children identified that those with higher independence tend to know more people in their area. (Their level of independence was measured in terms of the extent to which they were permitted to visit places in
Overall finding/ significance	their neighbourhood unaccompanied by an adult.) In their conclusion, the author argues that children "have the right" (p. 174) to participate in their communities and benefit from daily, independent social interaction with people in their neighbourhoods.

Source https://library.aut.ac.nz/doing-assignments/annotated-bibliographies

Annotated bibliographies will save you time and effort on re-examination. If you are using MS Word, by maintaining a database of sources (Reference > Insert Bibliography >), the List of References can be automatically retrieved at the end of the paper. That tool not only provides the List of References automatically, but it would also allow you to change the referencing format with a single click, for example, changing APA format to Harvard or Chicago formats. As different journals use different formats, changing these 'List of References' and 'citations' to different formats can be a tedious task. This task can also be accomplished by installing various software such as EndNote.

1.3   Positive and Flexible Attitudes

Ensuring a strong focus and determination from the beginning on positive and flexible attitudes is also a very important precaution to take. A research process is an exercise in practising patience. It is not uncommon for inexperienced students, as well as mature research students, to have conflicts with their supervisors and to sabotage research as well as their entire career. It is essential to control any anger and frustration felt, knowing that cases of over-tolerance are also common, such as those illustrated in the following conversations:

"I'm done with my supervisor."

"Why, what happened?"

"Last time he said that my grounded theory research design was not right. It was also we decided after many rounds of discussions. Then he suggested case study methodology and advised to read more on that and gave me two books written by Robert Yin."

"So?"

"So I made numerous reviews of related literature during the last two weeks, including these two books, made a case study research design, wrote a draft and went for today's discussion."

"So?"

"So, now he says my original grounded theory design is better and blah blah blah."

"Huh huh ... you just started to learn how to deal with supervisors – don't be surprised if he asks you to change the questionnaire, even after the field work, after collecting the data!"

Delays and dissatisfaction are common problems among doctoral students. A study at a university in the Netherlands has revealed that the relationship between satisfaction, progress and (desperate) abandonment, and the nature of socio-psychological and research activities, were significant, as seen in the quote below.

High dropout rates, delay, and dissatisfaction among Ph.D. students are common problems in doctoral education. In this study, we investigate which supervision, psychosocial, and project characteristics are related to satisfaction, progress, and quit intentions in a sample of 839 PhD candidates at a university in the Netherlands. Results of regression analyses show that experienced workload was negatively related to satisfaction and progress and positively to quit intentions. The quality of the supervisor-PhD candidate relationship, the PhD candidate's sense of belonging, the amount of freedom in the project, and working on a project closely related to the supervisor's research were positively related to satisfaction and negatively to quit intentions (van Rooij et al ., 2019, 1).

The summary of these findings can be listed as follows:

- Abundance of experience negatively affects the progress and satisfaction of research students and motivates them to drop out of school.
- The better the relationship between supervisors and candidates, the higher the candidate's sense of entitlement and broader their freedom in the research process; this will promote positive encouragement effects against dropping out of school.

- In addition, the 'match' between Ph.D. candidate and supervisor is crucial, the PhD candidate's.
- research topic should be closely related to the supervisor's field of research (van Rooij et al., 2019).

1.4 Research

Research is defined as a systematic inquiry on a specific topic. Research is all about the world we live in and how we understand that world. We all have an instinct and the intelligence to investigate when we are faced with something we do not know. This inquisitiveness is the mother of all knowledge and man's pursuit of knowledge of anything unknown is called research. Methodological investigation, inquiry, study can be defined as 'research'—a systematic inquiry, study on a specific topic. It is, in fact, a never-ending journey of discovery, because the 'knowledge' that is presented today as 'new' can be nullified by another 'new' discovery tomorrow. Such negatives occur frequently and inevitably, reminding us of the impermanent nature of the world.

Scientific knowledge is an outcome of the critical dialogue in which individuals and groups holding different points of view engage with each other. It is constructed, not by individuals, but by an interactive dialogic community (Longino, 1990). Scholar Karl Popper argues that the very possibility of refuting scientific research findings is an essential feature of scientific methodology.

Research is an academic activity and the term should be used in a technical sense. Therefore, academic research consists of several steps. These steps include defining and articulating research problems, gathering data, analysing data, organising information, interpreting results and drawing conclusions.

Scientific knowledge is the result of critical dialogue and consensus generated by different individuals and groups from different perspectives. This 'scientificness' seems to depend on certain attributes and assumptions of the researcher, research methods, and methodologies; they are diverse, varying in context. These attributes and assumptions are multifaceted. It is common for many scholars of research methods to divide these traits into two extremes (a dichotomy) called quantitative/positivists and qualitative/interpretivists & critical theorists. But such extremism is not appropriate for academic researchers and falls within a narrow framework. Contemporary scholars are of the opinion that the successful researcher should enjoy the freedom to use the acceptances and assumptions of both these extremes appropriately in his/her research work. These divisions are summarised under various factors as shown in Tables in Chap. 3.

The Falsification Principle, proposed by philosopher of science Karl Popper, is a way of demarcating science from non-science. It suggests that, for a theory to be considered scientific, it must be able to be tested and conceivably proven false. Ever since Popper's seminal work, for a scientific theory to be worthy of its name, it has to be falsifiable by experiments or observations. This requirement has become the foundation of the 'scientific method' (Mario Livio, 2013). For example, the hypothesis that "all swans are white," can be falsified by observing a black swan. The need of critical dialogue and consensus has become the basis of the 'scientific method' today.

In his book, *Brilliant Blunders*, astrophysicist Mario (2013) states that, in order to be scientific, a theory must be able to be falsified by experiments or observations, as suggested by Karl Popper. But one could argue that this argument, too, is an aspect of radical fundamentalism itself, which represents a certain extreme. Accordingly, more modern scholars, such as Sinham (2020), suggest that the idea that a scientific theory can be 'false' is a myth and we have now abandoned it.

It should be emphasized here that laboratory tests/experiments, observations and conclusions are only different stages of scientific research that we use in common practice (natural sciences; biology, chemistry, mathematics, etc.). They are only stages of data collection and analysis. Similarly, research in the social sciences, such as economics, political science, accounting, history, and aesthetics, are scientific because those research methods are logical and methodological. Scientific status cannot be confined to only certain subjects or methods. What is really needed for a scientific inquiry, or research is a systematic approach. It can be called academic research. Interpretation exists in all types of scientific studies, be they quantitative or qualitative (Gummeson, 2003).

Depending on the nature of the data, the method of analysis, and the method of drawing conclusions, academic research is basically of two types, quantitative and qualitative. Academic research can be done using data from large samples (surveys) using statistical methods as well as using phenomenology, observations, in-depth discourse, narrative/thematic analysis or using both methods (mixed methods). Regardless of the method used, systematic methodologies must be used to conduct research scientifically.

Until recently, there was a commonly held myth that only quantitative research using numbers were scientific, and that qualitative research was not scientific. Today, it is widely accepted that going beyond dealing with numbers is also considered as valid and dependable and is therefore credible.

1.5 Importance and Objectives of the Research Work: Publishing

To be official and world-recognized, the 'knowledge' discovered through academic research, the research results, should be published as articles in international peer-reviewed journals (IPRAJs) or as book chapters in edited books published by reputed publishers or presented at academic conferences. There is a well-known saying among scholars (within academia) that such discoveries are wasted; "Publish or perish."

"Publish or perish" is an aphorism describing the pressure to publish academic work in order to succeed in an academic career. Such institutional pressure on academics is generally strongest at research-oriented universities. Successful publications also intensify the focus on the authors as well as the commercial sponsors, which in turn helps to attract more funding.

Successful publications bring attention to scholars and their sponsoring institutions, which can help continued funding and their careers. In popular academic perception, scholars who publish infrequently, or who focus on activities that do not result in publications, such as instructing undergraduates, may lose ground in the competition for available tenure-track positions. The value of published work is often determined by the prestige of the academic journal in which it is published. Journals are often measured by their impact factor (IF), which is the average number of citations to articles published in a particular journal.

The value of a published work often depends on the ranking/indexing of the academic journal or publisher's recognition and reputation. The chances of losing eligibility for permanent appointments or promotions are high for academics who rarely publish research papers, who focus only on activities such as teaching and advising undergraduates.

The number of research papers and citations published by a scholar can be easily found on the internet. All you have to do is go to the Google scholar page of the researcher you are looking for—you get all the information, if those scholars have uploaded their profiles.

Over the years, many other popular indexation services have developed. Econtent Pro International provides the following list as prominent indices (https://www.econtentpro.com/blog/understanding-scholarly-indexing-and-what-it-means-for-you/45).

- **Scopus**: An Elsevier-maintained citation and abstract database, Scopus has an international listing of interdisciplinary scientific information, from medicine, to computer science, to the social sciences, and humanities.
- **Clarivate**: Formerly Thompson Reuters, this index is recognized for its highly selective Web of Science databases (formerly known as ISI Web of Knowledge) and provides a platform for researchers to access content in the sciences, social sciences, arts, and humanities.
- **Compendex**: This is known as Elsevier's Engineering Village and comprises a mass of carefully selected engineering research articles and publications.
- **ProQuest**: A well-recognized database with multiple subject-specific indices encompassing the arts, humanities, and the social sciences.
- **EBSCO Information Services**: EBSCO provides research databases, e-journals, magazine subscriptions, and e-books to libraries and academic institutions all over the world.
- **DOAJ (Directory of Open Access Journals)**: This is a web directory indexing entirely open access journal content.
- **BioOne**: This index focuses on listing scholarly publications in the biological, ecological, and environmental science fields.
- **ACM Digital Library**: One of the leading databases covering full-text articles and bibliographic literature on computing and information science.

- **ERIC**: One of the most well-known online databases of education research and information.
- **PubMed**: A bibliographic database of biomedical literature covering emerging research in journals and books.

1.6 Summary

- Even though writing in English is not mandatory, reading in English is essential.
- Acceptance of a research paper depends on the depth of the existing knowledge studied.
- Objectives cannot be achieved by studying your native language alone. It should be a practice to present it back again in English. However, it is also possible to write your paper in your own language and it can be translated into academic English professionally, by a qualified professional.
- There are many translating tools/websites available in internet, but translations do not take the context into account and, therefore, may not be credible. Therefore, one must learn adaptation rather than translation.
- What is more important is not to find or create words of your native language related to Western terminology, but to try to understand those concepts and to sense the meaning that is conveyed through those words.
- Reading patiently in English is an essential part of this process.
- Care should be taken to keep a back-up of the data collected during the research process, such as sketches, and to avoid any damage to the work done so far.
- Annotated bibliography entails keeping track of references and citations in every important source to which you refer.
- It may also be important to keep hard copies of some frequently used documents.
- Having a positive and flexible attitude and persistent determination is also very important preparation.
- There is nothing wrong with recognizing the research process as an exercise in practising patience.
- A strong, cordial relationship between the research student and the supervisor depends on maintaining personal friendliness and keeping your research topic close to the supervisor's field of research.
- Research is an academic activity; It involves defining problems, gathering data, analysing and interpreting, and then drawing conclusions.
- These findings can be a solution to a research problem, constructing a theory, a generalization or an expansion of understanding, or an initiative for social progress.
- Scientific knowledge is the result of critical dialogue from different perspectives.
- This 'scientificness' varies and may do so depending on the characteristics and assumptions of the researcher and the research methodology.
- For a theory to be scientific according to Karl Popper's falsification principle, it must be able to be refuted by experiment or observation.

- According to some scholars, the theory that a scientific theory can be 'falsified' is a myth.
- Experiments, observations and research results to be recognised, the research results must be published in an internationally recognised medium (publish or perish).
- Academic publications have a critical impact on the professional development of researchers as well as the organization's international ranking (world ranking).
- The value of a publication is determined by the impact factor (IF), which is a measure of the ranking/indexing of a journal or publisher's reputation.

References/Further Reading

Gummesson, E. (2003). All research is interpretive! *Journal of Business & Industrial Marketing, 18*(6/7), 482–492. https://doi.org/10.1108/08858620310492365.

Longino, Helen E. (1990). Science as Social Knowledge: Values and Objectivity in Scientific Inquiry. Princeton: Princeton University Press.

Mario, L. (2013). *Brilliant blunders: From darwin to einstein-colossal mistakes by great scientists that changed our understanding of life and the universe.* Simon & Schuster.

Singham, M. (2020). The idea that a scientific theory can be 'falsified' is a myth. It's time we abandoned the notion. Policy & Ethics. https://www.scientificamerican.com/article/the-idea-that-a-scientific-theory-can-be-falsified-is-a-myth.

van Rooij, E., Fokkens-Bruinsma, M., & Jansen, E. (2019). Factors that influence PhD candidates' success: The importance of PhD project characteristics. *Studies in Continuing Education, 43*(1), 48–67. https://doi.org/10.1080/0158037x.2019.1652158.

Chapter 2
Evolution of Western Research Methodology

2.1 Historical Development

Social research has been widely used throughout history for over 2500 years. The origins of Western philosophy go back to the ancient Greek philosophers such as Plato (424–348 BC) and Aristotle (384–322 BC). Research and knowledge were highly valued by these philosophers; Socrates (470–399 BC) has said that "there is only one good; knowledge, and one evil; ignorance" and Aristotle has mentioned that "the educated differ from the uneducated, as the living from the dead". In fact, this seems to be in line with the teachings in the Eastern philosophers such as ancient Hindu scholars and Gautama Buddha (563–483 BC) etc. who called this evil of ignorance "avijja" many more years before Aristotle.

This process of knowledge accumulation evolved gradually, and Western scholars believe that Bacon's (1561–1626) introduction of the scientific approach to research was a historic milestone. Various scholars' analyses of the evolutionary process of the leading schools of thought in social inquiry from the early Bacon era to modern times are updated and presented in Fig. 2.1.

The researcher's philosophical stance and strategic approach are crucial in presenting arguments effectively and drawing convincing conclusions. Factors such as the researcher's personal characteristics, the resources available, access to data, expertise, and ethical considerations should be carefully considered and selected. It is also important to justify those choices regarding this historical evolution and state it clear in the research report.

An understanding of this historical evolution is useful in justifying your choices and presenting them clearly in the research report. Awareness and understanding of the different paradigms, perspectives, worldviews and theories affect many aspects of research, and it is important to mention that in a research report. This is due to the recognition that interpretations and conclusions of the research results may be

© The Author(s), under exclusive license to Springer Nature Singapore Pte Ltd. 2022
C. A. Saliya, *Doing Social Research and Publishing Results*,
https://doi.org/10.1007/978-981-19-3780-4_2

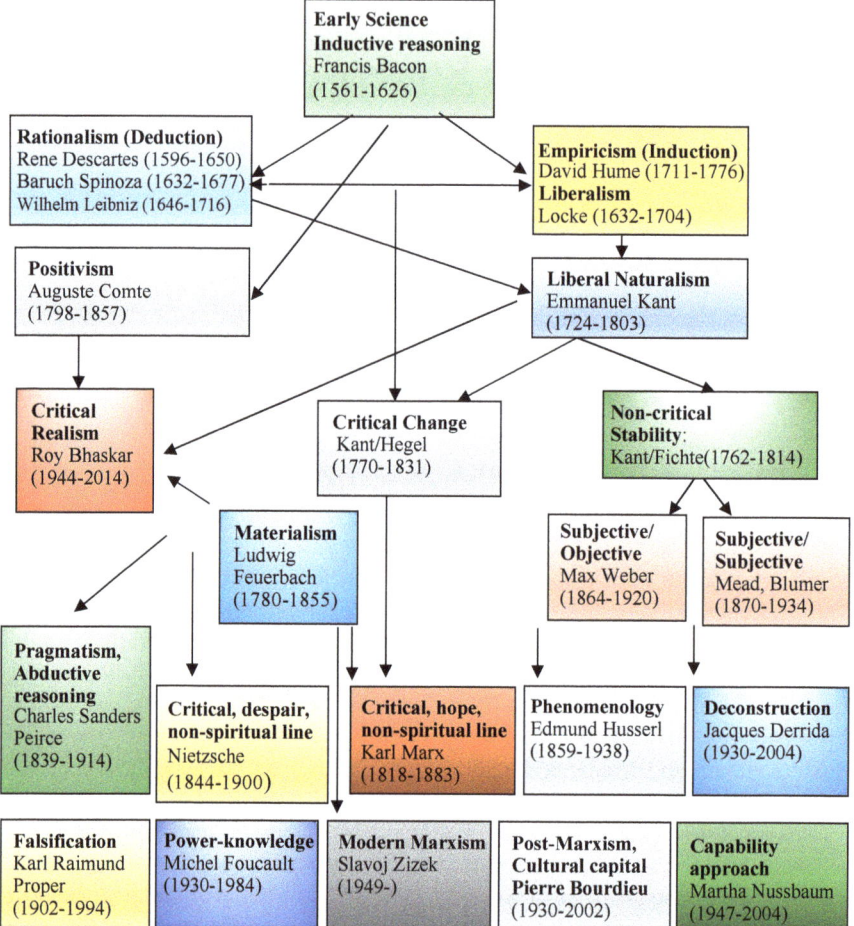

Fig. 2.1 Evolution of research methods and ideologies. *Sources* Laughlin (1995), Saliya (2017)

significantly influenced by the socio-economic and cultural background to which the author belongs. In fact, researcher's stances with regard to paradigms, perspectives, worldviews and preferred theories critically depend on his/her socio-economic and psychosocial, sociocultural and economic background.

2.2 How Do We Know the World?

"Do you really know what you think you know?" and if so,

"How do you know what you know?" and.

"What is the relationship between the inquirer and the known?"

Answers to these questions are heavily influenced by the paradigm to which the researcher belongs.

Denzin and Lincoln (2005) posed the above questions to explain epistemology. That understanding is as important to the reader as it is to the researcher, to understand the researcher's epistemological position and to assess the how independent the researcher is and whether the research results are interpreted in an unbiased way.

2.3 Scientific Knowledge and Positivism

Positivism is a philosophical theory that states that "genuine" knowledge is exclusively derived from experience of natural phenomena and their properties and relations. Thus, information derived from sensory experience, as interpreted through reason and logic and verified data (positive facts) received from the senses is known as empirical evidence; thus positivism is based on empiricism.

2.3.1 Empiricism and Rationalism

In philosophy, empiricism is a theory that states that knowledge comes primarily from sensory experience and inductive reasoning. It is another method of acquiring knowledge in epistemology such as rationalism and skepticism. Empirical ideas emphasise the importance of empirical evidence in gaining knowledge rather than from innate ideas or traditions.

It is a basic element of primitive methodology, and all assumptions and theories must be tested against observations of the natural world, relying solely on primitive reasoning, intuition, or revelation. Francis Bacon, of England, advocated empiricism in 1620, and Rene Descartes of France confirmed rationalism in about 1640. In the eighteenth century, both George Berkeley of England and David Hume of Scotland became pioneers of empiricism.

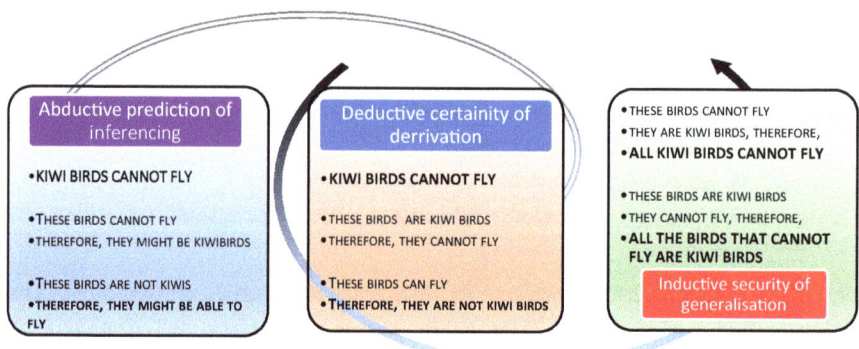

Fig. 2.2 This is a comparative representation of the threefold conclusions, with examples. *Source* Saliya (2021a)

Thinkers such as Henry de Saint-Simon (1760–1825), Pierre-Simon Laplace (1749–1827) and Auguste Comte (1798–1857) accepted only the principle of positivism as scientific. Emil Durkheim (1858–1917) reformed sociological positivism as the basis of social research. In the early part of the twentieth century, a new paradigm of critical thought emerged, despite the failure of some Marxist predictions regarding capitalism.

2.3.2 Liberalism and Naturalism/Realism

Liberalism is a political and moral philosophy based on liberty, consent of the governed and equality before the law. These ideas were first drawn together and systematised as a distinct ideology by the English philosopher, John Locke, generally regarded as the father of modern liberalism (see Fig. 2.1).

John Locke (1632–1704), in an essay on human understanding (1689), strongly established the idea that empiricism is the only knowledge that people can have based on experience. Locke's fame is attributed to the suggestion that human knowledge is the experience of a person's life experiences accumulated on a blank sheet of paper. This implies that liberalism and realism are based on empiricist principles.

Around 1780, the German Emmanuel Kant redefined the distinction between the empiricism of Francis Bacon and the rationalism of Rene Descartes. Kant argues that conscious experience recognises not only the way things are in objects but also the way we see them under the conditions of our senses. This means that objects are just 'looks' and we cannot know their nature as they are. His theory is known as liberal naturalism.

Rationalism versus Empiricism

Rationalists, such as Rene Descartes, thought that reason could explain the working of the world; without reference to sense experience.

Conversely John Locke's empiricism argued that the mind was like a tabula raza (blank sheet of paper) which was informed by the world of experience.

Kant rejected Locke's empiricism, arguing that the rational mind is capable of structuring and interpreting sense experience.

Rene Descartes

John Locke

Sources https://www.quora.com/Is-Descartes-a-rationalist-or-empiricist
https://qph.fs.quoracdn.net/main-qimg-0cd8ef1caeee168c53073ac37158c1d3-lq

At the beginning of the twentieth century, the first wave of German sociologists, Max Weber (1864–1920) and George Siemmel (1858–1918) rejected positivism and laid the foundation to post-positivism in sociology. Subsequent post-positivists and critical theorists have suggested that positivism as both scientism and science as an ideology. In the early twentieth century, several forms of practical philosophy emerged, combining the basic understanding of empirical (experience-based induction) and rationalist (concept-based deduction) thought. Charles Pierce's (1839–1914) major contributions were the addition of the concept of abduction to the logic of deduction and induction. These three forms of reasoning serve as the basic conceptual basis for empirically based scientific methodology.

2.4 Pragmatism

At the beginning of the twentieth century, the German sociologists Max Weber and George Simmel rejected positivism and several forms of empirical and rational thought emerged. Charles Sanders Peirce, father of pragmatism, introduced the concept of abduction inferencing (abduction reasoning) as an alternative to David Hume but complementary to (in parallel with) the induction and deduction logics.

These three forms of reasoning were established as the primary conceptual basis for empirically based scientific methodology, and the ontological position of the abduction method clearly differs from that by suggesting that the world exists as multiple realities (instead of one reality). Wilhelm Dilthey (1725), who fought hard against the argument that only the "explanations from science" were valid, argued that scientific explanations did not reach the inner nature of phenomena, and that the most plausible argument for understanding thoughts, feelings, and desires was humanist knowledge which is a confirmation of G. B. Vico's argument presented far back in 1725. Vico said that "anyone who intends to pursue a career in court, senate or life should also be taught to defend themselves on both sides of the controversy. Because man has a free and brilliant style of expression, they are able to build arguments that are more logical and of the most optimal accuracy". But Vico also condemned the defence of both sides in the controversy as bogus rhetoric.

Title page of *Principj di Scienza Nuova* (1744 ed.)

Source https://www.space.com/17661-theory-general-relativity.html

Raoul Nakhmanson (2003) argued that the analogy between evolution and quantum mechanics and sociology gives rise to the idea that quantum mechanics has an individual and a social consciousness for material objects.

Marx's critical theories of society, economics, and politics, collectively understood as Marxism (Karl Marx, 1818–1883), suggest that human societies develop through class conflict. In the capitalist mode of production, this shows the conflict between the ruling classes (known as the bourgeoisie) who control the means of production and the working class (known as the proletariat) who activate these means by selling their labour power for wages.

Using a critical approach known as historical materialism, Marx predicted that capitalism, like previous socio-economic systems, would cause internal tensions and cause these systems to self-destruct and be replaced by a new system known as the socialist mode of production.

2.5 Critical Education

Today theories are not rejected just because of allegations such as "not based on the senses or experience", "not rigorous", "non-scientific" or "not scientifically convincing" etc. Anthony Giddens (1938-) argues that, because mankind is constantly using science to discover and research new things, mankind will not move beyond the second metaphysical phase.

The Right Honourable
The Lord Giddens
MAE

Paulo Reglus Neves Freire (1921–1997) was a Brazilian educator and philosopher who was a leading consultant in critical education. He is best known for his book, *Pedagogy of the Oppressed*, which is considered as one of the books on which his critical educational movement is based. "Critical teaching" is a philosophy of education and a social movement. It developed concepts from critical theory and related traditions and is applied to the field of education and the study of culture.

Charles Sanders Peirce, who is considered the father of pragmatism and abduction inferencing or reasoning and the theory of conjecture, presents an alternative to the induction and deduction logic of David Hume, however this is complementary (in parallel) to induction and deduction logics.

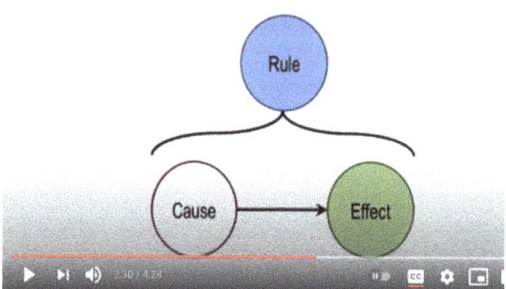

Deduction, Induction, Abduction - Georgia Tech - KBAI: Part 5

Source https://www.youtube.com/watch?v=-nn3XMoPC7s

How to Argue - Induction & Abduction: Crash Course Philosophy #3

Source https://www.youtube.com/watch?v=-wrCpLJ1XAw

2.6 Newtonian and Quantum Mechanics

Newton's theory of gravitation and Einstein's theory of relativity are not sensitive to the five senses. Although gravity or any other attraction is felt towards the Earth, we do not feel the smell, taste, touch, to see with our eyes or to hear. According to Raoul Nakhmanson (2003), the wave function is purely a mental construct. Einstein called these so-called non-material 'matter waves', 'Gespenster Felders' (demon fields). However, they control the behaviour of material objects. They argue that there is no sensory evidence of gravity or any other attraction toward the Earth. It may not be a pull in the centre of the Earth, but, as Einstein put it, and which is acknowledged by modern physicists, special relativity argues that space and time are inextricably connected, and these scientists do not acknowledge the existence of gravity.

Quantum mechanics
as a
sociology of matter[1]

Raoul Nakhmanson[2]

Source https://arxiv.org/pdf/quant-ph/0303162.pdf

Nakmanson argues that the evolution theory and analogy between quantum mechanics and sociology led to the idea that material objects in quantum mechanics have an individual and a social consciousness too. Such a hypothesis explains the essence of wave function and its collapse as well as "mysterious" experiments known as "two-slits", "delayed-choice", "EPR", "Aharonov-Bohm", and"interaction-free measurement".

The wave-particle duality is a mind–body one. In the real 3D-space there exists only the particle, the wave exists in its consciousness. If there are many particles, their distribution in accordance with the wave function represents a real wave in real space. "Many worlds", "Schrödinger's cat", "Great Smoky Dragon", etc. exist only as virtual mental constructions (Nakhmanson, 2003).

You can get a clear idea of this by watching the amazing videos below.

Double Slit Experiment explained! by Jim Al-Khalili

2,226,029 views · Feb 2, 2013   👍 24K   👎 1K   ➤ SHARE   ☲₊ SAVE

Sources https://www.youtube.com/watch?v=A9tKncAdlHQ; https://www.youtube.com/watch?v=0ui9ovrQuKE; https://www.youtube.com/watch?v=5hVmeOCJjOU

He goes on to say that the wave-particle duality is the same as the mind–body duality. Only the particle exists in real, three-dimensional space. The wave exists in consciousness. If there are many particles, their distribution represents a real wave in real space according to the wave function.

2.7   Paradigm

Choosing a paradigm is important because it allows you to start forming research questions (RQs). This is useful, even if your questions start off vague (Gournelos et al., 2019, p. 8).

A paradigm is the entire sets of beliefs, values, techniques that are shared by members of a community (Kuhn, 2012). Guba and Lincoln (1994), who are leaders in the field, define a paradigm as a basic set of beliefs or worldview that guides research action or an investigation. Paradigms are thus important because they provide beliefs and dictates which, for scholars in a particular discipline, influence what should be studied, how it should be studied, and how the results of the study should be interpreted. The paradigm defines a researcher's philosophical orientation and, as we shall see in the conclusion to this paper, this has significant implications for every decision made in the research process, including choice of methodology and methods. And so a paradigm tells us how meaning will be constructed from the data we shall gather, based on our individual experiences, (i.e., where we are coming from). It is, therefore, very important that when you write your research proposal, you clearly state the paradigm in which you are locating your research.

Scholars have identified three main paradigms: positivism, interpretivism/constructivism and critical thinking. Pragmatism is also cited by some scholars as the latest paradigm shift. Positivists aim to explore, explain, evaluate, predict, or test theories. The goal of interpretivists and constructivists is to understand human behaviour. Critical theorists aim to critique social reality, liberate, empower people, and propose solutions to social problems.

The pragmatic paradigm refers to a worldview that focuses on 'what works' rather than what might be considered absolutely and objectively 'true' or 'real.' Early pragmatists rejected the idea that social inquiry using a single scientific method could access truths regarding the real world.

Critical realism is a branch of philosophy that distinguishes between the 'real' world and the 'observable' world. The basis of this theory is that 'reality' cannot exist or be excluded from human cognitions, theories and creations. Interpreting the world scientifically through cause-and-effect mechanisms contradicts/alternates the methods of empiricism and fundamentalism.

2.8 Pseudoscience and Social Constructivism

Born in 1984 and still in his early career, Maarten Boudry is best known for his skepticism and critical attitude towards pseudoscience. He studies the errors of human reasoning, which can be subject to pseudoscience and irrationality. Pseudoscience is described as an 'imitation of real science'. Social reformism provides a quantitative answer to this confusion

Maarten Boudry

Born	15 August 1984 (age 36)
	Moorslede, Belgium
Nationality	Belgian
Alma mater	Ghent University
Occupation	Philosopher
School	Scientific skepticism

Source https://en.wikipedia.org/wiki/Maarten_Boudry

Social constructivism is a sociological theory of knowledge in which human development is socially situated and knowledge is built by interacting with others. Like constructism, social constructivism states that people work together to create something.

Social constructism focuses on the objects created through the social interaction of a group, while the worldview of social reformism focuses on the learning of an individual as a result of his or her interactions within a group. Strong social reformism as a philosophical approach suggests that the natural world has little or no role in the construction of scientific knowledge.

2.9 Paradigm Comparison

Objectivity, reliability, validity and generalizability are the keywords used by positivists in their vocabulary whereas non-positivists, often guide qualitative research, may employ terms such as 'credibility', 'transferability', 'dependability' and 'conformability' (Guba and Lincoln, 2005).

The basis of the positivist paradigm is positivism. Positivism was a social research philosophy introduced in 1848 by Auguste Comte. According to this view, the researcher begins the process of discovery with theory and tests hypotheses using deductive reasoning in accordance with scientific methods.

Their ontological belief is that an objective reality exists and can be known through research in contrast to post-positivists who concede that "we might never know reality perfectly but … accumulated efforts will move us toward discovering what is real" (Bailey, 2007, p. 52).

Positivists believe that truth or fact is independent of any theory or human observation. On the contrary, interpreters and reformers argue that 'truth' is 'creation' and that it is created in the minds of individuals and among people in a culture. Therefore, the various socio-economic and religious backgrounds in these two extremes are innumerable, both within the broader philosophical and more narrow frameworks, such as tribalism.

According to Denzin and Lincoln, "The interpretivist/constructivist paradigm assumes a relativist ontology (there are multiple realities), a subjectivist epistemology (knower and subject create understandings), and a naturalistic (in the natural world) set of methodological procedure" (2005, p. 27).

According to Denzin and Lincoln (2005), research methodology involves selection, justification and sequential arranging of activities, procedures, and tasks in a research project.

The epistemological stance of positivists is that the available knowledge does not depend on the researcher. Realist fundamentalists believe that research and data must be objective or empirical. That is, the five senses must be based on sensible observations and experience. It should also be unbiased. This means that the researcher's feelings or values should have no bearing on the research results.

Does this mean that the senses do not play any role in the creation of knowledge beyond the empirical evidence built on the five senses? In fact, it is important to examine in depth whether observation or experience can be obtained only by the five senses without the sense of mind.

The existentialist position of the interpretive/reformist paradigm is the belief in the existence of the world, that reality is relative/subjective, that it does not exist independently, but as multiple collectively created realities in the human mind. Their epistemological stance is that the available knowledge depends on the researcher's thoughts and desires. That is, 'knowledge' is created by the seeker and what is sought by a specific methodology.

The epistemological position in critical reality is that the researcher is not independent of what he is researching and discusses the findings of the research through his or her values. Bailey argues that one of the most important values often included in such (critical; critical) research is the desire to eliminate social injustice. In that sense, fundamentalists and interpretation/reformists are mere observers and judges.

The prevailing general understanding is that researchers in the critical paradigm often want to describe, educate, understand, and change how powerful groups oppress the weak, the poor, the discriminated. They seem to hold opportunistic interpretations of reality, the existence of 'truth', the fact that truth is absolute and relative, that there is only one truth or that there are multiple truths, and that 'truth' can be found or not.

Differences in these assumptions cannot be ruled out as mere philosophical differences. The philosophical positions of these researchers are critical to the practical investigation of research problems as well as the interpretation of findings and policy choices.

2.10 Induction, Deduction and Abduction

Based on the results of academic research, three methods have been identified to reach final conclusions. These reasoning methods are called induction, deduction and abduction.

> Deduction proves that, for logical reasons, something must be the case; induction demonstrates that there is empirical evidence that something is truly so; abduction, by contrast, merely supposes that something might be the case. It therefore abandons the solid ground of prediction and testing in order to introduce a new idea or to understand a new phenomenon (Bude, 2004, p. 322).

Inductive reasoning is based on the assumption that laws or generalisations can be created by the accumulation of observations and events. In stark contrast, deductive reasoning is based on the assumption that theories should be tested and derived using only empirical methods and strict adherence to principles.

Some scholars today acknowledge that both of these theories are unsatisfactory in the design and communication of ideas, and that it is a matter of urgency to break free from the narrow frames imposed by traditional logic. Therefore, such scholars are of the opinion that abductive inferencing is more appropriate for qualitative inquiries that present an open-minded intellectual approach. However, when generalising conclusions from empirical evidence, inductive generalisation is used to answer the 'how' question; deductive derivation is best suited for hypothesis testing.

Deduction, for logical reasons, seeks to prove something; Inference seeks to show that there is empirical evidence to prove that something really is so; In contrast, hypothetical logic seeks to infer that something is simply possible.

Deduction	Abduction
Rule: All the beans in this bag are white.	*Rule*: All the beans in this bag are white.
Case: These beans are from this bag.	*Result*:These beans are white.
Result:These beans are white.	*Case*: These beans are from this bag.

Induction

Case: These beans are from this bag.
Result:These beans are white.
Rule: All the beans in this bag are white.

Source Peirce (1878)

Charles Sanders Peace (1839–1914) thus argued that speculative abductive reasoning, which extends beyond the derivation of deductive certainty and the generalisation of inductive security, can lead to introduction of a new idea or the realisation of a new phenomenon, in which case one has to abandon specific expectations, such as predicting and checking for accuracy.

In other words, Charles Sanders Peace emphasizes that deduction is a mathematical derivation or some other logical method that is intended to prove certain conclusions.

Also, using the security of generalization (as an alternative to the certainty of deduction) from the induction process, one seeks to present empirical evidence to prove something as truly so. In contrast, the abduction method introduces a discovery by conjecture, which goes beyond these derivations and inferences, such as looking at right and wrong in theories and/or creating or predicting theories. What is happening in this conjecture is simply arguing that something might (might be). Therefore, in order to introduce a new idea or to understand a new phenomenon, abductive conjecture rejects the theory of rigid, stubborn, positivist, deductive certainty and the flexible humanistic interpretive judgments of induction methods.

A concise comparison of positivists, interpretivists, and critical thinkers' assumptions is provided in Tables 2.1 and 2.2.

O'Leary (2014) explains that the purpose of the knowledge varies from just "building understanding" to "action change within a system" to "emancipate through action" or further to "expose the systems". Therefore, the research methodologies could vary from

Table 2.1 Assumptions' comparison

Assumptions \newline Factor	(Typical)		(Modern)
Identification	Positivism	Interpretivism, Constructivism	Critical approach
Ontological belief	An objective absolute realities exist outside the observer.	A subjective concept, multiple realities constructed in human minds.	Reality is historically determined by power relations.
Epistemology	Independent and value free	Could be biased and value laden	
Axiology	Neutral. Disengaged. Logical. Absolute truth	Rational, humane, contextual	Emancipation. Social justice and Transformation/ improvement
Aims, purpose	Make predictions, Generalisation	Broadening understanding	Discuss social injustice and stimulate change
Data types	Numbers. Surveys, large samples	Phenomenon, perceptions. Case studies, interviews, observations	
Methods	Quantitative dominance, Statistical estimations	Qualitative dominance, Narrative/ discourse/ thematic analyses. Logical inferences	
Methodology	Hypothesis/Theory testing	Interpreting/ Constructing theories	Critical evaluation
Making conclusions	Deductive derivation	Inductive generalisation \newline Abductive inferencing/assertion	
Verification, Justification	Empiricist.	Phenomenalist	Critical thought/dialogue

"basic" to "applied/evaluative" to "participatory" or further to "critical/radical ethnography" accordingly.

Therefore, it is not so enlightening or logical for academic researchers to be trapped in quantitative or qualitative boxes. Thus, the mixed methods school emerged, combining the different elements of these two traditions in different ways facilitating many different strategies for exploring the vast world around us.

2.11 Perspectives and Theories

Perspectives are a set of rules or theories that apply to the interpretation of a phenomenon and a framework of its own. For example, in a car accident, the driver of one car may have one view, another driver or passenger may have another, and each observer may have slightly different perspectives on their location and distances.

Table 2.2 Alternative Inquiries Comparison of paradigm fundamental beliefs (metaphysics)

Paradigm	Positivist/Post-positivist	Interpretivist/Social constructivist	Critical
Researcher	Detached observer	Attached participant	Transformative intellectual
Purpose	To discover the laws governing the universe	To understand and describe human nature	To destroy myths and change society
Objectives	Explore, explain, evaluate, predict and to develop/test theories	Understand human behaviour	Criticize social reality, liberate people, and propose solutions to social problems
Ontology	Reality or 'truth' can be known and independently exists outside of perceptions. Post-positivists concede that reality can never be known perfectly.	Reality or 'truth' is unknown and constructed within the minds of individuals. Multiple realities exist. No direct access to the real world.	Reality is created and shaped by social, political, cultural, economic forces that have been historically crystalized over time. Relativism.
Epistemology	Objective knowledge does not depend on the researcher and value-free.	Subjective perceived knowledge and value-bound/neutral; no value is wrong.	The objective-subjective label is socially contrived. Value-mediated Findings. Some value positions are wrong and some are right.
Key concepts	Scientific, Experimental, Objectivity, Reliability, Validity and Generalizability.	Credibility, Transferability, Dependability and Conformability.	Virtual-reality shaped by social, political, cultural, economic, ethnic, and gender values.
Methods	Quantitative methods are preferred. Deductive logic. Begins with theory.	Qualitative methods are dominant. Inductive generalization inferencing/reasoning. Hermeneutical, dialogical and dialectical Abductive-dialectical reasoning.	
Theories/ Perspectives and Contributors	Rationalism: Francis Bacon (1561-1626), Rene Descartes (1596-1650). Empiricism: John Locke (1632-1704). Positivism: Auguste Comte (1798-1857). Post-positivism; Falsificationism: Karl Popper (1902-1994).	Idealism: Johann Fichte (1762-1814). Protestantism: Max Webber (1864-1920). Pragmatism: G. H. Mead (1863-1931). Symbolic interactionism: Herbert Blumer (1900-1987).	Materialism: Georg Hegel (1770-1831). Marxism: Karl Marx (1818-1883). Friedrich Nietzsche (1844-1900). Power-knowledge: Michael Foucault (1926-84). Cultural capital: Pierre Bourdieu (1930-2002). Deconstructivism: Jacques Derrida (1930-2004).
	Liberal Naturalism: Emmanuel Kant (1724-1804), Middle-range thinking: Robert Merton (1910-2003)		

Source Bailey (2007), Creswell (2007), Denzin and Lincoln (2005, pp. 193–196), Saliya (2017, 2021)

Holding a single, strict worldview may enable an American (for instance) to disregard other perspectives, and may be unnecessarily confusing, confusing, and perhaps adaptable by worldviews.

Snyder (2015) asserts that low wage labour or sweatshops are often described as self-evidently exploitative and immoral. But defenders of sweatshops might describe these as the first rung on a ladder toward greater economic development. On the other hand, in another perspective, it could be described as enhancing productivity or wealth maximisation and such excessive explanations would lead to too much noise to the reader.

Emerging worldviews from the eighteenth century onwards can be divided into four basic categories according to their basic positions and/or periods of introduction:

1. Classic worldviews (eighteenth century)
2. Modern worldviews (1800–1950)
3. Post-modern worldviews (1900–1990)
4. Dynamic worldviews (since 1980).

Classical worldviews consist of research approaches based on extremely abstract theoretical creations, such as idealism, materialism, rationalism, and positivism. Modern classical worldviews consist of research approaches based on theories such as Marxism, naturalism/realism, and symbolic interactionism. Postmodern worldviews, identified as theories that evolved in the post-industrialization period, consist of major research strategies such as phenomenology, critical sociology, ethnography, and neo-pragmatism. Among these worldviews, dynamic worldviews have their place because they are separated from dual or bipolar interpretations such as idealism and materialism, quantitative and qualitative research, and free market economic policies and centrally planned economic policies. Theories related to paradigms such as critical thinking and pragmatism belong to this set of dynamic worldviews (Saliya, 2010, 2017).

2.12 Summary

- Awareness and understanding of the different paradigms, perspectives, worldviews and theories affect many aspects of research, and it is important to mention that in the research report.
- "How do you know what you know?" is the question to be answered to determine your epistemological standpoint.
- A paradigm is the entire sets of beliefs, values, techniques that are shared by members of a community.
- Positivism, based on empiricism, assumes that all knowledge is based on experience derived from the senses.
- Empiricism is a theory that states that knowledge comes primarily from sensory experience and inductive reasoning.

- Phenomenalism is the view that physical objects cannot justifiably be said to exist in themselves, but only as perceptual phenomena or sensory stimuli situated in time and space.
- Critical thinking is the analysis of facts to form a judgment. A Critical Dialogue is a conversation that inspires insight and wisdom on a particular topic, both for the individuals as well as the collective thinking of a group.
- Emmanuel Kant argues that conscious experience recognizes, not only the way things are in objects, but also the way we see them under the conditions of our senses. This means that objects are just 'looks' and we cannot know their nature as they are. His theory is liberal naturalism.
- Charles Sanders Peirce introduced the theory of conjecture as an alternative to the induction and deduction logic.
- Researchers must seek knowledge beyond the scientific method in the second metaphysical phase so that mankind can constantly discover new things.
- Critical teaching is a philosophy of education and a social movement. It is said to develop concepts from critical theory and related traditions and apply them to the field of education and the study of culture.
- The three forms of reasoning, deduction, induction and abduction, were established as the basic conceptual basis for empirically based scientific methodology; the ontological position of the abduction method is clearly different from that which suggests that the world exists beyond the dual, that is, multidimensional realities (instead of one reality).
- Humanistic knowledge is the more plausible argument than scientific explanations and more helpful in understanding thoughts, feelings and desires.
- Because humans have a free and brilliant style of expression, they are able to build arguments that are more logical and of the most optimal accuracy.
- Evolution, and the analogy between quantum mechanics and sociology, have led to the idea that quantum mechanics has an individual and a social consciousness for the material objects.
- According to social constructivism, human development is socially situated, and knowledge is built by interacting with others.
- Like constructism, constructivism states that people work together to create something.
- Social constructivism focuses on the worldview of individual interaction within a person's learning.
- Strong social constructivism suggests that the natural world has a role to play in building scientific knowledge.
- Emerging worldviews from the eighteenth century onwards can be divided into four basic categories according to their basic positions and/or periods of introduction; Classic worldviews (eighteenth century); Modern worldviews (1800–1950); Post-modern worldviews (1900–1990); Dynamic worldviews (since 1980).

References/Further Reading

Bailey, C. A. (2007). *A Guide to qualitative field research*. Thousand Oaks, CA: Sage.

Bude, H. (2004). The art of interpretation. In *A companion to qualitative research*, eds. Uwe Flick, Ernst von Kardorff and Ines Steinke, London: Sage.

Czarniawska-Joerges, B. (1992). *Exploring complex organizations*. London: Sage.

Creswell, J. W. (2007). *Qualitative inquiry and research design; Choosing among five traditions*. Thousand Oaks, CA: Sage.

Czarniawska, B. (2004). Narratives in finance science research: Introducing qualitative methods. *London. Sage.* https://doi.org/10.4135/9781849209502.n5.

Denzin, N. K., & Lincoln, Y. S. (2005). *The landscape of qualitative research: Theories and issues.* London: Sage.

Gournelos, T., Hammonds, J. R. & Wilson, M. A. (2019). *Doing Academic Research: A Practical Guide to Research Methods and Analysis.* London: Sage.

Guba, E. G. & Lincoln, Y. S. (1994). Competing paradigms in qualitative research. In Denzin, N.K. & Lincoln, Y.S. *The Handbook of Qualitative Research*, 3(105–117). California: Sage.

Guba, E. G. & Lincoln, Y. S. (2005). Paradigmatic controversies, contradictions, and emerging confluences. *The SAGE handbook of qualitative research.* N. K. Denzin and Y. S. Lincoln. 191–216. Thousand Oaks, CA: Sage.

Kuhn, T. S. (2012). *The structure of scientific revolutions, Chicago.* IL: University of Chicago Press.

Laughlin, R. (1995). Empirical research in accounting: alternative approaches and a case for middle range thinking. *Accounting, Auditing and Accountability Journal, 8*(1), 63–87.

Nakhmanson, R. (2003). Quantum mechanics as a sociology of matter. https://doi.org/10.48550/arXiv.quant-ph/0303162

O'Leary, Z. (2005). *Researching real-world problems.* London: Sage.

O'Leary, Z. (2014). *The essential guide to doing research.* London: Sage.

Ontology versus Epistemology: https://study.com/academy/lesson/ontology-vs-epistemology-differences-examples.html

Peirce, C. S. (1878). Deduction, induction, and hypothesis. *Popular Science Monthly, 13,* 470–482.

Saliya, C. A. (2010). *The Role of bank lending in sustaining income/wealth inequality in Sri Lanka.* PhD Thesis, Auckland University of Technology, New Zealand. https://openrepository.aut.ac.nz/handle/10292/824

Saliya, C. A. (2017). Doing Qualitative Case Study Research in Business Management. *International Journal of Case studies.* 6(12), 96–111. http://www.casestudiesjournal.com/Volume%206%20Issue%2012%20Paper%2010.pdf

Saliya, C.A. (2021a). Conducting case study research: Practical guidance formanagement students. *International Journal of KIU, 2*(1), 1–13. https://ij.kiu.ac.lk/article/read/9.

Snyder, J. (2015). Exploitation and Sweatshop Labour: Perspectives and Issues. *Society for Business Ethics, 20*(2), 187–213. https://doi.org/10.5840/beq201020215

Youtube Link

Constructivism—Research Paradigm: https://www.youtube.com/watch?v=EDEXXvpbOIM

Critical paradigm: https://www.youtube.com/watch?v=Xx9JM1gcc3E

Interpretivism: https://www.youtube.com/watch?v=FybkUMplAlI

Ontology, epistemology, and axiology (2021): https://www.youtube.com/watch?v=AhdZOsBps5o

Paradigms: https://www.youtube.com/watch?v=URWcOJWfSnI

Positivism: https://www.youtube.com/watch?v=8QkVqT3EPyk

What is Interpretivism: https://www.youtube.com/watch?v=FybkUMplAlI

Chapter 3
Qualitative Versus Quantitative

Experts who use qualitative methods allege that qualitative research methods have been criticized, rejected, or ignored by scholars who prefer quantitative methods for various reasons. The following are some of those criticisms.

- It is similar to 'soft science' or journalism.
- It was simply 'humanism' in disguise.
- It is 'non-scientific' and subjective/personal.
- It breaks down the notion of 'non-attachment' in scientific research.
- It cannot make statements that can be proven.
- It cannot make statistically generalizable findings.
- It is free from strict rules, not rigorous.

Scholars who advocate qualitative methods have sought to refute these criticisms as described below;

- Labels such as 'soft', 'weak' and 'humanitarian' are arbitrary misclassifications that serve no effective purpose to the researcher or critic.
- The accusation of subjectivity/personalized indicates that their *presumption* is that the world is a notion of a reality that is completely conceivable and independent of the observer, instead.
- The qualitative researcher investigates a notion of a reality which is complex, multifaceted, with intangible relationships, meanings and insights that cannot exist independent of participants and researchers.

However, the main criticism of quantitative positivists is that qualitative results or conclusions are emotional and cannot be verified. Post-positivists, who deny this allegation, argue:

- Qualitative data is auditable and therefore dependable.
- The opposition argues that quantitative research is based on more rigorous, reliable data and results, and that the argument that quality research is truly scientific is not.

C. A. Saliya, *Doing Social Research and Publishing Results*,
https://doi.org/10.1007/978-981-19-3780-4_3

The qualitative opposition argues that the claim of quantitative research is based on more rigorous, reliable data and results are truly scientific, and that the qualitative research would not is baseless.

- They argue that qualitative research can also be truly scientific, and that quantitative research, on the other hand, uses quantitative data that can be too weak and cannot be guaranteed the adequacy of sufficient evidence.

On verifiability and statistically generalizability, qualitative scholars argue that:

- In the sense of trying to identify, penetrate, understand and express relationships, qualitative strategies such as contextual processing and theoretical depth, data richness (thick descriptions) are second to none in methods such as estimation/computation in quantitative approach.
- The critique of quantitative schools, such as 'lack of rigor' reflects the critics lack of knowledge of the basic methods used in the various strategic traditions of qualitative research.
- More structural qualitative methods also have their own methods and tools such as 'validity', 'dependability' and 'triangulation' second to none to compared to positivist methods.
- Structurally soft, qualitative methods reject the rigid rules, calculations and estimates of positivists, and instead rely on flexibility, creativity, and other inaccessible insights (alternative approaches) and broad and in-depth explanations such as dialogues.

The more structured quantitative methodologies have their own equivalents to positivist method concerns such as validity, dependability and triangulation. the less structured qualitative methodologies reject many of the positivists' constructions of what constitutes rigour, favouring instead the flexibility, creativity and otherwise inaccessible insights afforded by alternative routes of inquiry that embrace storytelling, recollection, and dialogue (Parker, 2003).

3.1 Quantitative or Qualitative?

Quantitative or qualitative? Is a question that is often asked by beginners. Some scholars point to it as an attempt to put the "strategy cart" before the "content horse" (Gournelos et al., 2019). Before focusing on tools and techniques for data collection, it is important to focus on a wide range of topics, such as research issues, research design, and the philosophical ideas behind them. This book does not discuss the selection of strategies but gives an overview of the most used strategies. Once you chose a strategy, you need to study that strategy in depth to work with it. Selected you tube links are provided wherever necessary and 'Further readings' are also provided at the end of each chapter for that purpose.

The differences between quantitative and qualitative research

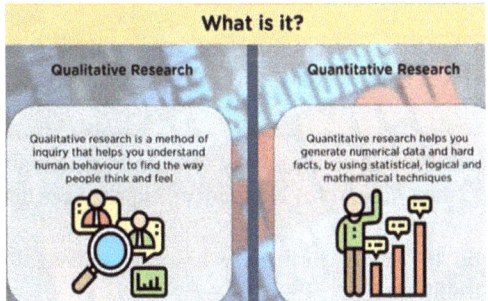

Sources https://www.scribbr.com/methodology/qualitative-quantitative-research/ AND https://www.simplilearn.com/tutorials/data-analytics-tutorial/qualitative-vs-quantitative-research

3.2   From Positivism to Post-positivism

O'Leary says that "'quantitative' and 'qualitative', however, have come to represent a whole set of assumptions that dichotomize the world of methods and limits the potential of researchers to build their methodological designs from their questions" (O'Leary, 2004, p. 99). Assumptions of quantitative and qualitative traditions are summarised in Table 3.1.

3.3   Summary

• The researcher's philosophical position and strategic approach are very important in presenting cogent arguments and convincing conclusions.

Table 3.1 Assumptions related to quantitative and qualitative methods

Tradition factors	Quantitative	Qualitative
Paradigms	Positivism	Post-positivism
Theoretical beliefs	Empirical evidence, world is knowable and predictable, single-reality	Intuitive, holistic interpretations/constructions, world is changeable/ambiguous, multiple-reality
Terminology	Scientific, deduction, hypothesis-driven, reliable, generalisability, objectivist, significant	Credible, induction, abduction, data-driven, exploratory, dependable and auditable, subjectivist, useful/valuable
Strategies	Statistical inferencing and estimations, experiments	Ethnography, biography, phenomenology, case study, grounded theory
Data	Survey based, large-scale, numbers	Observations, interviews, small-scale, words and symbols
Analysis	Statistical techniques	Thematic/narrative

Sources Bailey (2007), Denzin and Lincoln (2005), O'Leary (2004), Creswell (2007), Saliya (2017)

- Research topic should be selected carefully and consciously considering the availability of resources, accessibility to data, researcher's expertise, experiences and skills, and ethical considerations.
- The research methodology can no longer be limited to a set of universally applicable laws, conventions and traditions.
- Realist fundamentalists also evaluate or formulate theories and make predictions. The purpose of interpreters is to understand human behaviour.
- Critical theorists aim to critique social reality, discuss liberation, and social issues.
- Beyond the empirical evidence built on the five senses, the mind also plays a role in the creation of knowledge. In fact, it is important to examine in depth whether observation or experience can be obtained only by the five senses without the sense of the mind.
- The basis of critical realism is that 'reality' cannot exist outside of human cognitions, theories and creations. Interpreting the world through cause-and-effect mechanisms contradicts the assumptions of empiricism and anti-fundamentalism.
- Induction is the creation of theories by accumulation of observations and events. Empirical research on deduction tests theories. Qualitative inquiries made to suggest hypotheses are called abductions.
- Deductive methods attempt to prove something logically; induction is proving something empirically; Abduction makes many possible inferences.
- Qualitative methods also have their own methods second to none. Instead of the rigorous calculations of the fundamentalists, quality methods involve in-depth narratives such as flexible storytelling and dialogue.

References/Further Reading

Bailey, C. A. (2007). *A Guide to qualitative field research.* Thousand Oaks, CA, Sage.
Creswell, J. W. (2007). *Qualitative inquiry and research design; Choosing among five traditions.* Thousand Oaks, CA, Sage.
Czarniawska, B. (2004). *Narratives in finance science research: Introducing qualitative methods.* London. Sage. https://doi.org/10.4135/9781849209502.n5
Denzin, N. K., & Lincoln, Y. S. (2005). *The landscape of qualitative Research: Theories and issues.* Sage.
Gournelos, T., Hammonds, J. R., & Wilson, M. A. (2019). *Doing Academic Research: A Practical Guide to Research Methods and Analysis.* London: Sage.
Kuhn, T. S. (2012). *The structure of scientific revolutions,* Chicago, IL: University of Chicago Press.
Laughlin, R. (1995). Empirical research in accounting: Alternative approaches and a case for "middle-range" thinking. *Accounting, Auditing & Accountability Journal, 8*(1), 63–87. https://doi.org/10.1108/09513579510146707
Ontology Vs Epistemology: https://study.com/academy/lesson/ontology-vs-epistemology-differ ences-examples.html
O'Leary, Z. (2005). *Researching real-world problems.* Sage.
O'Leary, Z. (2004). *The essential guide to doing research.* London: Sage.
Parker, L. D. (2003). Qualitative Research in Accounting and Management: the Emerging Agenda. *Journal of Accounting and Finance 2,* 15–30.
Saliya, C. A. (2017). Doing qualitative case study research in business management. *International Journal of Case studies. 6* (12), 96–111. http://www.casestudiesjournal.com/Volume%206%20I ssue%2012%20Paper%2010.pdf.
Snyder, J. (2015). Exploitation and sweatshop labour: Perspectives and issues. *Society for Business Ethics, 20*(2), 187–213. https://doi.org/10.5840/beq201020215

Chapter 4
Research Types and Approaches

There are many ways to classify different types of academic research. The words you use to describe your research depend on your goals, approach, intended methods, strategies, and your field of research. In general, your approaches are shaped by considering different aspects, including the following:

- Type of knowledge you intend to present: testing the validity of existing theories, creating new knowledge, expanding understanding, and participating in policy making/social advancement, etc.
- The nature of the data you collect and analyse: quantitative numbers and qualitative descriptions.
- Sample selection methods, time frame, analysis methods, strategy and research location: surveys, case studies, predictive associations, estimates, comparisons, experiments, action research and organizations or associations of different communities.

4.1 Academic Research Types

The following Table 4.1 is an attempt to illustrate the diversity of academic research types. It is an attempt to freeze the strategies broadly.

Brew (2007) classifies academic research into four groups according to their realities and interpretations, as shown in Table 4.2.

The 'domino approach' is considered to be a synthesis of several elements such as tools, experiments and readings to solve a research problem. The 'layer approach' is the researcher's focus on expanding understanding: uncovering the underlying meaning; the 'trade approach' is research that focuses on product exchanges such as grants, publications, benefits and personal recognition. In the 'travel approach', research is seen as a personal journey of discovery that can lead to personal transformation. More than one approach to any research, several approaches may be common, for example a doctoral dissertation by a researcher may be linked to the

© The Author(s), under exclusive license to Springer Nature Singapore Pte Ltd. 2022   39
C. A. Saliya, *Doing Social Research and Publishing Results*,
https://doi.org/10.1007/978-981-19-3780-4_4

Table 4.1 Wide range of academic research types

Aspect		Field				
		Lab/closed studies	Environment based	Social structure	Human studies	Products & processes
Purpose	Discovery	Double slit experiment	Flora & fauna species	Cause & effect relationships		The wheel
	Innovations				Coronary-artery stents	Information technology
	Developments	Quantum physics	Artificial rain	Policy formulations		
	Explanations			Social injustice	Biographies	Cancerous effects
	Predictions	Comets & asteroids	Climate change	Millennium goals		Hazards
Approach	Observations		Direct observations	Ethnography	Surveys/Interviews	Market surveys
	Experiments	Robotics	Satellites		Thought experiments	Electronics and mechanicals
	Implementations	Green houses		Action research		
	Historical		Excavations & longitudinal studies			Effects of chemicals
Technique	Statistical analyses	Descriptive analysis		Regression analysis	Multivariate analysis	Significance Tests
	Logical arguments			Grand & Conspiracy theories		Complementy & substitute goods
	IT based	Software developments, simulations and algorithm etc.				
Reasoning	Deductive	Proving or falsifying		Confirming		Proving or falsifying
	Inductive			Generalising		
	Abductive		Conjecturing and inferencing			

Table 4.2 Four groups academic research

Conception	What is in the foreground	Research is interpreted as
Domino conception	Separate elements, techniques, problems, etc. are viewed as linking together in a linear fashion	A process of synthesising separate elements so that problems are solved, questions are answered or open up
Layer conception	Data containing ideas together with (linked to) hidden meaning	A process of discovering, uncovering or creating underlying meaning
Trading conception	Products, publications, grants and social network. These are linked together in relationships of personal recognition and reward	A kind of social market place where the exchange of products takes place
Journey conception	Personal existential and dilemmas. They are linked through as awareness of the career of the researcher and viewed as having been explored for a long time	A personal journey of discovery, possibly leading to transformation

Source Brew (2007, p. 52)

researcher's profession and may be understood as research with a tour approach or a layered approach. Another example is, according to the research focus, that it can also be seen as a layer approach to uncovering the motives behind an approval of credit application, but it is also a commercial approach when the researcher publishes the results of the research.

4.2   Research Designs

The research design refers to the way that you choose to integrate the different components of the study in a coherent and logical way, thereby, ensuring you will effectively address the research problem; it constitutes the blueprint for the collection, measurement, and analysis of data.

In sociological research, it is usually the research problem that determines the type of evidence needed to test a theory, evaluate a program, or accurately describe a phenomenon. However, they can begin their research well in advance of critical thinking about what information is needed to answer the study's research questions. The researcher does not have to be confined to a rigid framework that specifies a specific starting point where academic research should begin. Thus, the depth and complexity of research plans can vary significantly, but the following procedures are followed in any research methodology. This does not mean that you have to complete one task at a time and do the other. For example, the literature review may be followed by reconsideration of the research problem or vice versa. These tasks should be repeated to select the optimal method.

Different academic scholars have identified different research designs. These designs are basically grouped into two categories for research purposes, namely basic or fundamental design designs and applied research designs. Sacred Heart University (2021) provides very useful list of research designs which can be downloaded from this link: https://library.sacredheart.edu/c.php?g=29803&p=185902.

For the convenience of further study, all these research plans can be further categorized into three main categories according to the strategies and methods used, namely, qualitative, quantitative and mixed methods. The chapters in Part III and IV are devoted to introducing the most commonly used research strategies used today under each category. In order to make good use of those research strategies, it is essential to study the particular strategy in depth and reviewing the relevant literature using the same strategy.

4.3 Ethical Issues

Ethical issues are the ethical principles and a code of conduct that guide research. Ethical principles emphasis the need for an ethical approach that goes beyond conducting research more efficiently or appropriately. Much attention has been paid to this ethical code of conduct due to the research done in Nazi concentration camps in Germany during World War II at the cost of human lives. As a result, the Nuremberg Code 1947, a ten-point charter on human research, was released.

The basic premise here is to ensure that human research does not cause any harm to society in general and to the privacy and security of research participants. Investigations, etc., should be done by scientifically qualified persons and should be aimed at social progress with good intentions. This charter can be found by doing a Google search.

Principles of ethics are basically in four parts. They are (a) that participants in the research should be prevented from being harmed, (b) that they should be properly informed and give their consent, (c) that their privacy should be respected, and (d) that they should not be deceived.

4.3.1 Anonymity and Confidentiality

Anonymity means that even the researcher does not know the participant's details. Confidentiality means that the researcher knows the participant's information but does not disclose it. What often happens in journalism is that confidentiality is maintained, but they are compelled to disclose certain information when required by law, such as in court. The research proposal should explain in detail how the researcher maintains this anonymity and/or confidentiality to get the approval from the ethics committee of the relevant institute for almost all academic research.

4.3.2 *Universalistic and Contingent/Relativistic*

Morally questionable acts (by universalistic standards, e.g., lying, treating some people inhumanely) are justified if they produce 'good' consequences (Gray, 2018).

Sovereignty and utilitarianism are two opposing ideologies. Those following sovereignty (e.g., Kantianism) insist that ethics should never be violated. They strongly believe that to do so is immoral and harmful to society (deontological). But relativists (e.g., utilitarianism) argue that some factors, such as 'falsehood' and 'inhumanity', which are immoral from a universal point of view, can be justified if they are used with good intentions to bring 'better' results.

4.3.3 *Whistle Blowing*

Whistle blowing has to do with ethics because it represents a person's understanding, at a deep level, that an action his or her organization is taking is harmful—that it interferes with people's rights or is unfair or detracts from the common good. Whistle blowing also calls upon the virtues, especially courage, as standing up for principles can be a punishing experience. (Nadler & Schulman, 2015).

Whistle blowing is an alarm bell. In academic research context, whistle blowing is the publication of any sensitive information that the researcher investigates and finds for the benefit of society in general. As an example, if you are involved in a case study of the organization you work for, you need to get ethical approval from the heads of the institutions, but you may not be able to get that approval because of the confidentiality/sensitivity of the information. But if that information is critical to society in general and revealing it helps to prevent bigger loss or damage, disclosing that information would not be against ethical principles or violate the code of conduct, according to the relativists.

4.4 Summary

- The terminology of research depends on the methods, strategies, and field of research intended to be used.
- The research design depends on the type of knowledge intended to be presented; the nature of the data collected and analysed; the sampling methods, time frame, analysis methods, etc.
- Research is grouped into four categories according to their perspectives and interpretations: the 'domino approach' (problem solving, experimentation, etc.); 'layer access' (to expand understanding); 'trade access' (to a product exchange); 'travel access' (which can lead to personal transformation).

- Research design is the whole process of choosing and combining different elements in a compact and logical way to conduct research effectively.
- The researcher should not be confined to a rigid framework in which a specific place or procedure is to be used to initiate academic research.
- Research consists of several functions. This does not mean that you have to complete one task at a time. The optimal method should be selected from continuous repetitions.
- Research strategies can be classified into three main categories: Qualitative, quantitative and mixed methods.
- It is more important to conduct research ethically than efficiently.
- The Nuremberg Code 1947 is a 10-point charter for research was introduced after WW II, because humans had been used for research purposes, thus risking their lives.
- Unethical things like 'falsehood' and 'inhumanity' can be justified if the research is done with 'good' intentions to bring benefits to the larger society.
- The whistle blowing is the publication of sensitive information found by a researcher for the benefit of society in general.

References/Further Reading

Bovaird, J. A., & Kevin A. K. (2010). Sequential design. In N. J. Salkind (Ed.), *Encyclopaedia of research design*. Thousand Oaks, CA: Sage.

Brew, A. (2007). Academic autonomy and research decision-making: The researcher's view. In C. Kayrooz, G. S. Akerlind, & M. Tight (Eds.), *Autonomy in social science research:* The view from United Kingdom and Australian Universities. UK: Oxford.

Gray, E. D. (2018). *Doing research in the real world* (4t ed.). Sage.

Greer, S. (2005). *Encyclopaedia of social measuremen*. Sacred Heart University. https://library.sac redheart.edu/c.php?g=29803&p=185902.

Nadler, J., & Schulman, M. (2015). Whistle Blowing in the Public Sector. Markkula centre for applied ethics program in government ethics. Santa Claire University.

Chapter 5
The 10 Elements of Academic Research (10P$_\mathrm{S}$)

The focus of this book is not only on doing research as course work, but also to organise and conduct the whole research process with a view to publishing important results. For this purpose, we have to deviate from certain traditional procedures. There are many models suggested on this topic of designing and conducting academic research—such as Sander's Research Onion (see Fig. 5.1), the Research Methodology Tree (see Fig. 5.2). The author's own 7Ps model (Saliya, 2021), which was developed by encompassing seven essential activities that are common to most research methods, has been improved by adding three more essential elements to match the purpose of this book to include activities such as Planning (research design), Programming (time frames, co-ordinating and budgeting) and Publishing (Fig. 5.3: The 10Ps of Academic Research).

Guidelines are scarce on how to conduct research in alignment with ontological and epistemological hypotheses. This book describes a holistic approach that encompasses those philosophical elements of the research throughout the process, including publishing results. The process focuses on formulating a philosophical framework, articulating research problems, combining them with paradigms, perspectives, worldviews and theoretical issues, choosing methods, and developing convincing, cogent arguments based on research results. The 10 elements are identified as **P**aradigm, **P**erspective, **P**urpose, **P**revious knowledge (literature review), **P**lanning and **P**rogramming, **P**lot (strategies), **P**rocedures (methods), **P**ersuasion (drawing conclusions) and finally, **P**ublishing.

This 10Ps structure projects an overall picture of the research process in different contexts. This exercise is not a smooth process that moves from one stage to another, or a sequential process like peeling an onion, layer by layer (Fig. 5.1); nor is it a process that starts from roots of ontology and epistemology and spreads like a tree growing upwards for methodology, but it is an interactive, dynamic and flexible operation. The researcher's ontological and epistemological standpoints can be (or should be) changed from research to research—it is not a straitjacket where the researcher is confined to a particular box.

© The Author(s), under exclusive license to Springer Nature Singapore Pte Ltd. 2022   45
C. A. Saliya, *Doing Social Research and Publishing Results*,
https://doi.org/10.1007/978-981-19-3780-4_5

Fig. 5.1 Research onion. *Source* https://gradcoach.com/saunders-research-onion/

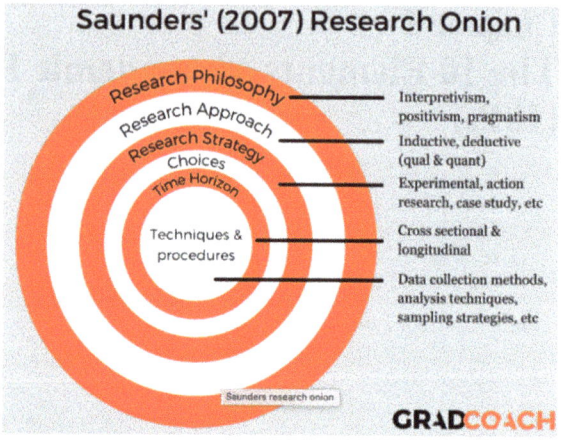

Fig. 5.2 Research methodology tree. *Source* https://www.youtube.com/watch?v=X5GF3GYBOOE

5.1   Is the Research Process Linear and Sequential?

According to Denzin and Lincoln (2005), research methodology involves the *sequencing of the activities*, procedures, and selection and justification of a research project. Traditionally, academic research is a linear, sequential operation which begins with the articulation of the research problem, followed by a review of existing research papers (the literature review), a design process which includes data collection, analysis, interpretation and finally, the drawing of conclusions. But, especially when amateur researchers are doing research, it is not practical to implement this process in such a linear or sequential manner. For example, it may be necessary

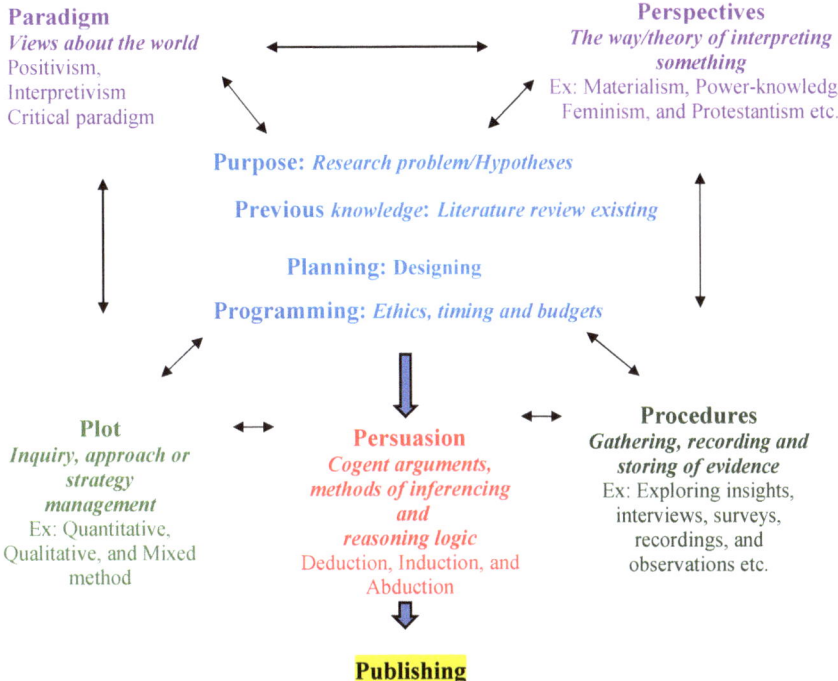

Fig. 5.3   Ten elements of the research process: The 10Ps model

to redesign the research problem (the first step 'already decided') after reviewing the relevant literature which is traditionally considered as the *second* step in this process. Similarly, other steps should also be revisited back and forth as the among research lenses research proceeds, if the results are to be extracted, arguments are to be convincing and conclusions are to be useful.

Therefore, it is more pragmatic to follow a repetitive interaction method between these research activities by revisiting, reconsidering and revising supposedly 'completed' tasks—instead of a sequential chain of tasks performed, one after another. On the other hand, in post-positivist paradigms, it is of great value to select and justify the methodology of a research report in a pragmatic and philosophical way, because the paradigm which the researcher follows may influence and obscure the observations and conclusions. The following graphic, Fig. 5.4, illustrates how the 10Ps model can be presented as lenses in conducting research and how the 30 chapters have been allocated to explain them. It is important to understand that these lenses are not in an unalterable alignment but signify a flexible programme that will, arguably, lead to more successful results.

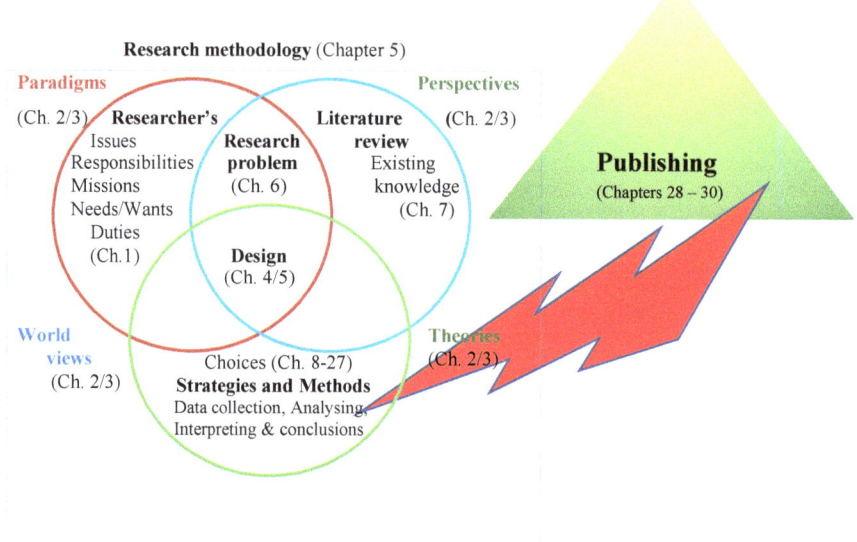

Fig. 5.4 Distribution of 30 chapters

5.2 Evolution and Composition of Research

Many researchers insist that research methods and knowledge are co-evolving (Varga, 2018).

In the early stages of scientific research, idealists suggest that the absolute truth can be discovered irrespective of the observer's standpoints, but more materialist-empiricists and naturalists argue that we have no ideas other than the ideas that come through us. French philosopher, Auguste Comte (1798–1857), who is generally recognised as the founder of both positivism and sociology, presented his thesis based on some key features, such as:

- reality consists in what is available to the senses;
- philosophy is parasitic on the findings of science;
- there is a basic difference between fact and value; and
- science deals with the facts while values belong to an entirely different order of discourse.

Anti-positivists believe that human actions are complex and have multiple meanings and argue that the concept of *variable* used in modern quantitative analysis can only register quantifiable change, not its cause. Therefore, anti-positivists rely on intensive studies of a small number of cases rather than large amounts of survey-based data (Alvesson & Sköldberg, 2009). Therefore, instead of a large amount of data ($n = 1000+$) based on surveys, post-anti-positivists rely on in-depth studies of a few cases. Critical dialectics such as Karl Marx (1818–1883) and non-critical interpreters such as Max Weber (1864–1920) did not believe in a definite set of rules governing social research, instead flexibility in application of research methods and critical thought.

5.3 What is the Starting Point?

Traditionally, identifying a research problem is the starting point of research. Such an initial position is valid for some project-based research (such as seeking funding), but does not appear to be valid for academic or course-related activities. Candidates preparing for a general, postgraduate or doctoral degree should be more careful about what kind of research will help them to successfully complete their dissertation and achieve high marks. On the other hand, a scholar who is aiming for publication should be more concerned about the importance of publishing her paper in an international academic journal. For example, taking into account:

(1) the researcher's passion for the particular field of study;
(2) interesting or burning problems in society;
(3) special skills and experience possessed by the researcher;
(4) access to data; ability to influence, socio-economic power, etc.;
(5) the strength and nature of the relationship with the supervisor;
(6) alignment with the supervisor's field of research;
(7) time required to complete the research work; and
(8) other physical resources such as funds and equipment, etc.

Research conducted by a team of scholars in Netherlands, using regression analyses with a sample of 839 Ph.D. candidates, has revealed that:

> …experienced workload was negatively related to satisfaction and progress and positively to quit intentions. The quality of the supervisor-PhD candidate relationship, the PhD candidate's sense of belonging, the amount of freedom in the project, and working on a project closely related to the supervisor's research were positively related to satisfaction and negatively to quit intentions. The high workload of PhD candidates should be a major point of attention for universities who wish to increase their rates of PhD completion and PhD candidates' satisfaction. In addition, the "match" between PhD candidate and supervisor is crucial, both personally – a good relationship – and academically, i.e. that the PhD candidate works on a topic closely related to the supervisor's research. (van Rooij et al., 2019, p. 1).

Table 5.1 shows a different, logical presentation of the overall 10Ps research process presented in earlier Fig. 5.3.

Table 5.1 Research methodology and the process

Research methodology[a]					Research design[b] (Plan-P8)		
Selection and justification							
Paradigms (P1)	Perspectives (P2)	Strategies, approaches, traditions (Plot-P6)			Activities (P3, P4 & P5)	Methods[c] procedures (P7)	Programming (P9)
		Quantitative statistical techniques based	Qualitative cognitive process based	Mixed methods of Quan and Qual			
Assumptions Epistemology Ontology Axiology	World views Theoretical frameworks				Articulation of problem statements (Prob. P3) Literature review (Past-P4)	Data collection Analysis Presentation	Approvals Collaborations Timeframes Budgets (P5)

Drawing conclusions[d] (Persuation-P7)		
Production of research papers	Reflections and future research	Publishing (P10)

[a] Research Methodology is an arrangement of chosen research design and its justification influenced by researchers' orientation in conducting research. Researchers' orientation is shaped by his/her views and values about existence of reality (ontology), the way of acquiring knowledge (epistemology) and preferred theoretical stances (Marxism, Feminism, Queer theory, etc.) and the design

[b] Research design consists of chosen methods and strategies for data collection, data analysis and framework of doing the research such as the resolving ethical issues, the timeframe and the budget etc

[c] The research methods are the tools and techniques of data collection and analysis that will be discussed in the next section. Research strategies, quantitative, qualitative and mixed methods are discussed in Parts III and IV

[d] The drawing of conclusions using deduction, induction or abduction reasoning

5.4   Summary

- Designing and conducting a research project is a flexible, dynamic process; the essential elements are identified and grouped into 10 elements (10P).
- These 10 elements are: **P**aradigm, **P**erspective, **P**urpose and **P**revious research, **P**lanning & designing and **P**rogramming, **P**lot-strategies, **P**rocedures—methods, **P**ersuasion—drawing and **P**ublishing.
- This dynamic exercise is not a smooth process that moves from one stage to another, nor is it a process of sequential exploration like a peeling an onion or like a tree grounded on rigid roots.
- Research methodology is a bundle of activities, procedures and selection and justification of activities in a research project.
- Experience has shown that continuous interaction leads to more credible outcomes by revisiting, reconsidering and revising activities in order to get final results, convincing cogent arguments and presentation of useful conclusions.
- The research problem has traditionally been accepted as the starting point for research but, more important in academic or course related work, is what kind of approach contributes to successful completion and achievement of a higher grade or improves the possibility of publishing research results.
- Personal (positive) relationship, and academic 'compatibility' (closeness) between the candidate and the supervisor are crucial when choosing a research topic.
- Research methodology is a research design chosen to conduct research that is justified under the influence of researchers' own research philosophy.
- Research design consists of methods and strategies selected such as data collection, data analysis, ethical problem solving, timetabling, and budgeting.
- Methods are the topics covered in research design such as data collection and analysing tools and techniques discussed in Part II.

References/Further Reading

Alvesson, M., & Sköldberg, K. (2009). *Reflexive methodology: New vistas for qualitative research.* Sage.

Denzin, N. K. & Lincoln, Y. S. (2005). Introduction: The discipline and practice of qualitative research. In: N. K. Denzin, Y. S. Lincoln (Eds.), *The landscape of qualitative Research: theories and issues.* London. Sage, 1–6.

Gournelos, T., Hammonds, J. R., & Wilson, M. A. (2019). *Doing academic research: A practical guide to research methods and analysis.* Thousand Oak, Taylor & Francis Group/Routledge.

Gray, E. D. (2018). *Doing research in the real world* (4th ed.). Sage.

van Rooij, E., Fokkens-Bruinsma, M., & Jansen, E. (2019). Factors that influence PhD candidates' success: The importance of PhD project characteristics. *Studies in Continuing Education.* https://doi.org/10.1080/0158037X.2019.1652158.

Saliya, C. A. (2021a). Conducting case study research: Practical guidance for management students. *International Journal of KIU,* 2(1), 1–13. https://ij.kiu.ac.lk/article/read/9.

Varga, L. (2018). Mixed methods research: A method for complex systems. In E. Mitleton-Kelly, A. Paraskevas, & C. Day (Eds.), *Edward Elgar handbook of research methods in complexity science* (pp. 34–39). Edward Elgar Publishing.

Part II
Methods 1: Purpose and Approach Articulation of Research Problem, Literature Review and Data Collection

Chapter 6
Articulation of Research Problem

6.1 Introduction

Often, the researcher's insight and experience direct the researcher towards a problem that needs to be researched (O'Leary, 2005). However, figuring out a problem depends on the researcher's ontological and epistemological stance; therefore, chosen paradigm and theoretical perspective provide the necessary guidance to articulate a research problem (Saliya, 2021a, pp. 4–5). Often, it is the researcher's insight and experience that direct the researcher towards a problem that needs to be researched (O'Leary, 2005). However, figuring out a problem depends on ontological and epistemological stance of the researcher therefore, chosen paradigm and theoretical perspective provide the necessary guidance to articulate a research problem (Saliya, 2021a, pp. 4–5).

The researcher's attitudes, beliefs, and perceptions are influenced and shaped differently by the paradigm, worldviews, perspectives, and theories that the researcher belongs to. Accordingly, the 'burning' problems that a social researcher are different from each other. As a social researcher, the 'pain' of finding solutions to your 'burning' problems makes you more passionate about research. Therefore, it is inappropriate and ineffective at the outset to review existing literature, which is the tradition of the general research methodology, as explained in the quotation above.

The problems we face in the academic world are vast. They are not questions but broad topics or disciplines. Academic research involves narrowing these broad topics as much as possible, exploring them narrowly and deeply, and finding the root causes. Sometimes the exploration stops just by abduction because of data saturation. Nevertheless, paving the way for further research due to that disclosure is a good yardstick for assessing the importance of research results. The question is how to narrow down this broad field, topic or question and focus on a specific goal.

At this juncture, literature can assist with your burning issue. The most innovative approach is to start searching through databases (databases; EBSCO, Science Direct, Pro-Quest etc.), drawing more attention to the recent publications.

C. A. Saliya, *Doing Social Research and Publishing Results*,
https://doi.org/10.1007/978-981-19-3780-4_6

Research methods, statistical techniques, and grand theories such as Marxism, Bourdieu's concepts of capital and Foucault's concept of Power Knowledge etc., are generally accepted for use in research, even though they are very old. However, reviewing research papers older than ten years may not be very useful if it is not about a methodology, theory, statistical technique or method. This does not mean not referring to pioneering, seminal and pioneering works or original sources.

Another thing to be careful of is that the paper you are reviewing, whether a book or a recently published source, may be based on older sources. Therefore, it is important to check the list of references or bibliography provided at the end of the article you review and be satisfied that recent publications have been used to construct that research paper.

O'Leary (2014) suggests that finding an 'angle' on the topic would help move from the general to the specific. For example, taking a topic but looking at what contemporary commentators are saying about it, or what 'hot issues' generate argument and debate. Another apparent angle is identifying a gap in the literature where a theme has been ignored or where a researcher has recommended other themes worthy of research at the end of an article.

The selection of a tentative research topic will guide future research and will be helpful in the tasks such as choosing appropriate theories, relevant literature and necessary research methods relevant to the study. However, for deductive approaches, a research methodology cannot be chosen without a research question that is clearly stated at the outset.

6.2 Articulation of Research Problem

The next stage is to write a research question that is concise and unambiguous. This, however, is easier than done. It is astonishing how difficult most students find the formulation of research questions. Even when they have identified a focus for their research, locating one or more researchable questions seems to pose a Herculean task (Gray, 2018).

As previously discussed, research approaches vary. They can be either in-depth exploratory, descriptive, explanatory or interpretive. The nature of the problems that build up also varies according to the research approaches. Consider the following problem.

Is mobile phone abuse at work a problem?

This issue needs to be redesigned and clarified to a considerable extent. What is meant by 'abuse'? Who cares? To an individual, to a company, to both parties? Also, not choosing 'research questions' that you may have an emotional connection with will help you maintain your neutrality and demonstrate your independence.

On the question of phone abuse, a researcher who grew up in a family with parents addicted to phones may find it difficult to formulate a neutral, unbiased question. The

Table 6.1 Examples of research problems and explanations

Research question	Approach	Fucus
What changes in workplace Mobile phone use have taken place over the last ten years?	Descriptive	The research could explore changes in Mobile phone usage levels and/or changing levels across different business sectors or professional groups
Do high levels of Phone use lead to low productivity at work?	Explanatory	Seeks to explore a relationship between two variables: Phone usage time and Output. It is relatively easy to convert this question into a hypothesis
Why is Mobile phone usage at work on the increase?	Explanatory	A question that seeks to identify the factors behind a phenomenon
What is the scale and cause of Phone abuse amongst older employees?	Exploratory	A question that seeks to identify themes when little is currently known about the subject
What is the impact of rising Phone usage on workplace performance/accidents?	Interpretive	Seeks to uncover people's views and perspectives—a valid question for exploratory, largely qualitative studies

Source Adopted from Gray (2018)

complexity of data collection, the time it can take, the ability to reach respondents, and ethical issues must be considered. Table 6.1 presents some examples of solving this Mobile use problem and turning it into a research problem.

It is worth noting that not all researchers would accept that formulating research questions is actually necessary. Some qualitative researchers argue that their approach is so inductive and emergent that the application of research questions is superfluous and inappropriate. However, most qualitative researchers are prepared to formulate research questions, even if they regard these as tentative and subject to change during the research process (Gray, 2018).

It should be noted that not all researchers agree that formulating a research problem is essential at the beginning of a research. Some qualitative researchers argue that building research problems at the outset is unnecessary and inappropriate, as their approach is strongly motivated and arises from the research itself. However, many qualitative researchers consider research questions casual/tentative, even though they are considered temporary and subject to change in the research process.

The emergence of a researcher's research problems out of curiosity intensifies the researcher's enthusiasm. Such a choice minimizes the barriers such as fatigue, frustration, or stress that often occur in long-term research programs. That curiosity, if it can be reconciled with the researcher's intentions, makes him more eager to work tirelessly. Research work can be easily facilitated by creating research problems consistent with the resources available to the researcher (subject knowledge, expertise, experience, power relations, social networks and opportunities). This process is illustrated by Saliya (2021a) in the following example. This illustration is based on a real-world experience encountered in a workplace by a researcher.

Example 6.1 Conducting Case Study Research: A Concise Practical Guidance for Management Students (Saliya, 2021a)

Box 6.1 Articulating a research problem

Curiosity: Poverty,   high-rate   unemployment/underemployment, income/wealth inequality in an emerging economy

Aspiration: Fair and just income/wealth distribution system.
This curiosity and aspiration bear a subjectivist ontological and value-mediated epistemological stance stemming from a historically crystallised situation; hence dialectical type methodology is preferred. Therefore, the suitable paradigm would be, critical paradigm.

Critical paradigm: Social justice and emancipation.
Capacity of the inquirer in gathering and analysing data depends largely on the expertise and exposure to the research field, for example;

Expertise: Accounting, finance and strategic management.

Experience: Auditing, accounting, management, and banking (Corporate planning, treasury and credit management).
Meanwhile, a review of the literature reveals that;
"SMEs make diverse contributions to economic and social well-being, which could be further enhanced SMEs play a key role in national economies around the world, generating employment and value-added…provide the primary source of employment, accounting for about 70% of jobs on average…" (OECD, 2017, p. 6).

The Topic: Small and medium-sized enterprises (SMEs) and employment.
According to the researchers' expertise, experience and accessibility to data;

The context: Financial capital; loans and advances to SMEs by commercial banks in the country.

6.3 Articulation of Research Problems for Qualitative Studies

When research is mainly exploratory or qualitative, it may be sufficient to create a problem rather than to build a hypothesis. Of course, hypotheses are not suitable for qualitative research. For qualitative research, let us consider step by step the articulation of a problem statement. The vital issues addressed in this process are;

curiosity, the aspirations and/or goals, appropriate research paradigms/perspectives, researchers' expertise and experience, data accessibility, the topic, the context, the issues/questions, the motivation and potential relationship that could be explored. Articulating a reasonably acceptable research problem could stem from curiosity such as 'a burning issue', 'a mystery' or 'nice to know passion' etc.; however, the lucidity of the problem and objectives largely depends on the process; an interactive exercise through initial investigations, preliminary review of literature, matching with underpinning philosophical stances, and with continuous cross-reference to the vital issues (Saliya, 2021a).

Important points to focus on this process are the researcher's paradigms, perspectives, etc., capability (to persuade participants, etc.), power relationships, network connections, topic, context, subjective bonds, and interest in exploring. Expressing a reasonably acceptable research problem will arise out of curiosity, such as a 'burning question', a 'mystery' or a 'desire to know'.

However, the clarity of the problem and the objectives largely depend on the process. That is, an ongoing interactive exercise between preliminary search, literature review, adapting to philosophical standpoints, and focusing on the nature of the problem.

This process is described in Box 6.1. Box 6.2 explains how to guide that problem-building process into a more clear, unambiguous proper researchable problem. Although this example was built for case study research, this problem can be used for any qualitative study with minor modifications.

Example 6.2 Conducting Case Study Research: A Concise Practical Guidance for Management Students (Saliya, 2021a)

Box 6.2 Articulating a Case study (qualitative) research problem

In this example, the preliminary investigations and a large amount of literature revealed that there is a disadvantaged group of entrepreneurs who are deprived of credit capital;
"Access to the appropriate finance is one of the most crucial resources for business survival, development and growth. The literature on this is expansive and suggests that disadvantaged entrepreneurs may experience specific challenges to gaining external finance for a variety of reasons…limited know-how and network connectivity—this is compounded by relatively low levels of finance capital—… disadvantaged groups may be less likely to have a track record of running a business…" (Blackburn & Smallbone, 2014, p. 7).

The questions: How and why SMEs are financed. Are certain credit applicants treated favourably while some other applicants are discriminated against?
The potential associations are identified in two aspects: (i) Social power and favourable credit decisions, and (ii) poor-powerless and denial of credit.

The motivation: Analyse and document the factors driving discriminatory credit decisions made by credit officers in a lending institution.
Therefore, the final research problem is arrived at as follow;
Is social power and credit approval a mutually reinforcing function while poor-powerless and denial of credit creates a vicious cycle in society?

In other words, the purpose of the above case study research is to explore the role of credit capital concerning power relations and access to business loans. In this study, the researcher seeks to explain how certain credit decisions are made, their impact on society/country, and whether such credit decisions create specific vicious cycles.

The form of qualitative questions involves a central question followed by sub-questions. The central question is the most general question that could be asked about a phenomenon. It typically begins with the words how, why or what. It also focuses on a central phenomenon or idea that the researcher wishes to explore (e.g. What does it mean to wait for a kidney transplant?).

When phrasing the qualitative question, the researcher also uses action-oriented exploratory verbs, such as investigate, understand, explain, or describe. These questions often change during data collection as the researcher learns how to collect data in the field best. Using a specific type of qualitative design may influence the wording of the question as well. A grounded theory question might be, "What theory explains why migrants feel isolated?" Whereas a narrative research question might be "What stories of migrants in New Zealand have?".

Therefore, when formulating qualitative questions, research-based exploratory verbs such as 'to discover', 'to understand', 'to describe' or 'to report' are often used as research objectives. For example, when a research methodology such as grounded theory research is used in research: 'What is the theory that explains why migrants in New Zealand feel isolated?' What kind of stories do migrants have to tell about their migration? This is presented as a further small question as shown in Box 6.3: Research questions below as a guide to review the academic knowledge in the above example research paper.

Box 6.3 Research questions

The following questions could also be raised as guidance for the literature review:

Are credit decisions made in favour of influential businesspeople?
Are certain demographic groups at a disadvantage in obtaining credit?
As a result of favourable credit decisions, could influential groups get more affluent and more influential?

Are "ability to obtain credit" and "becoming more influential" mutually reinforcing?

Such multifaceted research questions can provide helpful insight into where and when to look for relevant literature and evidence. Such research issues are instrumental when the researcher's approach is critical and structural changes are expected for a more equitable financial capital mobility system as the ultimate goal. Therefore, such research questions can provide a solid basis for theorizing research findings more effectively and meaningfully. How this research belongs to the critical paradigm is further explained in Box 6.4.

Box 6.4 Discriminatory credit and critical paradigm

This particular example research problem belongs to the critical paradigm because it focuses on critique and transformation and the issues addressed are on social power relations and inequality. Therefore, it can guide the researcher to aim at documenting, understanding and even suggesting changing the negative implications of unequal power relationships and promoting justice. The questions asked in this example research are 'why do certain bank lending processes appear discriminatory?' and 'what methods are used by the decision makers to make preferential or discriminatory credit decisions?'; therefore, the answers (and potential relationships such as 'reinforcing?' and 'co-integrated?') could be inferred from the views of research participants' experience and values.

The Fig. 6.1 shows a wholistic picture of the research problem and potential relationships with their interconnectivity to the variables identified through the literature review. This approach would help the researcher to review further literature, identify appropriate research fields, data gathering methods

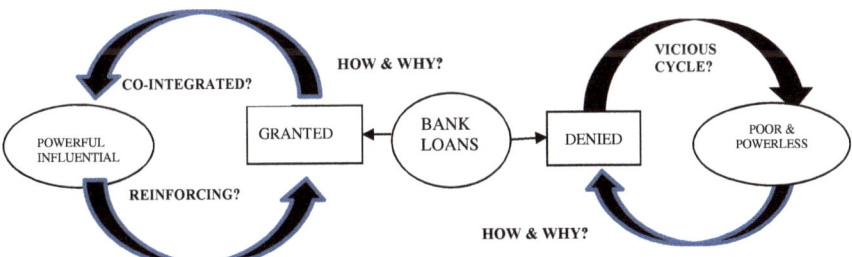

Fig. 6.1 Illustrated integrated research questions and proposition. *Source* Saliya (2021a)

and focus on data description, analysis and interpretation (D-A-I formula of Wolcott, 2008) in developing cogent arguments and plausible conclusions.

This research problem can be illustrated by presenting a diagram as illustrated in Fig. 6.1.

6.4   Hypothesis Testing

The hypothetical investigation can be thought of as a court case. In a court case, a jury must decide whether a person is guilty or not based on the evidence presented to them. Watch the Youtube video below.

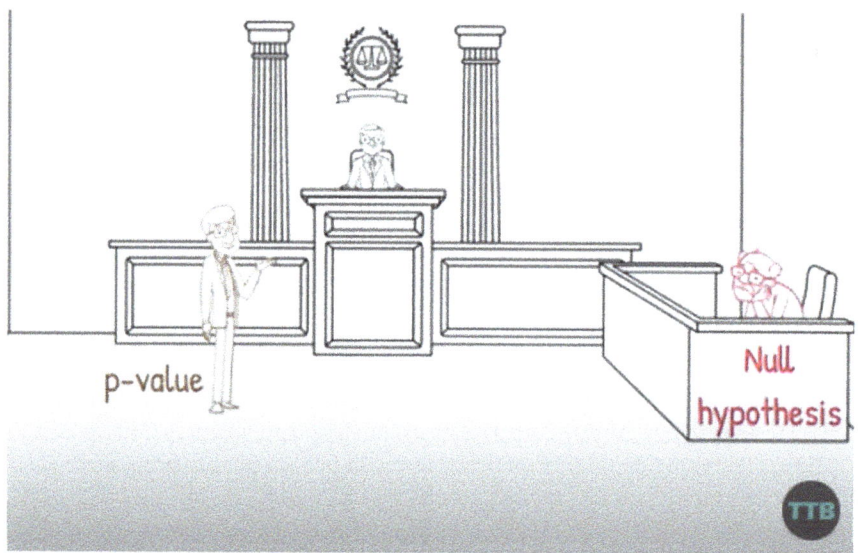

Source What Is A P-Value?: https://www.youtube.com/watch?v=ukcFrzt6cHk

H_0: Null hypothesis: The hypothesis that the researcher expects to REJECT

This hypothesis assumes that the sample data do not provide evidence of an abnormality. That is, the sample represents the normal condition; That the difference is zero ($\bar{x} - \mu = 0$); That there is no effect; That the evidence is not statistically significant. The researcher hopes to reject the null. Therefore, the researcher expects a low p (Sig.) value from the statistical test. Statistics are less than the critical values.

In a court case, this hypothesis can be formulated as follows.

H_0: Null hypothesis: Defendant is innocent

An alternative hypothesis, **H1: The accused is guilty**.

The defendant is presumed innocent until proven guilty with credible evidence. The notion of 'innocence' that is expected to be refuted can be refuted only if it is accepted by the jury with sufficient evidence to substantiate the expected hypothesis (the charges, in this court case). If a court case is a hypothetical investigation, the jury will consider the likelihood that the accused would be guilty based on the evidence presented. This will be discussed in the next section with further examples.

6.5 Construction of Quantitative Research Hypothesis

A quantitative research model describes how a hypothesis (research problem) will be tested and what predictions it can make. A hypothesis can also be presented as a directional statement, such as 'null hypothesis that there is no difference between X and Y' or 'it happens when it happens' (positive) or 'it does not happen when it happens' (negative).

Punch (2006) suggests that in determining whether a hypothesis is appropriate, a researcher needs to reflect on the following two questions, and if the answer to both questions is 'YES', then hypotheses can be formulated, otherwise to work with a set of research questions:

For each research question, is it possible to predict what findings will emerge in advance of the research?

Is the basis for this prediction a rationale, a set of propositions, or a 'theory' that explains the hypothesis?

Otherwise, you will have to work with a series of research questions rather than hypothetical tests.

Hypotheses are used when existing knowledge and theory make predictions about a relationship between variables. Hypotheses are predictions of outcomes based on the literature or theories. They can be stated in a null form ("there is no significance between…") or in a directional form, "Higher motivation leads to higher achievement". Hypotheses are a formal way of writing questions, and they are typically found in the experimental research components of a mixed-methods study. An alternative to constructing hypotheses would be to state research questions; 'Is higher motivation related to higher achievement?'.

Further, hypothetical testing is appropriate if existing knowledge and theories facilitate predicting a relationship between variables. Hypotheses are predictions of results based on currently published knowledge or theories. They can be expressed in the form of ambiguity (H0: no significant significance among variables).

6.6 Basic Guidelines for Building Quantitative Hypotheses or Questions

There are some fundamental guidelines in writing quantitative hypotheses or questions.

- First, you need to identify your variables, typically the major independent variables that influence your dependent variables or outcomes in a study.
- Second, the most rigorous quantitative studies base their hypotheses or questions on a theory that explains or predicts the relationship between the independent and dependent variables.
- Third, researchers must select either hypotheses or research questions; typically, both are not used in a single mixed methods study.
- Fourth, be clear about the variables and their intent. The two most important variables are the independent and dependent variables-probably indicating cause and effect. Following these are other variables such as mediating variables (those that stand between the independent variables and the dependent variables as a means of influence), moderating variables (which combine with the independent variables to influence the outcome, e.g. age x motivation influences achievement); and covariates that are controlled in a study for their impact, such as demographics like social-economic status, years of education, gender, and so forth.

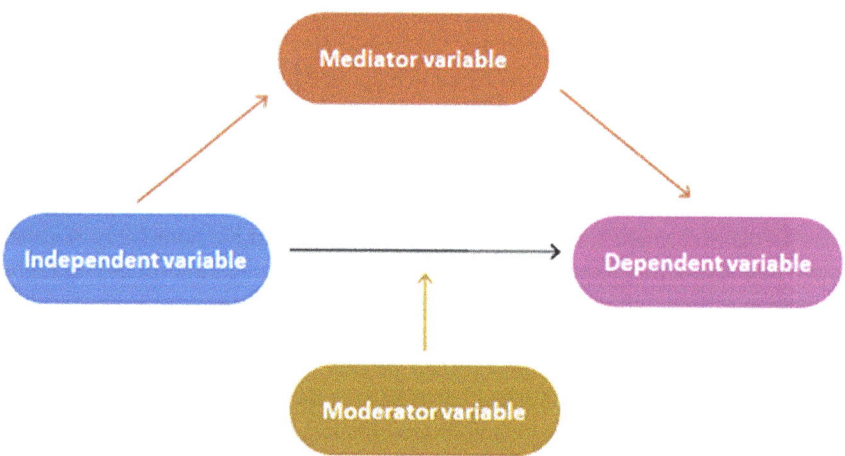

Mediator and moderator variables

- Fifth, to assist readers, it is helpful to make the word order of variables—from independent to dependent—consistent in each research question or hypothesis.

Here is an example of parallel word order:

Does home resident location influence the choice of a medical clinic?
Does input from family members influence the choice of a medical clinic? (Gray,
2018).

Example 6.3 Role of Enterprising personality in motivating stock investors: Case
of Fiji, (Saliya, 2022a).

Box 6.5 Hypothesis 1

Enterprising, as a personality trait, is defined as good at thinking of and doing
new and difficult things, especially things that will make money (Cambridge
2020). Using Holland's (1997) SOS enterprising (E) scale, Zulaifah (2005)
showed that ENP is significantly related to uncertainty tolerance. In this light,
and because studies in behavioural finance have shown that personal resources
have an effect on investment decisions (Durand et al., 2008), we contend that
enterprising personality (ENP) is a broad construct of positive psychosocial
feelings with various dimensions motivating individuals towards stock market
participation (SMP).

Recently, Martinez-Loredo et al. (2018) have introduced a multifactor implicit-
measure model to assess ENP dimensions enfolding eight traits (with associated stimuli words); achievement motivation (persistent), autonomy (initiative), innovativeness (creative), self-efficacy (competent), locus of control
(responsible), optimism (positive), stress tolerance (stable, calm) and risk-
taking (courageous, daring).

In addition, Vasile (2018) reveals that persons who have an ENP make active
decisions and were actively involved in taking control of their lives more
than others. Thus, consistent with the behavioral financial models previously adopted, we hypothesize that: **Within our sample of potential stock
investors (employed university graduates), ENP will positively influence
their inclination towards SMP** (H_1).

Box 6.6 Hypothesis 2

Intuition means an accumulation of attitudes triggering inclinations to believe
(Wilder 1967; Earlenbaugh & Molyneux, 2009). Intuitive thinking is defined
as automatic, fast, effortless, unconscious, and based on vast amounts of prior
experience (Hogarth 2001), and demonstrates an integration of information and
feelings in an cumulative manner (Hogarth, 2001; Glöckner & Betsch, 2008).

Intuitive processes have little or no information-processing costs (Hogarth & Karelaia, 2007) and empower individuals to justify their behavior quickly and rationally (Glöckner & Betsch, 2008; Saliya, 2019a).

Finally, intuition is typically contrasted with deliberation which describes slow, effortful, stepwise, and mostly rule-governed processes (Horstmann et al., 2009). According to Hunjra et al. (2016), the main determinants of choice of investment are propensity of risk, framing of problem, asymmetry of information, and perception of risk. Connecting intuition to common sense, Yurttadur and Ozcelik (2019) reveal that the investment behavior of people is directed by common sense but with an overconfidence tendency. Based on these findings, we anticipate that intuition, as captured by the instinctive feelings of respondents, will be positively associated with Inclination to invest (IIN) directly and/or indirectly affecting SMP.

As Martínez-Loredo et al. (2018) point out, because the trait of 'risk-taking' contributes to the development of ENP, we hypothesize that: **The individual's intuitive characteristics (which are loaded with risk-related attitudes) influence ENP** (H_2).

Thus the hypothetical construction from the literature review must be presented logically. Hypotheses 3, 4 and 5 are presented in the same way as can be seen from the Hypothesis 1 and 2. As another example, the 'Age', which acts as a moderating variable, is presented below.

Box 6.7 Hypothesis 6

It is argued that elderly individuals typically show more maturity with longer social experiences and are cautious about their savings and investments. Previous studies have shown that associations between psychosocial resources and IIN differ across age (Beatty et al. 2012). Further, there is evidence that self-esteem is positively associated with age (Vasile 2018). As a result, the age of an individual may influence his/her self-evaluations (Rosenberg, 1979) and could directly influence IIN. Yurttadur and Ozcelik (2019) show that an overconfidence-tendency is observed in middle-age. Thus, we hypothesize that: **Individuals' age will have a positive impact on IIN** (H_6).

It is also important to summarize and present these 6 hypotheses together as follows.

> **Box 6.8 Summery of the Hypotheses**
>
> In sum, we tested whether ENP plays a mediating role in relation to the association between three individual characteristics (Intuition, Knowledge and education, and Sociocultural norms) and IIN, and the Age factor directly towards IIN for SMP as a causal structural equation model (SEM). So we endeavored to analyse and estimate how individuals are motivated to invest in the SPX. As depicted in Fig. 6.1, the specific hypotheses are as follows:
>
> H1. Intuition (H1a), Knowledge and education (H1b), and sociocultural norms (H1c) influence enterprise personality (ENP) which influence SMP.
>
> H2. Maturity (Age) positively influences investment inclination.
>
> H3. ENP mediates the relationship between Intuition, Knowledge, Sociocultural norms and investment inclination (IIN).
>
> H4. IIN positively influences stock market participation (SMP).
>
> Figure 6.2 presents the theoretical framework for the study. In this framework ENP, as the key mediating variable, links individual psychosocial and cultural characteristics, namely, Intuition, Knowledge and education, and Socio-cultural norms to inclination to invest (IIN). In addition, maturity (Age) is hypothesized to directly influence IIN as a moderating variable between ENP and SMP through IIN. We will discuss each of these constructs in the paragraphs that follow.

6.7   Articulation of Research Problems for Mixed Methods Approach

It is important to acknowledge that in published journal articles reporting mixed methods research, purpose statements (or study aims) and research questions are typically not both reported. More often than not, only purpose statements are reported. The role of research questions or hypotheses is to narrow down the purpose statement to questions or statements that will be specifically addressed in a project (Creswell, 2015, p. 69).

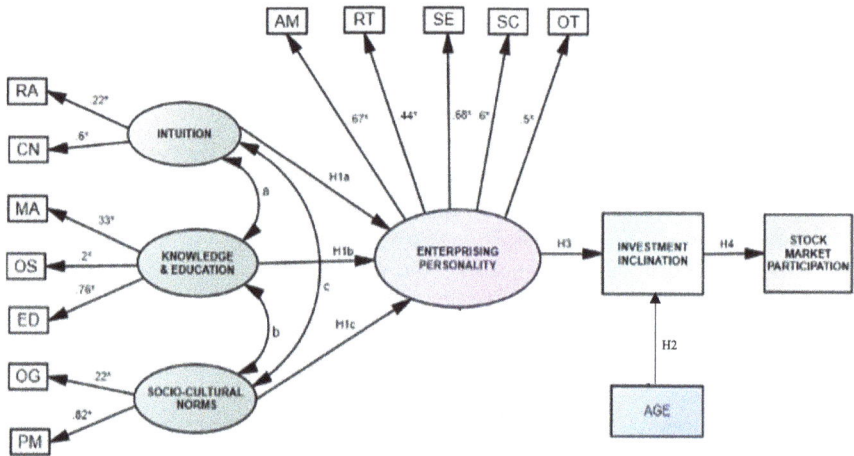

Fig. 6.2 The theoretical framework and hypothesized pathway analysis. *Source* Saliya (2022a)

Formulating quantitative and qualitative questions leads to the question of the mixed method—a question that is new to most researchers and not found to date in research methods textbooks. Creswell (2015), with his colleagues, has developed this question because a question was being asked in mixed methods that were beyond the quantitative or qualitative questions. That "something beyond" can be represented by the intent of the mixed methods design.

It is helpful to have three questions in a mixed-method approach: quantitative questions or hypotheses, qualitative questions, and mixed-method questions. Quantitative and qualitative questions lead to a mixed-method problem. This is new to many researchers, and research methods have not yet been presented in textbooks.

What additional information do you expect from having a plan that combines both quantitative and qualitative research results? They say that 'something beyond that' should be represented by the intention of designing mixed methods. Knowing the mixed method plan makes it possible to think of research questions intended to answer the plan. The following list describes the most commonly used mixed method questions related to each design (basic and more advanced).

Convergent: *To what extent do the qualitative results confirm the quantitative results?*

Explanatory: *How do the qualitative data explain the quantitative results?*

Exploratory: *To what extent do the qualitative findings generalize to a specified population?*

Intervention: *How do the qualitative findings enhance the interpretation of the experimental outcome?*

Social Justice: *How do the qualitative findings enhance understanding of the quantitative results and lead to identifications of inequalities?*

Multistage: *A combination of the previous questions for the different phases in the project to address the overall research g. These mixed methodological questions seem to have been expressed with a focus on data analysis, both quantitative and qualitative. In other words, these mixed methods can be written from a question orientation.*

Finally, probably the best possible mixed methods question is featured in both the methods and content. This is called a "hybrid" mixed methods question, and again, it needs to reflect the type of design being used.

Alternatively, they can be expressed from a content-centric perspective. Consider the following problem as an example:

How do the ideas of boys support their perspective on self-esteem in school?

In this example, the term '*ideas*' symbolizes the study's qualitative part, and the term '*perspective on self-esteem*' symbolizes the quantitative part. Therefore, the mixed method question can be articulated as follows:

What results emerge from comparing the exploratory qualitative data about self-esteem of boys with outcome quantitative instrument data measured on a self-esteem instrument?

In this example, we can easily determine the types of data to be collected (qualitative and measurement data) and focus on the content results of the study. That is, comparing 'self-esteem' measured (quantitatively) by the scale and extracted (qualitatively) by interviews.

Example 6.4 aims to quantitatively measure the effect of macroeconomic and institutional variables on the stock market's performance using the statistical methods of reflection and use qualitative strategies to illustrate those results further.

Example 6.4 Stock Market Development and Nexus of Market Liquidity (Saliya, 2020a)

Box 6.9 Articulating mixed-method problem statement 1

The purpose of this paper is two-fold: First, to explore the influencing factors of stock market development (SMD) in Fiji by examining the impact of macroeconomic and institutional determinants during the period 1996–2019. Second, to explore the causes of the prolonged sluggish status of the South Pacific Stock Exchange (SPX) in Fiji in the context of market liquidity, financial liberalisation and stock market integration.

Box 6.10 Articulating mixed-method problem statement 2

The current study will add to the literature as a comparative discussion to examine whether the impact of the structural determinants of SMD identified in other emerging economies is also applicable to the SPX in Fiji. Thus, in this study, using data on stock market behaviour and macroeconomic and institutional determinants in Fiji from 1996 to 2018, we will examine their influences on the development of the SPX in Fiji. This analysis will be followed by a discussion on the theoretical and practical implications of the study findings together with supplementary qualitative information to arrive at conclusions towards policy recommendations.

6.8 Summary

- Creating a research problem depends on the researcher's insight, experience, ontological and epistemological position.
- The research strategies that the researcher has also mastered significantly impact the selection of the problem and the clear presentation.
- The 'pain' of finding solutions to a 'burning' problem that a social researcher may have makes you more passionate about research.
- Researcher's aspirations and/or objectives with researcher's paradigms, perspectives, etc. Factors such as 'desire to know' also help to create a good research problem.
- Researcher's aspirations and/or objectives, researcher's expert skills and experience, along with researcher's paradigms, perspectives, etc. alone with the factors such as the ability to access data, the ability to respond (to participate), power relationships, network connections, subjective relationships, 'mystery' or 'desire to know' also help to create a good research problem.
- Literary review as the traditional approach to research methodology at the outset is inappropriate and ineffective. More attention should be paid to more recent publications.
- The emergence of research problems out of curiosity intensifies the researcher's enthusiasm and minimizes barriers such as fatigue, frustration or stress.
- Quality questions include a central question and sub-questions, and research purposes often use exploratory verbs such as 'discover', 'understand', 'describe' or 'report'.
- Is it possible to make predictions before research, for research questions, in determining whether a hypothetical test is appropriate? Is the basis for this prediction a causal relationship, a series of propositions, or a 'theory' that explains the hypothesis? It should be asked whether the answer to both the questions is 'yes'.

- It is useful to have three questions in a mixed methodology investigation: quantitative/hypothetical, qualitative questions, and mixed methodological questions.
- Mixed method questions express and analyse data from both quantitative and qualitative aspects. Mixed methods can also be written from a question orientation.

References/Further Reading

Creswell, J. W. (2015). *A concise introduction to mixed methods research*. Thousand Oaks, CA: Sage.

Creswell, J. W., & Plano Clark, V. L. (2011). *Designing and conducting mixed methods research*. Thousand Oaks, CA: Sage.

Gray, D. E. (2018). *Doing research in the real world*. Thousand Oaks, CA: Sage.

Locke, L. F., Spirduso, W. W., & Silverman, S. J. (2013). *Proposals that work: A guide to planning dissertations and grant proposals*. Thousand Oaks, CA: Sage.

Maxwell, J. A. (2013). *Qualitative research design: An interactive approach*. Thousand Oaks, CA: Sage.

O'Leary, Z. (2005). *Researching real-world problems*. London: Sage.

O'Leary, Z. (2014). *The essential guide to doing research*. London: Sage.

Punch, K. (2006). *Developing effective research proposals*. London: Sage.

Saliya, C. A. (2019a), Credit capital, employment and poverty, *International Journal of Money, Banking and Finance, 8*(1), 4–12. https://search.proquest.com/docview/2249669069?pq-origsite=gscholar.

Saliya, C. A. (2020a). Stock market development and Nexus of market liquidity. *International Journal of Finance and Economics*. https://doi.org/10.1002/ijfe.2376.

Saliya, C. A. (2020b). Stock market development and Nexus of market liquidity. *International Journal of Finance and Economics, 25*(3), 23–45.

Saliya, C. A. (2021a). Conducting case study research: A concise practical guidance for management students. *International Journal of KIU, 2*(1), 1–13. https://doi.org/10.37966/ijkiu20210127.

Saliya, C. A. (2022a). Impact of Enterprising Personality on Individual Stock Investors in Fiji. International Conference on Sustainable and Digital Business - 1–2 December 2022, Sri Lanka Institute of Information Technology, Malabe, Colombo.

Wolcott, H. F. (2008). *Writing up qualitative research*. University of Oregon, USA.

Youtube Link

What is A P-Value? https://www.youtube.com/watch?v=ukcFrzt6cHk.

Chapter 7
Literature Review

7.1 Introduction

> *...the literature review should lead the reader to one or more research questions. Hence, there needs to be a tight connection between the literature reviewed and the research study that follows. The connection between the two is the formulation of research questions and/or hypotheses. (Gray, 2018, p. 58)*

The purpose of Literature Review: "What is known about the subject of study?"

Literary Review means studying the latest knowledge through credible sources, especially relevant research papers published in International Peer-Reviewed Academic Journal (IPRAJ). These sources must not be too old. Even though the source you review could be recent, its references could be obsolete. Therefore, the list of references or bibliography provided at the end of a source should be checked and be satisfied that recent publications have been used to develop that research paper.

A process of reviewing academic knowledge usually involves the following actions.

- Selecting a topic
- Search Searching and surveying the literature
- Crit Critique of literature
- Writing

C. A. Saliya, *Doing Social Research and Publishing Results*,
https://doi.org/10.1007/978-981-19-3780-4_7

Source https://www.lib.ncsu.edu/tutorials/lit-review

7.2 Selecting a Topic

The origin of research can be based on a social problem or a curiosity/excitement about society/people or a curiosity about something or a phenomenon. The illustration in the above chapter discusses how to use contemporary knowledge in constructing such problems.

This currently published knowledge has to be obtained from a preliminary level review. This process can be described as follows.

1. Identify personal needs (such as curiosity or excitement).
2. Translate that need into a researchable topic.
3. Identify the relevant academic field.
4. Broad touch on the knowledge expressed in that field.

Existing knowledge of a topic could be acquired through published sources such as any source (such as Wikipedia, newspapers, magazines, and professional publications) and from colleagues and your own experiences and one's own and others' experiences.

The literature review would expose you to essential theories, arguments, and controversial debates and highlight how others have researched the field. Some of the Academic Publications Review objectives are to identify knowledge gaps suitable for further investigation, challenge existing ideas, or apply an accepted theory to a new field.

7.3 Searching and Surveying the Literature

There are two types of literature reviews:

- Publications on the topic of your research problem (On the topic)
- Methods Statements related to choosing and justifying your method (On methods).

7.3.1 Choosing Search Terms

The best way to find publications is to use the online databases provided by your educational institution.

Databases: World Cat, EBSCO, JSTOR, Sage Premier, Gale, ProQuest, Science Direct, Google Scholar, PubMed, Web of Science, Zetoc, ERIC, SSRN, etc. (produced more than 10,000 hits for the term 'Inequality' or 'Credit').

Suppose you need to find the latest knowledge published on a topic discussed above, "the issue of inequality in bank loan approval". If you search for the word 'inequality' or 'credit' in the above database, you will find no <10,000 hits.

Nevertheless, if you combine the term 'inequality and credit', you will find about 6056 finds that are too broad and too much. However, if you search for the phrase 'co-integrated function of credit and inequality', you will find only eight too narrow findings. Alternatively, you can first search for the words 'inequality' or 'credit' only in research paper titles; if this is not enough, you can try only in the abstract; you should have at least 10–20 publications. If none is available, it can be searched in the body.

Most important to make sure no essential articles have been overlooked. This can be done by referring to the list of references of a high-quality article publisher in the top-ranked journal.

7.3.2 Accessing Articles

After selecting the relevant research papers, you should first read the summaries and first select some of the publications published in the most acclaimed journals (ranked in Scimago Q1 or Q2 etc.) that are most relevant to your issue. It is crucial to make sure that you do not miss an essential research papers. High-quality research paper publisher (Elsevier, Emerald, Sage, John Willey, Inderscience, Routledge and Taylor and Francis). Google search can sometimes find full-text papers, especially for older seminal works. However, Wikipedia should only be used as a medium for trusted sources and should never be used as your primary source. It can only be used to get a 'feel' about the subject, author profiles, etc.

Many publishers' websites now offer some degree of free support. For example, Sage Publishers, which provides the following link, provides very useful websites for free (Methodspace), www.methodology.co.uk. And https://study.sagepub.com/grayresearchworld4e for research methods.

In addition, there are open access publications and publishers available, and the following are some examples.

- Open access books: https://www.routledge.com/
- The most prominent Journal publishers' websites: https://www.elsevier.com/, https://www.emerald.com/insight/, https://www.inderscience.com/, https://journals.sagepub.com/, http://cesmaa.org/, https://www.ssrn.com/index.cfm/en/, https://www.wiley.com/en-us
- Social media networks: www.researchgate.net, www.academia.edu, https://www.mendeley.com/search/, https://publons.com/about/home/.

7.3.3   CRAP Test

The following test (CRAP) may help select research papers.

- Currency; how recent it is
- Relevance; how related to your topic and its particular discipline
- Authority; how credible the publisher or writer is
- Purpose; are they just trying to persuade you of their opinion, or do they want to inform you

In general, for secondary sources, you will want to stick with scholarly, peer-reviewed journals. No cherry-picking: Do not choose sources that agree with you. The more diversity you find, the better.

7.3.4   Screening: Scan, Skim and Map

Scan: Determining the criteria for extracting the relevant literature. For example; Abstracts with search words containing your search terms, avoid publications older than ten years on the subject, official websites (United Nations, World Bank, International Monetary Fund, Universities, Governments) and peer-reviewed journals.

Skim: Reducing the number of research papers that should be reviewed.

Map: Read the list of references and choose the most relevant and quality research papers from top journals.

7.3.5   *Storing*

There are several benefits to storing resources using software such as the MSWord
reference menu (shown in the figure below) or EndNote etc.

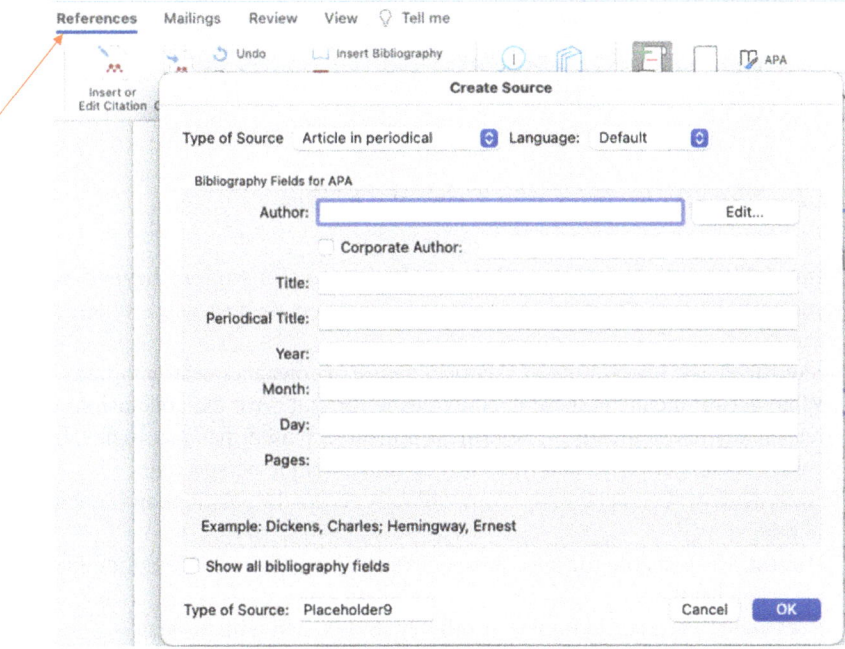

The advantages of maintaining such a database are as follows:

- This database facilitates one-click citing in the text when providing sources for
  quotes and references in your research paper.
- Provides the facility to automatically retrieve the source list of quotations and
  references in your research paper at the end (creating the list of references).
- Allows you to change the format of all sources at once (changing referencing styles
  to match the styles of different journals) when you need to provide a different list
  of sources for different journals. [You may have to submit your research paper to
  several journals. You will then need to customize your sources to the format that
  the journal follows (e.g. APA, Harvard, Chicago, etc.). This process is one when
  there are a large number of sources.]

It may also be useful to download the entire paper of the most valuable articles in
PDF or other digital format and store it on a computer or print it (hard copies).

7.4   Critical Review

Researchers have opinions about the problems in their field and often have pet viewpoints to which they are committed. These preconceptions and personal attachments are both strengths and weaknesses in a research effort. Personal attachment provides the passion and dedications but can carry bias and opinion causing premature conclusions. Rather than arriving at a conclusion based on methodological scholarly work. While bias and opinion can never be removed completely, they must be recognized and controlled. (Machi & McEvoy, 2016, p. 21)

7.4.1   The 10 Qualities of a Critical Thinker

- **Inquisitiveness**: Curiosity to learn and discover. To explore beyond what is currently known, the need to continue to question such as why? What? Is that true?
- **Skeptical**: The uncertainty or skeptical nature of constant questioning; say what? What does it mean? Is there good evidence for that? Are the conclusions drawn reasonably and logically? The critical thinker is biased, has ideologies, beliefs, values and experiences and tries to keep it in a unique perspective.
- **Independent**: Critical thinkers do not blindly accept the positions and conclusions of others.
- **Honest**: Responsible for bias, perspectives and conclusions. Weighing facts and ideas against them.
- **Persistent**: The critical thinker is diligent. Stay tuned with projects.
- **Patient**: It takes time for the critical thinker. Calmly and deliberately works for the task itself.
- **Deliberation**: The critical thinker is concerned with focusing on burning issues. They strive to be organized and organized.
- **Collegiality**: The critical thinker shares ideas with others. Engages in conversation to improve.
- **Rational and logical thinking**: Critical thinkers prefer logical thinking and check for the accuracy and value of the data. Search for evidence, Weigh pros and cons. Relying on logical thinking.
- **Circumspect thinking**: With an open mind, consider different perspectives. Trying to maintain flexibility is a constant reflection.

7.5   Structure of the Presentation of the Literature Review

The structure of the publication review presentation consists of the following elements.

Coverage: Justify the excludes and set limits.

Synthesis: A combination of components or elements to form a connected whole. Distinguish between what has been done and what needs to be done, absorb and improve the vocabulary, and articulate important variables and phenomena related to the topic.

Methodology: Identify the critical strategic and research methods used in the field and their advantages and disadvantages/weaknesses and identify areas for improvement.

Significance: Rationalize the practical significance of the research problem.

Rhetoric: Writing using rhetoric with a friendly and clear structure that supports the review.

7.5.1   Annotated Bibliography

It is important to distinguish this concept of annotated bibliography from the literature review. An annotated bibliography is a list of references that summarize your sources, keep them organized, and explain why that source is relevant to your topic. In contrast, a literature review identifies critical points in the reviewed research paper, explores themes, and initiates a narrative relevant to your research. It is a convincing, integrated presentation that each piece of writing frames on its own.

Extensive accumulation of findings: This means reviewing a large number of research papers and presenting a concise summary of existing views on a particular statement or argument. This is not a critical review.

Example 7.1 Stock Market Development and Nexus of Market Liquidity

The following is an example of a summarised presentation of a literature review. It summarizes recent macroeconomic criteria studies and their impact on stock market development. This summary is based on a review of more than 40 studies in more than 50 countries. This is not a critical review.

Table 7.1 summarizes the macroeconomic and institutional criteria and their impact on stock market development. The last column states whether their effects are positive, negative, or weak or insignificant.

Table 7.1 Summary of macroeconomic and institutional determinants of stock market development, the studies and their impacts

Determinant	Study and author	Impact
Economic growth	40 emerging economies (El-Wassal, 2005), Pakistan (Raza, 2015)	Positive
	41 countries (Levine & Zervos, 1996; Ho & Iyke, 2017)	Positive
	42 emerging countries (Yartey, 2010), India (Subathra & Mohideen, 2015)	Positive
	High and middle-income countries (Fufa & Kim, 2018)	Situational
	Cameroon (Jun et al., 2015), Jordon (El-Nader & Alraimony, 2013)	Negative
	China (Pan & Mishra, 2018)	Negative
Income level	42 emerging economies (Yartey, 2010)	Positive
	14 African countries (Adjasi & Biekpe, 2006)	Positive
	14 MENA countries (Cherif & Gazdar, 2010)	Positive
	Jordon (Al-Sharkas, 2004; Al-Tarawneh & Al-Assaf, 2018)	Positive
	Jordon (El-Nader & Alraimony, 2013)	Negative
Inflation	Turkey (Kutan & Aksoy, 2003)	No relationship
	Jordon (El-Nader & Alraimony, 2013)	Positive
	51 countries (Boyd et al., 1996), 48 countries (Boyd et al., 2001)	Negative
	19 European countries (Sukruoglu & Nalin, 2014)	Negative
	12 Middle Eastern countries (Naceur et al., 2007)	Negative
Banking sector development	13 African countries (Yartey, 2007), 42 Emerging economies (Yartey, 2010), 14 MENA countries (Cherif & Gazdar, 2010), 15 countries (Garcia & Liu, 1999), Philippines (Ho & Odhiambo, 2018), Malaysia (Ho, 2019)	Positive Positive Positive
	Cameroon (Jun et al., 2015), Mauritius (Matadeen, 2019)	Negative
Stock market liquidity	12 Middle Eastern countries (Naceur et al., 2007) 13 African countries (Yartey, 2007), 14 MENA countries (Cherif & Gazdar, 2010), 15 countries (Garcia & Liu, 1999), BRICS (Guru & Yadav, 2019)	Positive Positive Positive
FDI	Latin America (Fernández-Ariasa & Hausmann, 2001). Indonesia (Rheea & Wang, 2009)	Negative
	16 African countries (Agbloyora et al., 2013), Pakistan (Ihtisham & Shehla, 2013; Raza, 2015). 6 GCC countries (Al Samman & Jamil, 2018)	Positive
	Ghana (Acquah-Sam, 2016), Greece (Tsagkanos et al., 2019)	Insignificant/weak
Investments	Mauritius (Matadeen, 2019), South Africa (Yartey, 2008)	Positive

(continued)

Table 7.1 (continued)

Determinant	Study and author	Impact
Rule of Law	47 countries (Minier, 2003), MENA (Cherif & Gazdar, 2010)	Positive
	49 countries (La Porta et al., 1997)	Negative
	Origin of Law (La Porta et al., 2000; Sarkar, 2010)	Positive/negative
Government effectiveness	Mauritius (Matadeen, 2019), South Africa (Yartey, 2008)	Positive
Voice and accountability	Mauritius (Matadeen, 2019), South Africa (Yartey, 2008)	Positive
Control of corruption	MENA countries (Cherif & Gazdar, 2010), Mauritius (Matadeen, 2019), South Africa (Yartey, 2008)	Positive
Regulatory quality	Stock market integration (Bekaert & Harvey, 2000; Obstfeld, 1994) Financial liberalisation: 12 emerging countries; (Henry, 2000a) Financial liberalisation: 37 emerging countries (Ashraf, 2018), Taiwan (Cheng, 2012), Vietnam (Vo, 2019), Eastern Europe (Paun et al., 2019) 40 emerging countries (El-Wassal, 2005), 16 emerging markets (Levine & Zervos, 1996), 13 African countries (Yartey, 2007), MENA countries (Naceur et al., 2007), Financial openness (Ashraf, 2018) Privatisation: (Perottia & Oijen, 2001; Yartey, 2008) Trade openness: 41 countries (Braun and Raddatz, 2008) Malaysia (Ho, 2019b)	Positive Positive Positive Positive Positive Positive Positive Positive Positive Positive
	Philippines (Ho & Odhiambo, 2018)	Negative

Source Saliya (2020)

7.6 Development Arguments

Argument: The following four questions provide a handy guide for checking the validity of an argument:

- What is the stated conclusion?
- What are the reasons that support the conclusions?
- Do the reasons stated have convincing data to support them?
- Does the conclusion logically follow from those reasons?

7.6.1 Simple Argument

Developed by philosopher Stephen E. Toulmin, the Toulmin method is a style of argumentation that breaks arguments into six parts: claim, grounds, warrant, qualifier, rebuttal, and backing. In Toulmin's method, every argument begins with three fundamental parts: the claim, the grounds, and the warrant (https://owl.purdue.edu/).

A simple persuasive argument has three essential elements; evidence, warrant and claim. Claims are declarations of a proposed truth. The evidence consists of data that define and support the claim. At the intersection of evidence and claim is the warrant which represents the logical formation of the claims and evidence and is the glue that holds claims and evidence together. The warrant is because of the statement (Gray, 2018). Including a qualifier or a rebuttal in an argument helps build your ethos or credibility (Fig. 7.1). When you acknowledge that your view is not always accurate or provide multiple views of a situation, you build an image of a careful, unbiased thinker rather than of someone blindly pushing for a single interpretation of the situation (https://owl.purdue.edu/).

Example: See Fig. 7.2.

Recognizing the possibility of a condition or other opinion in an argument helps to build your values or credibility. When you admit that your vision is not always true, or when you offer different opinions about a situation, you build an image of an impartial thinker, rather than being blindly pushed for a single interpretation of the situation. Accordingly, the argument for the slowdown in world economic growth in the above Covid example (Fig. 7.2), if presented subject to conditions, can be worded as follows.

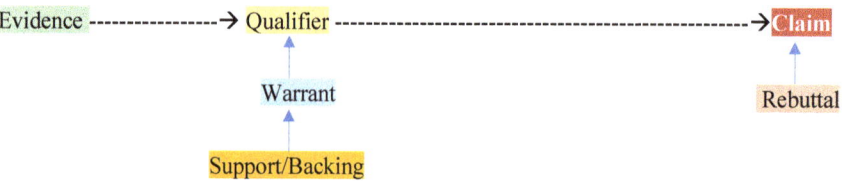

Fig. 7.1 Structure of an argument

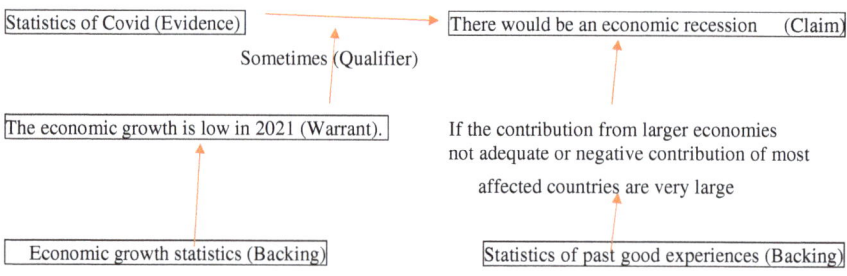

Fig. 7.2 Covid example

Table 7.2 Types of arguments or statements

Claim	Example	Evidence
Fact: Statements related to a person, place or something	The Industrial Revolution overcame human values, and capital emerged	Compare populations
Worth: Opinion on action, behaviour or stance	The Industrial Revolution overcame human values, and capital emerged	Statements endorsed by scholars
Policy: Statements on Criteria or Standards	Optimal democracy is to decentralize governance as much as possible and work locally	Comparing countries and providing written evidence. Statements endorsed by scholars
Concept: Comments on definitions, suggestions or phenomena	Group-thinking is the authoritarian subjugation of the ideas of powerful individuals to the opinions of weak group members	Statements endorsed by scholars
Interpretation: Statements that provide a source framework for understanding an idea	Marxism suggests that capitalists exploit workers	Evidence such as generally accepted, scholarly statements, statistical data

Some argue that the contribution of countries with large economies, where the impact of the Covid is minimal, would be adequate enough to negate the negative impacts of other economies, and therefore, the global economic situation as a whole will not be affected.

7.6.2   Type of Arguments, Claims and Assertions

The claim is the argument's assertion. According to Hart (2001a) there are five types of arguments or statements: fact; An opinion about a deed, behaviour, or stance (worth); Policy; Concepts; Interpretation of phenomena/interpretations.

The five types of arguments with examples are given in Table 7.2.

Four factors are considered to prove that an argument or statement is acceptable.

- On point—Relates directly to the argument
- Strong—Compelling reason
- Supportable—Evidence is available to justify
- Understandable—Specific. Clearly stated.

7.6.3   Complex Claims

A single argument is a single claim, its evidence, and its warrant. Most arguments are complex. Complex arguments are constructed using multiple simple claims. These

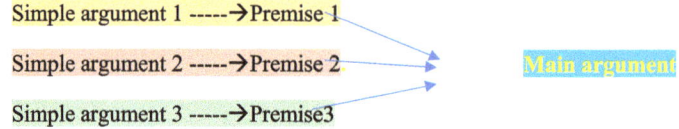

Fig. 7.3 Complex claim

Young graduates >>	for further education	>> migration	
Experienced scholars >>	For children's education	>> migration	**Brain drain**
Professional >>	For better earnings	>> migration	

Fig. 7.4 Complex claim Example 7.1

Rich-powerful	can influence to get financial capital as bank loans	hence can get richer and more powerful.	**Bank lendings contribute to sustaining income and wealth inequality**
Poor-Powerless	are unable to influence to get bank loans	hence remain poor and powerless	

Fig. 7.5 Complex claim Example 7.2

simple claims serve as the premises of the major argument. A premise is a previous statement of fact or assertion (claim) that serves as the evidence for warranting the claim of a major argument (Figs. 7.3, 7.4 and 7.5).

Example 7.2 The Role of Bank Lending in Sustaining Income/Wealth Inequality in Sri Lanka

The data from this research show that the 'powerful' is the prominent businessmen, and the 'weak' are the small businessmen who are less socially powerful. It also proves that weak small businesses do not have the strategies used by powerful businessmen to approve bank loans, such as accounting technology and social-power networks. Therefore, conclusions are drawn using the hypothetical method with the condition of 'possible'. Since a case study did this research, a problem of generalization may arise. However, if the evidence is convincing that these studied cases are typical, the induction method can justify the 'can' condition by removing the 'can' condition.

7.6.4 Organizing the Premises and Reasoning Patterns

Whether unraveling the plot or of a good detective novel or assembling a jigsaw puzzle, the reasoning is the same as that used to create the literature review. This description also helps in constructing theories. Successful convictions can be made by following such patterns in reaching conclusions.

7.6.5 Simple Reasoning

One-on-one reasoning: There is **Reason (R)**, therefore **Conclusion (C). R** causes **C.**

Example: The noon bell rang. So now it must be lunchtime.

Side-by-side reasoning: R_1, R_2, R_3, R_4....R_n, Therefore **C.**

Example: The noon bell rang. The train timed at 12 is running. The sun is right over the head. Therefore, now it must be lunchtime.

This is the most common pattern used by social researchers to give logical reasons. The authors generally use several theories to support this argument pattern; expert opinions, research studies, expert evidence and other data all lead to the same conclusion. The result is an enormous amount of evidence needed for a conclusion.

Chain reasoning: This is the most common pattern used by social researchers to give logical reasons. This pattern is also widely used by researchers. Following the natural sequence, it begins by quoting one or more reasons that justify the conclusion. As its basis, it uses a pattern of arguing one by one. The conclusion of the first pattern becomes evidence for the second conclusion. This logic continues until the conclusion is confirmed.

$$(R, :. C1) + (C1 :. C2) + (C2 :. C3 + \cdots : Cn).$$

Example: The engine of a vehicle burns less fuel when running at low speed, so low speed means less fuel combustion; low fuel combustion is a low toxic fuel release; reduces toxic emissions and reduces air pollution; so slowing down a vehicle means reducing air pollution.

This pattern builds logic as a chain, with A being B, B being C, C being D, etc. Each conclusion becomes the reason for building the following conclusion.

Joint reasoning: In this case, the reasons given may not stand alone, but together they may present the reasons necessary to confirm the conclusion.

$$[(R1 + R2) :. C].$$

This thinking pattern is represented as follows: "If x exists and y exists, then z exists". If one of the partial causes (x or y) is absent, there is no reasonable conclusion.

Example: Strong winds cause rain to flood the balcony.

This logic is inherently addictive. The argument is based on both 'rain' and 'strong winds'. Having only one factor is not enough to fill the **balcony** with rainwater. Individual data representing each approach cannot justify the conclusion on their own merits. The conclusion can only be drawn when the entries are combined. Data are the parts that put a theory or place together.

7.6.6 Complex Reasoning

Complex logic helps to build a complex argument by combining two or more simple logic patterns. Here is an analysis of simple arguments to present new ideas, profoundly explaining what is currently known about the subject. The analysis begins with a review of simple arguments to find a logical pattern. 'What does this data say?', 'What is the story?', 'How do the facts fit together?' etc. Critically analyzes evidence and arguments by asking questions.

Like the chief detective in a good mystery novel, the conspiracy must be revealed by examining the evidence to determine 'what happened?' Furthermore, 'who did it?' By combining arguments in some important way, by analyzing the evidence, it is possible to compose the story, that is, to present the final argument. An outline or planning of the argument can be done using the above reasoning methods as guidelines. The outline allows you to write an exploratory draft as a first attempt to tell the story of what you know about the research subject.

The basic simple reasoning patterns combine to form a complex pattern to organize the premise that forms the discovery argument. The basis for a complex argument is the combined claim of a set of simple arguments.

Divergent reasoning: This pattern represents an academic debate. Divergent logic is a logical approach to conflicting opinions.

(R1, R2.....∴. C1 against (Vs) R1, R2, ...∴. C2).

The basis of this complex logic is to present expert opinions, research studies, statistics, expert evidence and other data in a way that builds a clear pattern for one side of the question and to cite another similar set of data to show the opposite view.

The above divergent patterns are organized to facilitate the neutralization of opposite perspectives. This pattern represents the authors' positions, research findings, and theories in evidence with direct contradictions. By mapping the opposing data, it is possible to focus on the decisive factors for finding strong and weak points in each of the perspectives of the debate (Fig. 7.6).

Mapping opposing views

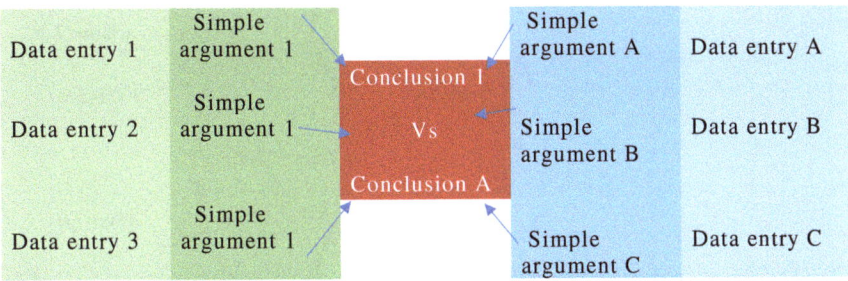

Fig. 7.6 Complex claim Example 7.2

Comparative reasoning: The Comparative Logic Scheme shows the relationships between data groups. Here we examine the similarities and differences of each group by comparing the evidence and arguments related to each, and this pattern of reasoning looks like the following formula:

$$R1, R2..... :. C1 \wedge R1, R2, ... :. C2.$$

Like multi-factor logic, the first argument cites expert opinions, research studies, statistics, expert evidence, and other data to build a clear pattern. The same set of data is presented for the second argument. The arguments of the two sides are compared to find differences and similarities between the presented data.

Comparison and contrast: Separate diagrams map the relationship between two or more data groups. They are commonly used to record theoretical data, contradictory locations, relationships between two compilations, or alternatives. Each circle in the Venn diagram is based on evidence. Once each argument has been described, the commonalities of the combined logic can be easily demonstrated by noting each of the arguments that fall into the circle. To see the differences, consider the sections that fall outside the circle intersection (Fig. 7.7).

The Discovery Argument

The purpose of the publication review is to collect the surveyed data to provide evidence and patterns for the premises of the main complex argument or to form small arguments and ask, "What is the currently published knowledge about the subject of study?" New knowledge is discovered by building a bridge to the knowledge gap discovered through the critical question.

Example: The following is an example of complex logic to formulate a discovery argument on a study topic. Suppose the subject of study is as follows; 'Definition of Human Intelligence in the 20th Century: From a Cognitive Perspective'.

Fig. 7.7 Each circle in the Venn diagram is based on evidence

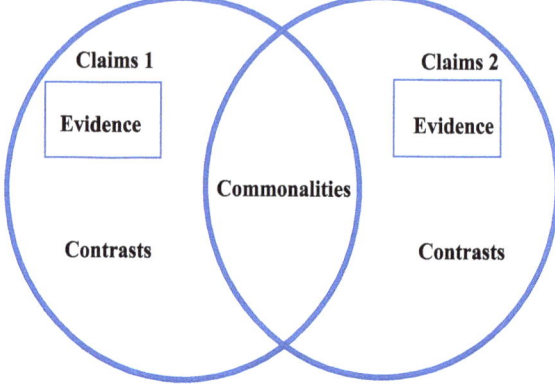

The Definition of Human Intelligence in the 20th Century: A Cognitive Perspective

The literature survey documents the seminal works on the subject. After completing the appraisal activities in surveying the literature, three positions emerge:

- Human intelligence consists of a single structure (Single, 2001), as opposed to the position that human intelligence consists of multiple structures of several domains or dimensions (Multiple, 2008).
- Human intelligence can be accurately measured (Accurate, 2004), as opposed to the position that human intelligence cannot be accurately measured (Guess, 2010).
- HI is inherited and static (Fixed, 2008), as opposed to the position that human intelligence is changeable and developmental (Develop, 2018).

You can add with simple arguments and using chain pattern that Single, Accurate and Fixed as a general theory of intelligence as oppose to theory of multiple intelligence by combining the works of Multi, Guess and Develop together.

What are the implications of the above discovery have to be critically reviewed asking the following questions;

Does what is known about the research subject answer your original inquiry?

Are there gaps, omissions, debates, and questions about topic that need further study?

Given what is known about the subject of the research, what can you conclude?

Source Gray (2018)

7.7   Synthesizing and Integration

A strong synthesis addresses issues such as 'What do we know about this problem (its seriousness, correlates, consequences) and its solutions (interventions tested to date), and how (by what methods and measures) do we know it?' (Kearney, 2016).

No matter how effectively the research is conducted, no matter how successfully the research is completed, no matter how impressive the results or innovations are revealed, your research paper would only be a dream without a powerful literature review unless it is integrated with the existing knowledge in the world. In this context, you will realize that what is meant by 'literature review' is not merely the 'in-depth study of existing knowledge' mentioned above.

Take this as an example, and you can get an idea of the 'context'. Accordingly, a 'literature review' is a scholarly, logical, 'integrated' presentation of your study in writing. Here you have to understand what 'integrated' means. You can get access to this by following the steps below.

Paraphrasing: Comparing and rewriting excerpts; There may be a critical review of the contents of the same research paper. There may be differences or similarities between two research papers by the same author or two authors (critical review of views). However, it is better to rewrite this writing to build a sharper argument and to write it in one's own words as much as possible (developing cogent arguments). If you are quoting direct quotes, you must present your opinion critically. These reviews may need to be re-interpreted and interpreted with your research results.

Given that all reviews should adopt a critical stance to the literature, this is also an opportunity to highlight the weaknesses of some of the studies described, particularly in terms of the validity and reliability of their results. The literature review hence becomes the basis for showing how the future research study will avoid these mistakes and produce more robust findings.

To see the quality of your critical review, check if you have identified several research questions. If none of the research questions arises and is explained, you should understand that you have often written only a commentary directory (Table 7.1).

7.7.1   Testing the Validity of the Arguments and Thesis

By legal logic, the legitimacy of a case presented depends entirely on the authority of the evidence. In descriptive arguments, validity depends on the powerful authority of the argument. Descriptive arguments are complex arguments that depend on the accuracy and adequacy of the evidence to construct. However, research in the social sciences seeks to develop conjecture. Uses consultative logic for hypothetical questions to conclude.

Example:

'If it is raining, then I should take an umbrella when walking to work.'

These two arguments, 'if it is raining' and 'take an umbrella', have two independent claims. Each should have separate proofs.

Once proven that rain is falling, it is necessary to propose that it would be reasonable to take an umbrella.

Taking an umbrella should logically advocate the implications of the facts to settle on a reasonable action.

The argument must show that, because it is raining, you will get wet walking outside. The implications of the first statement support the second one; "if you prefer to stay dry, and walk outside, then you should do something reasonable" Here, the reasonable action proposed is to take an umbrella.

Therefore,

"If it is raining" and "WE WANT TO REMAIN DRY", also "we want to walk outside in the rain", then suitable means to satisfy the end is using an umbrella. The success in supporting the thesis depends upon the IMPLIED LOGIC (we want to walk in the rain and remain dry) used in the argument.

The backing for the warrant's force is that previous experience has demonstrated that an umbrella can protect a person from rain. The warrant provided the logical bridge to justify the use of an umbrella as a reasonable conclusion.

Source Gray (2018)

As shown in Table 7.3, Grennan (1997) identified nine basic argument patterns. These nine patterns identify the relationship between the study question and the claim made by the disclosure or discovery argument. A good rule of reasoning provides the basis for each pattern. The key factor in choosing the correct pattern is identifying the type of logic that the research question seeks and the relationship between the evidence and the context developed by the literature survey.

Example research questions related to each pattern are suggested in the last column of Table 7.3 to understand these patterns.

7.7.2   Fallacies

Beware of false arguments or misconceptions that lead to erroneous or misleading conclusions. As we have seen, the lack of convincing data, inappropriate or irrelevant evidence, and unnecessary arguments can all lead to a false argument. The three most common mistakes in arguing are jumping to conclusions, justifying, and ignoring alternative explanations. The main mistakes that researchers should avoid are:

Table 7.3 Nine basic argument patterns

Argument pattern	Rule of logic	Prerequisite conditions "The researcher must show that…"	Example research questions
Cause and effect	For every cause, there is an effect	…the body of evidence identifies directly causal data	What are the determinants of stock market development in the CSE?
Effect to cause	Every effect has a cause	…the body of evidence contains the direct effects caused by the case defined in the research question	What are the effects of alcohol abuse in the workplace?
Sign	Identifiable symptoms, signals, or signs precede events and actions	…the data identified by the body of evidence are symptomatic of the action or event defined in the research question	What are the qualities of the dysfunctional group?
Sample to population	What is true of the sample is also true of the whole	…the sample identified in the body of evidence is truly representative of the population defined by the research question	Is a representative sample of graduates of government universities more likely to get MBA admittance at PIM
Population to sample	What is true of the population is also true of a representative part of that whole	…the pop identified in the body of evidence is truly representative of the sample defined by the research question	What leadership strategies can managers employ to foster teamwork
Parallel case	Where two cases are similar, the first case's true is also true of the second	…the case identified in the body of evidence is similar enough to the case defined by the research question to make them parallel	What teaching strategies employed in high-performing schools can be used by other high performing schools?
Analogy	Where two cases are alike, a conclusion drawn from one can be assumed to be a conclusion drawn about the other	…the case identified in the body of evidence contains qualities that provide explanation or clarity to similar qualities contained in the case defined by the research question	How can higher education institutions compare to the model of a living organism to explain their internal workings?

(continued)

Table 7.3 (continued)

Argument pattern	Rule of logic	Prerequisite conditions "The researcher must show that…"	Example research questions
Authority	The more a person knows about an issue, the more factual the claim about that issue	…the testimony presented in the body of evidence uses reliable expert testimony relevant to the case defined by the research question	What is the nature of human intelligence? What are the characteristics of effective leadership?
Ends-means	The result is directly attributable to performing a named action	…the action identified in the body of evidence of the literature survey will achieve the ends as identified by the research question	Are the results of this semester, based on the student cantered learning system, better than the last semester?

- Jumping to conclusions due to incomplete evaluation of evidence
- Jumping to conclusions with biased arguments without addressing other alternatives
- Name-calling attacks, a position, or an expert impugning the personal character of the author. "Nothing, my opponent, says is trustworthy…"
- Emotion-based arguments influenced by groupthink; "All good patriots must support my view…."
- Ignorance using logic that a claim must be true because it has not been proven false. "It is evident to all of us…."
- Misplaced causality often occurs when connections are refutable. "It is evident from the low test scores that teachers are incompetent."
- Begs the question occurs when the researcher asserts a claim and uses that claim as the evidence for the assertion. "There is. God because God said so."
- Disconnected conclusion without evidence supports it. "We must invade Iraq because they have weapons of mass destruction."
- "Everybody knows" those conclusions appear to make a case where none exists, ill-defined or vague notions.
- A loaded question has formed a research question that contains one or more false or questionable presuppositions. "When did you stop beating your wife?"

- Using controlling language. "This study examined the effects that the bureaucratic, authoritarian, and wasteful No Child Left Behind Act had on the reading achievement of third-grade inner-city children in California."

7.8 Using the Literature Review to Generalize the Cases

In order to justify case studies, the evidence must be provided that those case studies are fair representations of the field. Must appear as typical. Secondary data from the publication review suggest that the bank and loan applicants in the above examples may be role models. The following are examples of Box 7.1, Tables 7.4, 7.5 and 8.4.

Box 7.1 An assessment of representation of Sri Lankan credit applicants

Representation of the credit applicants

Twelve common characteristics of credit applicants were tested to assess the representativeness of the three credit applicants studied in this research, as tabulated in Table 7.4. All the bank officers contacted for this assessment have had branch manager positions in various parts of the country. They were of the view that ordinary credit applicants face problems with collateral, rigid credit evaluation rules and lack of support from accountants with regard to the accessibility, affordability and availability of credit. Also they agree that medium and large credit applicants are beyond their authority and are mostly accommodated with extra effort by the senior officers. Customer get-togethers are common in all Sri Lankan banks and all respondents agree that only large and medium sized clients are invited to such functions. Representation is ranked as high, average and low compared with other credit applicants the respondents had been aware of for the last 10–20 years.
(Table 7.4—*Source* Compiled by the author through annual reports and interviews)

Table 7.4 seeks to justify that the selected loan applicants described in Box 7.1 are typical credit applicants commonly found in Sri Lanka.

Table 7.4 An assessment of representation of Sri Lankan credit applicants in general by the credit applicants considered

Characteristic	Level of representation		
	Tony	Yusef	Silva
Legal form of the business entity	Average; with 2–5 private companies	High; with 1–3 private companies	High; sole proprietorship
Ownership	High; with 80% with owner-manager	High; with 80% with owner-manager	High; with 100% with owner-manager
Credit amount involved	Low; US$4 million is too large	Average; with US$500,000	High; with US$20,000
Type of business	High; labour intensive	Low; car manufacturing in Sri Lanka	High; labour intensive
Employment generation	Low (exceptionally high generation)	High	High
National interest	High with large exports	High with unique import substitution	High with high employment
Number of banks involved	Low (many banks)	High (a few banks)	High (only one bank)
Problems with collaterals	High	Average	High
Bank's authority level involved	Average with senior management	Average with senior management	High with middle management
Informal approach by the applicant	High	High	High
Informal methods by the decision-makers	High	High	Average
Success in credit approval	High	High	Low

Source Saliya (2010)

Box 7.2 An assessment of representation of private banks in Sri Lankan compared to the Soft Bank

Representation of the Soft Bank

Sri Lanka has a total of 22 commercial banks equally distributed between domestic and foreign (11 banks each) and, out of 11 domestic banks, nine private domestic banks, including the Soft Bank, share 40% of all banking business while the two state-owned banks share 57.5% in terms of branch distribution and approximately 41.1% in deposits, loans, assets and turnover in 2004 as well. That leaves only 2.5% of banking business to all of 11 foreign bank branches (The Central Bank of Sri Lanka, 2008). The Soft Bank alone enjoyed a 8–9% of market share in 2004.

A set of common characteristics were tested to establish the representativeness of the Soft Bank compared to other domestic private banks which share 41% of the banking business in Sri Lanka. In Table 7.5, fourteen such characteristics were tested with interviewees to assess the level of representativeness of the Soft Bank in the Sri Lankan private banking sector. The representation level is indicated by Yes/High, Average/Likely/Medium and No/Low. The banks numbered 1–5 represent listed privately owned commercial banks, number 5 being the Soft Bank and number 6 comprises all other small/unlisted/private commercial banks.

(Table 7.5—*Source* Compiled by the author through annual reports)

Table 7.5 An assessment of representation of private banks in Sri Lankan compared to the Soft Bank

Bank characteristic	Level of representation					
	1	2	3	4	5	6
A few large shareholders	Yes	Likely	Yes	Yes	Yes	Yes
Large number of shareholders	High	Likely	High	Likely	High	Low
Large shareholders interfere in the decision-making	Yes	Likely	Yes	Likely	Yes	Yes
Bureaucratic management style	Yes	Yes	Yes	Yes	Yes	Yes
Rigid management policies and procedures	Yes	Yes	Yes	Yes	Yes	Yes
Regulated closely by CBSL	Yes	Yes	Yes	Yes	Yes	Yes
The external auditors are from Big Four audit firms	Yes	Yes	Yes	Yes	Yes	Average
Positive perception of investors; Measured by the PE ratio	Medium	High	High	High	Low	N/A
Branches are distributed all over the island	Yes	Yes	Yes	Yes	Yes	No
Catering to the same market	Yes	Yes	Likely	Likely	Yes	Likely
Similar deposit base structure	Yes	High	High	High	Yes	Average
Catering to corporate clients	Average	High	Average	Average	Average	Low
Level of non-performing loans	High	Low	Medium	Low	High	Medium
Level of similarity in transaction processing	High	High	High	High	High	Average

Source Saliya (2016b)

Table 7.5 seeks to justify that the selected soft bank described in Box 7.2 is a typical commercial bank typical of Sri Lanka.

7.9   Positioning

The literature review placement depends on the nature of the project (dissertation, research paper, journal research paper, project proposals, etc.). However, literature reviews of research methodology are often presented under a separate heading, as is common in all research papers.

Literature review for a doctoral dissertation can be in one or several separate chapters. Example 7.1: The role of bank lending in maintaining income/asset inequality in Sri Lanka http://156.62.60.45/handle/10292/824.

In quantitative research papers, part of the literature review is presented under 'Introduction' and part under 'Hypothesis' as shown in Boxes 6.6 and 6.7, there is often a Literature review.

In quality research papers, part of the literature review should be presented under 'Introduction', and the main publication review should be presented under a separate 'Literature review'.

Literature review of research papers for a grounded theory is not presented under the heading 'Introduction' or a separate 'Literature review'. Here we compare the theories generated from the data in conceptualizing and constructing categories with the theories found in the literature review.

7.10   Summary

- 'Literature review' refers to the study of the latest knowledge that has been published so far through credible sources.
- The literature review objectives are: "What is the currently published academic knowledge about the subject of study?" Objectives include identifying knowledge gaps, challenging current ideas, or applying an accepted theory to a new field.
- Databases: WorldCat, EBSCO, JSTOR, Sage Premier, Gale, ProQuest, Science Direct, Google Scholar, PubMed, Web of Science, Zetoc, ERIC, SSRN, etc.
- It is important to follow the guidelines of CRAP: Currency/currentness, Relevance, Authority and Purpose.
- Ten qualities of a critical thinker: inquisitiveness, skepticalness, independence, honesty, persistence, patient patience, deliberation, collectiveness (collegiality), rational and logical thinking, circumspect thinking.

- Fundamentals of literature review; coverage, synthesis, methodology, significance, rhetoric.
- Annotated bibliography: A list of sources that summarize the sources and keep them organized, explaining why that source is relevant to your topic.
- Testing the validity of an argument: What is the stated conclusion? What are the reasons that support the conclusions? Is there any convincing data to support the stated reasons? Is the conclusion logical for those reasons?
- Claims are statements of the proposed truth. Evidence is supporting data. A warrant is a glue that holds logic and evidence together. A warrant is a statement of explanation as to 'why'. This can be further expanded to support/backing, qualifier or rebuttal steps.
- The basis for the complex argument that makes up the discovery argument is the combined plea of a set of simple arguments.
- 'Literature review' The end result is to present your study in a scholarly, logical, 'integrated' way.
- Beware of false arguments or misunderstandings. Lack of convincing data, inappropriate or irrelevant evidence and unnecessary arguments can all lead to a false argument. The three most common mistakes in arguing are making hasty conclusions, justifying, and ignoring alternative explanations.
- Literature review of research methodology is often presented under a separate heading, as is common in all research papers.
- Literature review for a doctoral dissertation can be a separate chapter or several chapters.
- In research papers, part of the publication review should be presented under introduction, part under hypothesis making and/or under a separate heading.
- In Grounded theory, the literature review is not presented under the introduction of research papers, but in the conceptualization of results, the theories generated from the data are compared with the theories found in the publication review.

References/Further Reading

Fisher, A. (2003). *The logic of real arguments*. Cambridge, UK: Cambridge University Press.
Fisher, A. (2004). *Critical thinking*. Cambridge, UK: Cambridge University Press.
Gournelos, T., Hammonds, J., & Wilson, M. A. (2019). *Doing academic research: A practical guide to research methods and analysis*. NY: Routledge.
Gray, D. E. (2018). *Doing research in the real world*. London: Sage.
Grennan, W. (1997). *Informal logic: Issues and techniques*. Montreal & Kinston, Canada: McGill-Queen's University Press.

Hart, C. (2001a). *Doing a literature review: Releasing the social science research imagination.* London, UK: Sage.
Hart, C. (2001b). *Doing a literature search: A comprehensive guide to the social science.* Thousand Oaks, CA: Sage.
Literature review. https://www.lib.ncsu.edu/tutorials/lit-review.
Machi, L. A., & McEvoy, B. T. (2016). *The literature review: Six steps to success.* Thousand Oaks, CA: Crown.

Youtube Link

How to do literature review. https://www.youtube.com/watch?v=lw8HPXJP1VA.

Chapter 8
Data Collection

8.1 Introduction

After reviewing the literature, you should have a clear idea of the research that other scholars have done so far on that field and topic and what methods they have followed to discuss your topic. The strategy you prefer; You may have thought of quantitative, qualitative or mixed methods. Now you have to collect the data, analyse and interpret it.

Smith and Hodkinson (2005) point out that "no special epistemic privilege can be attached to any particular method or set of methods" (p. 917).

If proper procedures are implemented, the subjectivities (e.g., ideologies, ideologies) of knowledge will be limited, and the arguments presented by the researcher will be plausible. Quantitative research allows for a 'correct' and 'objectivist' representation of reality in line with their ontological and epistemological positions.

According to Descartes' ontology there are substances, attributes, and modes. These are understood relative to one another in terms of ontological dependence. Modes depend on attributes and attributes depend on substances. The dependence relation is transitive; thus, modes depend ultimately on substances. No substances, no modes. https://plato.stanford.edu/entries/descartes-ideas/.

8.1.1 Validity/Conformability and Reliability/Credibility

Acceptance of research results is based on the validity/consistency, and reliability and methods used. There are four types of validity. Validity and reliability are the concepts of the quantitative tradition, while the qualitative concepts are consistency and credibility. The detailed study of these concepts should be done in accordance with the chosen research methodology, using specific sources.

© The Author(s), under exclusive license to Springer Nature Singapore Pte Ltd. 2022   99
C. A. Saliya, *Doing Social Research and Publishing Results*,
https://doi.org/10.1007/978-981-19-3780-4_8

- Construct validity: Does the test measure the concept that it has intended to measure?
- Content validity: Is the test fully representative of what it aims to measure?
- Face validity: Does the content of the test appear to be suitable to its aims?
- Criterion validity: Do the results correspond to a different test of the same thing?

Reliability is the consistency of a measure. Avoiding random errors or the noise increases reliability. Just because a measurement is reliable does not always mean it is a valid measurement. The reliability of acceptability of quality data can be enhanced by the same data set being verified by several different sources (data triangulation).

8.2 Data Collection

Czarniawska-Joerges (1992) uses the phrase 'insight gathering' instead of 'data collection' to encompass more comprehensive sources such as recalling memories and reconstructing experiences. Quantitative data collection methods are much more structured than qualitative data collection methods such as interviews, open-ended questionnaires, participant observations and reconstructions of experiences. Quantitative data collection methods include various forms of surveys—online surveys, paper surveys, mobile & kiosk surveys, longitudinal studies, website interceptors, online polls & forms, and systematic observations.

Data collection can be done through primary and secondary sources. The primary sources are the participants, including the researcher, and the secondary data are published documents and archival records. Data saturation is achieved by continuously collecting data and integrating and theorizing that data into analytical processes such as building relationships between events, finding standard features, patterns and relationships, and/or raising critical questions.

Saturation of data collection can be ensured by obtaining independent feedback from peer reviewers and colleagues when there is satisfaction or deadlock of posing further why and how questions. Now it is argued that gaining traditional 'rich/thick description' alone is not enough to ensure the validity and reliability of a case study research because, it may be limited to different levels of depth and detail (Woodside and Wilson, 2003). On the other hand whether the description is 'thick' or 'thin', if it provides adequate evidence to the claim, the description is considered as dependable. (Bailey, 2007).

8.3 Data Types

There are two types of data; statistical and non-statistical data. Statistical data are scales, counts, scores, etc. Non-statistical data includes words, pictures, clothing, accessories, locations or decorations, paintings, movies, etc. Quantitative data were discussed at length before. The following is a description of the qualitative data.

Textual data: Read one or more texts emphasizing their physical structure. This is a great help for other methods. Inherently cherry-picking and being limited to a few texts or textures is disadvantageous.

Content data: This usually means reading a large number of texts briefly. Common features and differences between all texts are usually examined and organized with the help of clear 'coding'. What happens here is a quantitative combination of qualitative data. Analyzing multiple texts helps to avoid arbitrary choices.

Discourse data: This usually means reading many different types of texts briefly to discuss the ideological issues surrounding a central topic. This is a time-consuming process.

Perception data: Data are collected through interviews and targeted group research.

Survey data: Data are collected through a questionnaire.

Observed data: These data are obtained by the researcher himself or by observation by the participants.

Experiments: In a controlled environment, data obtained by manipulating variables in one group to determine differences.

Archival records: Similar to conversational methods, archive research uses data from an organization's report on an event, a series.

8.3.1   Classification of Data According to Scales

Scales are created by assigning numerical values to an item or event. There are four types of such scale data; As ratios, intervals, ordinals and nominals. Many physical measurements, such as height, weight, blood pressure, distance, temperature, etc., are ratios or intervals, so they fall into numerical continuous scale variables. Ordinary data such as income levels and education levels and nominal data such as gender and preferred colour fall into categorical qualitative variables. Calibrated data can be obtained in large quantities and used as continuous numerical variables for analytical purposes, but the numerical values themselves have no real meaning. When data is processed for analysis, it is called a variable.

Unlike interval data such as temperature and scores, the differences between two ratio measurements have a meaning. For example, since weight measures are ratio variables, a weight of 4 g is twice as much as 2 g. Intermediate level data can be used in calculations but cannot be compared. For example, the temperature of 80 °C is not four times as hot as 20 °C. Zero in the interval data is just a point, not the absolute absence of a variable. Temperature zero (0-zero) is not a state without temperature, and it is the temperature at which water freezes (ice). Watch the Youtube video below for further clarifications.

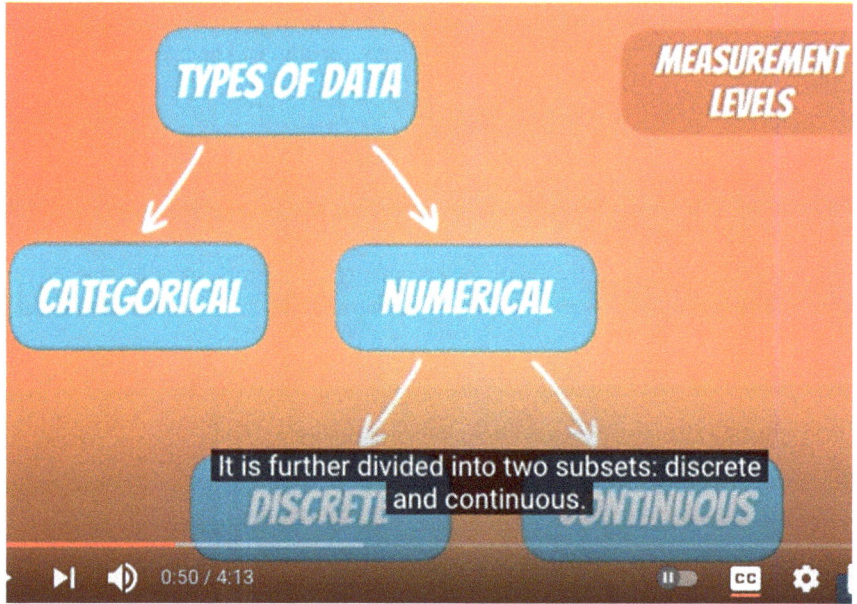

Sources Types of data: Categorical versus Numerical data. https://www.youtube.com/watch?v=DUcXZ08IdMo

8.4   Surveys

Surveys are a method of obtaining feedback on a topic from a given questionnaire. Samples are usually large, but the depth is often not as complex as focus groups or interviews. Common scales are scales such as Likert. The following are some examples of Likert-scale responses in Tables 8.1 and 8.2.

8.4.1   Sampling Techniques

Sample selections can be made using a variety of sampling methods. They range from random, purposive to convenient samples.

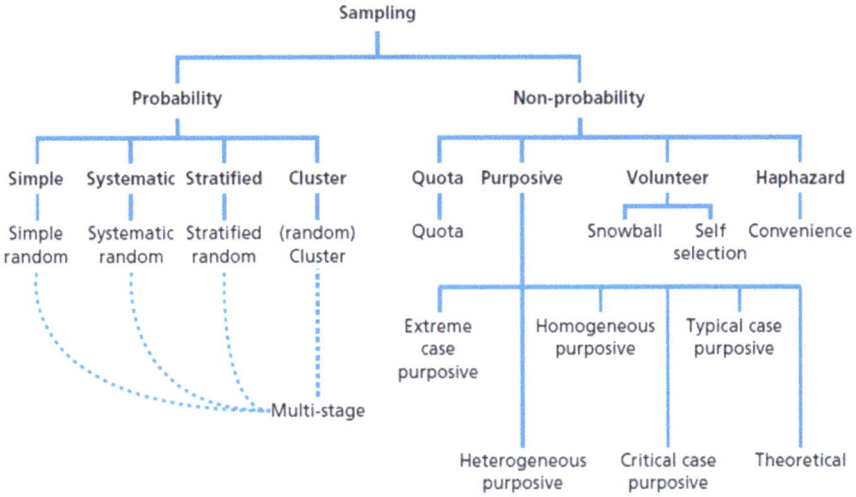

Source   BRM: https://research-methodology.net/sampling-in-primary-data-collection/

You can learn more about sampling by studying the following YouTube tutorials.

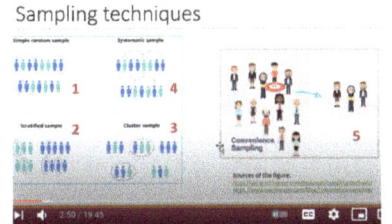

Source https://www.youtube.com/watch?v=6sk
CMCdh3FY

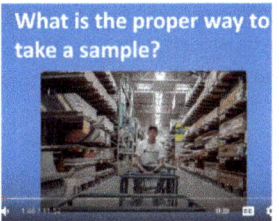

Source https://www.youtube.com/
watch?v=VoaQcff-uxQ

By submitting a limited number of answers to choose from, such as the Likert scale, participants' responses are limited to the options presented by the researcher. Therefore, survey questions are inherited with such disadvantages.

8.4.2 *Asking Questions that Make Survey Participants Feel Uncomfortable*

As a very extreme example, participants may be embarrassed by questions asked in a survey about the theft of books from libraries. Such inconveniences can be minimized if such issues are addressed as follows.

- Have you ever had to steal books from a library?
- As you know, why are people tempted to steal books from a library? Have you had a similar experience?
- Do you know people who stole books from a library? Has that happened to you?
- Which of the following suggestions is best for you?

  - Book theft can happen at the instigation of peers.
  - Never heard of book theft.
  - Book theft is a childish act.
  - An act of ignorance.
  - I am not so moved.

8.4.3 Type of Questions

Notice that the analysis potential is *cumulative*. That is, as you move up the nearby table to a higher type of data, you can perform more mathematical operations. This is important! If you want to trend average scores on some survey question(s), you have to choose question types that generate Interval or Ratio data. You cannot calculate means with ordinal and nominal data.

Note that the analytical potential is cumulative in nature. Table 8.1: Ratio and interval of the survey questions type table can be used for advanced mathematical operations. If you want to get average marks on some survey questions, you must select the types of questions that generate interval or ratio data. You cannot do calculations with calibrated and nominal data.

8.4.4 Response Categories

Below are some examples of Likert-scale responses in Table 8.2.

Example 8.1 Role of Enterprising personality in motivating stock investors: Case of Fiji, forthcoming.

Care should be taken to obtain similar responses to the same objective/concept for the same category (column) on the Likert scale. If it is difficult to construct such questions, the sequence of values that alternately alternate in the Likert answers should be carefully constructed. For example, when responding to the investment orientation in Fig. 8.1, consider the following questions 1, 5, 7 and 8. The same sequence is applied to answers 1, 5, 7 and 8 in the negative. The first question is the understanding of the Fiji Stock Exchange (South Pacific Stock Market—SPX).

For example. To the question "There are 25 real companies listed on SPX", Responding to 'Strongly disagree' implies an understanding of SPX. There are 21 real companies listed on SPX. Also, if the response to the second question, "I do not

Table 8.1 Types of survey questions

Question type	Data type	Analysis potential	Example question
Open-ended numerical	Ratio	Statistical tests	How old are you? How many students are in your class?
Scales	Interval	Many Statistical tests	Your opinion of your statistics teacher; very good; good; average; bad; very bad How do you recommend this course to a friend? How likely is it that you would recommend this course to a friend? Do not recommend 1 2 3 4 5 6 7 8 9 10 Recommend
Rank ordering	Ordinal	Cumulative frequency distributions	Mark all the following reasons as 1, 2, 3, 4, 5, 6, respectively, on the priority of the factors that led you to choose this institution Reputation__; Location__; Parents__; Partner__: Friends__; Employability__
Select all that apply	Categorical (Nominal)	Frequency distributions	The effort you put into education (Tick all applicable); An investment)__; Waste of money__; A correct effort__; A burden__; A gamble__; Waste of time__
Select one	Categorical (Nominal)	Frequency distributions	Gender: Male__; Female__; Other__ Family status: Married__; Never married__; Single__: Divorced__; Widow__
Yes/no	Categorical (Binary)	Frequency distributions	Age negatively affects education; Yes__; No__ Married: Yes__; No__. Children: Yes__; No__ This course fee is expensive: true__; False__
Open-ended comments	Text/words	Tally sheets/coding	What do you think of this course? What are your suggestions for improving this course?

know how to trade in the stock market," is 'Strongly disagree', it is also a correct understanding; the orientation is the same, and the magnitude should be the same.

Then the values for the answers can be obtained in the same direction. That is, answers to questions 3, 4, and 6 should have a high (5) value for 'Strongly agree' and a low (1) value for 'Strongly disagree', and a low (1) value for answers 1, 2, 5, 7 and 8 ('Strongly agree'). Substituting a value of (1) and a high (5) value for Strongly disagree (with a sequential scale). Then the scores would indicate that the higher the value, the greater the willingness to invest in the stock market.

The purpose of the research is to clearly state how long it will take to complete, anonymous or confidential, incentives for participation, and timing of feedback submissions. Also, the details of the questionnaire and instruction set should be provided. It is always effective to provide an incentive for the survey and the participants.

Table 8.2   Likert scale responses

Category	Quantifiers					
Opinions	Very satisfied	Satisfied	Average	Dissatisfied	Very dissatisfied	
	Very important	Important	No comments	Not too important	Not at all important	
	Strongly agree	Agree	Neither agree or disagree	Disagree	Disagree	Strongly disagree
	Strongly support	Support	Neither support nor oppose	Do not support	Oppose	Strongly oppose
Knowledge	Very familiar	Familiar	Not sure	Heard of it	Not aware	
	True	False				
	Yes	No				
Frequency of events	Always very often	Frequently often	Seldom sometimes	Never rarely	Never	
Ratings	Excellent high	Very good	Good	Fair medium	Poor	Very poor
Uncertainty	Very likely	Likely	Probably	May be	Unlikely	Very unlikely

Fig. 8.1   Response to investment orientation Likert scale

Writing an email quiz is relatively easy but less likely to provide visual stimulation or interaction. For example, skip techniques are difficult to use. If prospective participants have technical knowledge, today, the Google Form is a very effective method of surveying such as drop-down menus, pop-up instruction boxes and sophisticated skip patterns. Website-based questionnaires provide many facilities for developing a questionnaire, unlike in traditional paper-based forms. They can monitor the responses online-real time and download them directly to analytical tools such as Microsoft Excel Spreadsheet.

SurveyMonkey (www.surveymonkey.com) is a popular web-based research tool that provides a wide range of facilities for those who are recognized as researchers.

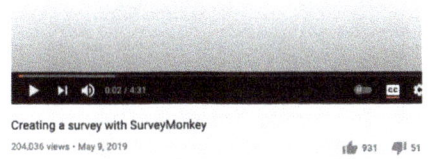

Creating a survey with SurveyMonkey
204,036 views · May 9, 2019   👍 931 👎 51

Source https://www.youtube.com/watch?v=7xdCDJxxoRk

How to use Google Forms - Tutorial for Beginners - YouTube
YouTube · Simpletivity

Sources https://www.youtube.com/watch?v=BtoOHhA3aPQ

Always do a pilot run with a few participants and test the questionnaire.

8.5 Interviews and Conversations

In-depth interviews: Structured, semi-structured or non-structured in-depth interviews with participants are usually 1–2 h long. Interviews try to tell participants stories and explore conceptual details in-depth, often with follow-up questions or examples.

At the root of interviewing is the intent to understand the lived experiences of other people, and the meaning they make of that experience. (Seidman, 2013).

Telephone interviews: Telephone interviews are becoming more and more common today as video conferencing is available for free or at a meagre cost. Today zoom, Skype, Viber, WhatsApp and Messenger (Zoom, Skype, Viber, WhatsApp and Messenger etc.) are widely used for interviews. Phone conversations are shorter than face-to-face interviews and may be at risk of getting shallower and more elaborate. You can learn more about this at the YouTube links below.

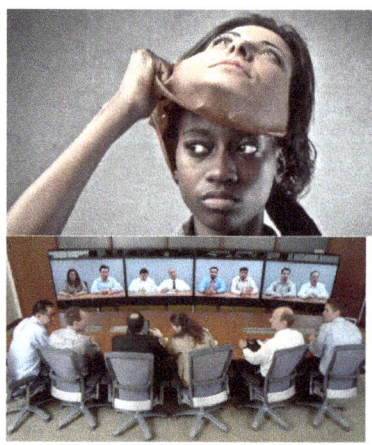

Sources http://www.discoversociology.co.uk/researchmethods/interviews-structured-and-unstructured

In addition, there are a variety of interviews, such as informal conversational interviews, problem-centred interviews, and non-direct interviews, depending on the research topic and the type of response expected.

Roulston (2010) links the type of interview used and the processes involved to the researchers' philosophical positions; Neo-positivist (skilled interviewer asking good questions, takes on a neutral role, avoid bias); Romantic (maintains a high level of rapport and an empathetic connection, generate self-disclosure) and; Constructivist (co-constructing the data through unstructured or semi-structured interviews).

Interviewing is a skill to be learned through experience and training. You need to be well prepared for that in advance. The respondents' consent, the dates and times, the time and place of the meeting, etc., should be decided in advance and agreed upon. The basics, such as explaining the purpose, maintaining confidentiality, using information, and requesting permission to record, must be organized first. The systematic support for exploring the inner feelings of the respondents must be built in a friendly and tactical manner. Finally, it is essential to pay attention to the language and focus strategically throughout the session to control the interview.

Interviews may take place face-to-face or at a distance (over the telephone or by email). It may be tightly structured, or open-ended or form of a discussion. It may involve two individuals or group events (often referred to as focus groups). It may be taped, or one might take notes (Blaxter et al., 2008).

From various questions, changing phrases is an essential feature of a good interviewer. It is a characteristic of a good interviewer to avoid the nature of an interview by asking follow-up questions, using exploratory questions, and allowing confidence to build. As well as listening, careful observation (such as body language) can help identify information about how the interview is progressing.

Example 8.2 Pink Unicorn and Gender Diversity in Fiji, Interview 8.1 shows an example of an interview with a semi-structured questionnaire via email. Using the same questionnaire, the researcher conducted a face-to-face interview with the other participants and took notes. Recorded with the permission of the participants. Table 8.3: Interviews summary shows an example of a summary of interview responses from a semi-structured questionnaire.

Example 8.3 The role of bank lending in sustaining income and wealth inequality, Table 8.4 Shows a summary of the nature of the interviews. By presenting in detail the number of discussions held, how the discussions took place, the time spent, the amount of data collected, etc., the fieldwork procedure of data collection can be built up, and the validity of the data can be highlighted.

Interview 8.1 Email interview with a semi-structured questionnaire

Interview 8.1 Gender diversity in Fijian Companies

Pink Unicorns and Reflecting Gender Diversity in Fijian Companies

Dear Officer, we are doing a study on gender inequality in Fijian companies. Tl questions are meant for interviews but due to current situation, we thought o responses by emails. We greatly appreciate your support by filling and sendin, ts.saliya@fnu.ac.fj **We expect short answers**

Questions for interviews

1. There is a trend of making female representation in corporate boards a requirement, now many European countries have made this. what is y it?
   It is a good thing that making females representation is a legal requirem shows that in past or currently there are less women in the boards thus become a legal requirement.

2. There is a compulsory 30% minimum representation of females in corp boards, do you think that is sufficient? Why?
   30% women representative is not enough in corporate board. At least th be 40 to 50% women in the board for better and equal decisions.

3. Do you think Fijian public companies also should follow a similar rule w listed companies have no female directors or only one? WHY?
   Yes the companies in Fiji should follow similar rules where women shoul included in the company's board for better understanding from both me women context.

4. What should be a legal requirement suitable to Fiji be: none, 10% 30%, 50%?
   At least 40%

5. Why/How do you think that female representation would make a diffe
   Females would be able to voice their opinion out in regards to the opera organization. The opinion from a male will vary from a female thus both opinion is needed to make better decision.

Interview 8.1 Gender diversity in Fijian Companies

6. Many companies appoint just one director to their boards. There are 1 Fijian companies have only one female director. Do you think this a ge to have gender diversity?
   Not really at least there should be more female directors.

7. This single female director in a board is referred to as 'pink unicorn' gi meaning that imaginary thing (an eye wash), Do you agree?
   No I don't agree

8. What kind of difference females can do in Fijian corporate boards, if tl appointed?
   They can raise their concerns in involving more women in leading roles : increase diversity in the professional life.

9. What are the barriers Fijian women face in their career advancement t boards?
   Not much opportunity; Culture barriers

10. Do you think that there is gender discrimination in Fiji, in general?
    To some extent... but it is seen that mostly the gender discrimination is i of vanishing.

11. Do you believe that women are at disadvantaged in career advanceme compared to men, in general?
    I don't believe this statement, as it is seen that in some organizations th females than the males now.

12. What is your perception about men's attitude towards gender diversit
    Man are now trying to change their attitude compared to the days befo

13. Are there different attitudes between Hindus, Muslims and Christians, toward female leadership in religion?
    To some extent as there are religions which restrict women from joining workforce.

14. Does a religious background condition women to accept male leaders!

15. Do you believe in glass ceiling? Why?
    To some extent yes, but now it is seen that the women and minorities a recognition.

Table 8.3 Summary of the interview responses classified into four themes

Item	Response type 1	Response type 2	Response type 3	Response type 4
Number of respondents	ℍℍ ℍℍ·ℍℍ	ℍℍ ℍℍ	ℍℍ	ℍℍ
1. View on female representation in corporate boards a legal requirement	A challenge in Fiji. We have different traditions and cultures; Women are expected to be 'Home-makers'	Good Favorable Diversity adds value Equality before the law is necessary	Females must be skilled in doing both or juggling	Proof of that women is at a disadvantage. Board members should be on merits only
2. Is 30% minimum representation of females in corporate boards, do you think that is sufficient?	Apart from sufficient, even the requirement that 30% has become a challenge	We cannot allocate a % when we consider equal treatment	Sufficient. Women have multiple roles to play	Not sufficient. However, the best person should be appointed to boards irrespective of gender
3. Fijian public companies also should follow a similar rule	A decent start but will not solve the problem	Yes, diversity enhances competitive advantage	30% is a bit much at this stage; first, try with 10%	Definitely, Fiji should follow this move Female or male, they should have the right people

(continued)

Table 8.3 (continued)

Item	Response type 1	Response type 2	Response type 3	Response type 4
Number of respondents	##### ##### ####	##### ####	####	####
4. What should be a legal requirement suitable to Fiji be: none, 10 30, 40 or 50%?	Not sure	We cannot promote any %. Maybe 30%	10% minimum but not out of concession	50% Right people. Selection should be based on merit

(continued)

Table 8.3 (continued)

Item	Response type 1	Response type 2	Response type 3	Response type 4
Number of respondents	‖‖‖ ‖‖‖-‖‖‖	‖‖‖ ‖‖‖	‖‖‖‖	‖‖‖‖
5. Why/How do you think that female representation would make a difference?	More than making a difference, it will be just to show that there is a gender balance and equality	Encourage diverse idea exchange (Multiple Perspectives), Different points of view and increase creativity and innovation. Enhanced Collaboration. to the decision-making process, which may not be possible in a board where they are all males	Because they are multi-skilled and can perform multi-task, they could bring a different perspective & innovation	Signals an attractive work environment for talent. Women have a more intuitive nature than men	Signals competent management for investors. Widen company Talent Pool

(continued)

Table 8.3 (continued)

Item	Response type 1	Response type 2	Response type 3	Response type 4
Number of respondents	‖‖ ‖‖ ‖‖‖	‖‖ ‖‖‖	‖‖‖	‖‖
6. Many companies appoint just one director to their boards. There are 10 leading Fijian companies that have only one female director. Do you think this is a genuine effort to have gender diversity?	It may not be an ideal number, but it is a good point to start from	No. Have no idea, but it could be	It should be more publicly advertised so more women could apply. I am sure more women could be there if they knew	Maybe there are no women there because no women applied or were seen fit It also needs to be merit-based

(continued)

Table 8.3 (continued)

Item	Response type 1	Response type 2	Response type 3	Response type 4
Number of respondents	|||| |||| ||||	|||| ||||	||||	||||
7. This single female director in a board is referred to as a 'pink unicorn', giving the meaning that imaginary thing (an eye-wash). Do you agree?	Not really. In Fiji, these most female directors have genuine interest/stake	Yes	No idea	No, females are becoming more prominent in the corporate landscape in Fiji

(continued)

Table 8.3 (continued)

Item	Response type 1	Response type 2	Response type 3	Response type 4
Number of respondents	ℍℍ ℍℍ-IIII	ℍℍ IIII	IIII	ℍℍ
8. How high is your family's religious attachment?	Very high, high	High	High, Average	Not much
9. How often do you participate in religious activities?	Every day	Almost every day	At least once a week	Occasionally
10. Are women and men equal?	No	No	No	Yes

(continued)

Table 8.3 (continued)

Item	Response type 1	Response type 2	Response type 3	Response type 4
Number of respondents	IIII IIII IIII	IIII IIII	IIII	IIII
11. Why are women not holding responsible positions as men do?	Religious principles treat women as not equal to men. Women are treated as weaker people. Islam promotes more protection for women. Traditionally men have been leading for a long period of time, but it is changing		In recent times the situation is changing, and women are getting more education	Men are at an advantage because they can devote more time to career related activities. Women lack experience
12. What are the main factors constructing these traditions?	Religious beliefs and conservative attitudes. Traditionally men have been leading for a long period of time, but it is changing		Women are more traditional, do not think out of the box. Women are reluctant to explore opportunities	
13. Are women discriminated against in Fijian society?	Yes	To some extent	No	

(continued)

Table 8.3 (continued)

Item	Response type 1	Response type 2	Response type 3	Response type 4
Number of respondents	‖‖ ‖‖-‖‖	‖‖ ‖‖	‖‖	‖‖
14. If so, why they are discriminated against?	Traditions, religious beliefs, cultural practices and male-dominant leaders. Opportunities are not the same for men and women		Women are treated equally in Fiji. Competent women have been promoted, given positions, and opportunities are the same for men and women	

Source Saliya and Hooper (2020)

Table 8.4 Primary data collection techniques & timing, nature & amount of data collected from different participants

Techniques item		Initial telephone interview (i)	Descriptive type questionnaire (1)	Series of short interviews (ii)			Analytical type questionnaire (2)	Further feedback Phone/e-mail (iii)
				E-mails	Chats	Phone		
When		June 2006	July–August 2006	September 2006			October–November 2006	December 2006
Amount of data	Case I	N/A	650 words	13	Once	3 Calls 45 min	3700 words	100 words
No. of minutes/words	Case II	25 min	680 words	14	Twice	2 calls 35 min	2400 words	150 words
	Case III	26 min	530 words	12	Twice	4 calls 45 min	2800 words	200 words
	Cross-checking; All cases	105 min	250 words	41	Nil	30 calls 62 min	3200 words	800 words
Nature of data		Background information of the bank/credit applicant, nature of involvement of the participant	Basic information on participants' experience on credit decisions considered	Friendly chat with the view to design the analytical type questionnaire to obtain more explanations on the credit decisions and to obtain critical analytical data through cross-questioning			Analytical and explanatory data of the credit application, approach, negotiation and decision-making	Verifications/cross-checking/reliability checks
Total duration (7 months)/ Words (16,260 words)		One month 156 min	Two months 2110 words	One month 600 words 137 min			Two months 12,100 words	One Month 1,050 words

Source Saliya (2010)

Table 8.5 DOs and DO NOTs of interviews

DOs	DO NOTs
Establish clearly what the interviewee thinks	Do not appear to judge or indicate that your meaning and understandings of their responses
Balance between open and closed questions	Do not ask leading questions, such as the answers would be 'yes', 'no' or 'agree' etc.
Listen carefully, Give plenty of time to respond	Do not rush
Get responses repeated for clarity	Don't repeat modified versions
If there are doubts or hesitations, probe them to share their thinking	Avoid showing your intentions
Be sensitive to possible misunderstandings	Do not make assumptions about interviewees thinking
Be aware that the respondents may make self-contradictory statements	Do not forget earlier responses
Try to establish an informal atmosphere	Do not interrogate the interviewee
Be prepared to abandon if not going well	Do not continue if respondents appear angry, agitated or withdrawn

Source Arksey and Knight (1999)

According to the emission methodology, this basic classification was summarized into four types of interview responses to facilitate data management and data analysis approach. Accordingly, out of the 32 interviewees, 14 were categorized as first-class, 9 as second class, 4 as third category and 5 as the fourth category.

Table 8.3 data analysis using the Grounded Theory coding strategies: Initial/Open, Focus/Axial and Selective Coding) presented in Part V: Data analysis.

The do's and don'ts of interviews are described in Table 8.5.

8.6 Focus Groups

Industry studies using Focus groups widely use semi-structured interview guidelines. This refers to groups that are usually small random groups of people or seek to get different opinions to obtain specific information on a topic. Such studies are based on the subjective responses of those who respond to a conscious state in which they are involved. The interviewer has prior knowledge of the situation and may respond again if respondents deviate from the theme. These interviews can be with one person or several respondents. They can be a single interview or multiple interviews. They can be with one researcher (interviewer) or with several interviewers.

Moderator must be appointed who fit the needs of the group taking part in the focus group, Moderator skilled for children is not suitable for a group of web designers (Krueger & Casey, 2009). An experienced facilitator not necessarily a subject matter expert (Langer, 2001). Also, ethnicity similarity can reduce communication barriers. (Halcomb et al., 2007).

8.7   Observation

There are many ways to collect observational data.

8.7.1   *Naturalistic (Non-participant or Non-reactive) Observation*

Such natural observations are often used in research, such as case studies, the study of a specific situation, often focusing on a business or individual, or explaining a broader trend or problem. The natural observation process allows natural observations to be made in a way that does not interfere with the behaviour of the observer as much as possible.

Observation is a complex combination of sensation (sight, sound, touch, smell, and even taste) and perception, through which we develop schemes, the mental structures we use to organize and simplify our knowledge of the world around us.

8.7.2 Overt and Covert Observation

Open observation is the method of observation in which the observer knows that the observation is taking place. The participation of the researchers is explained to everyone, and their support is requested. Confidential surveillance means that the observer is unaware that the observation is taking place. One of the arguments favouring covert surveillance is that people may change their behaviour when they find out an observation has been made, thus reducing the chances of threatening the validity of the results. The problem with covert surveillance is that it can be built unethically. But if it is no secret, the real behaviour of the people may not be observable. It highlights many of the overlapping benefits of covert participant surveillance: access to institutions, data collection as an insider, and no impact on the general behaviour of the subjects being monitored.

According to Berg (2006), it might be impossible to observe the real behaviour other than covertly. Meanwhile, Roulet et al. (2017) outline many overlapping benefits of covert participant observation, such as access to institutions, collecting data as insiders, and having no influence on the normal behaviour of the subjects being observed.

Figure 8.2 shows a classification of observers according to their openness or confidentiality and non-participation. Accordingly, there are four types of observers. The announced participant researcher knows that he, the observer population, and

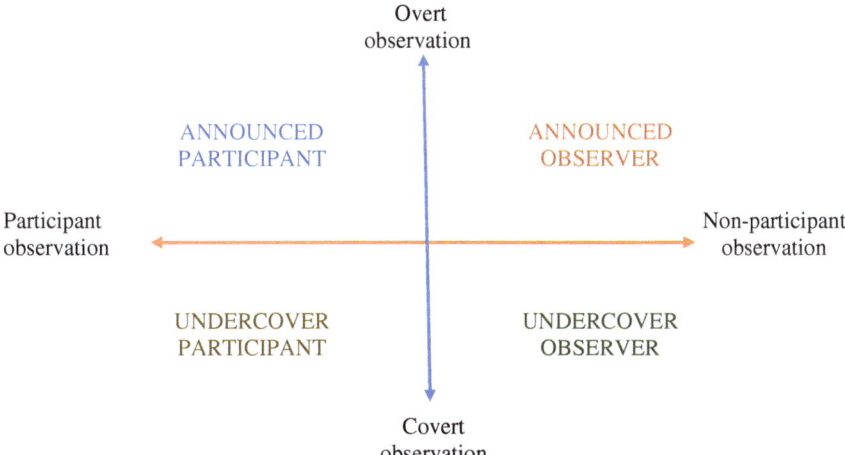

Fig. 8.2 Classify observers by open confidentiality and non participation. *Source* Gray (2018, p. 408)

other observers are being monitored. The unannounced participant, the researcher, is in the crowd he is observing, but the other observers are unaware that they (the researcher) are being observed. The announced observer is not in the crowd he is watching, and the other observers know that they are being watched. The undercover observer is not in the crowd he is watching, and the other observers are unaware that they are being watched.

8.7.3 Participant Observation

Participant observation involves investigating conversations, events, customs, and communications within a community or observing a group of people being a member of that community/group. Self-ethnography is definitely valid as a methodology, and it must be carefully selected and carefully considered to be the appropriate methodology for the impact and evidence needed for the project. Among the important factors to consider when collecting anthropological data is the selection of a field, setting a time frame, gaining access, obtaining informed consent (invisibility, building support and identity identification) and designing an exit mechanism.

**Participant observation in a
Tibetan Buddhist monastery**

8.8 Example 8.3: The Role of Bank Lending in Sustaining Income and Wealth Inequality

Below is a participatory observation made for a case study selected in association with a private bank; **Participant Observations** 8.1, 8.2, 8.3 and 8.4.

Participant Observation 8.1 The approach:

One Friday, while the Chairperson was conducting one of the review meetings of the Soft Bank, Mr. Tony, the owner of Tony Group, contacted the Chairperson for an immediate appointment to discuss an urgent matter. The Chairperson immediately made arrangements for a quick meeting showing all the respect due to another business tycoon of the country.

Tony was adjudged the Entrepreneur of the Year in 1995. His group had accounted for a major share of textiles and garments exports, which was the largest foreign exchange earner of the island. The Tony Group expanded very fast, especially during the previous regime of the Republican Government, with huge loan facilities from the government-owned banks, for the setting up of factories in rural areas. The Tony Group had a very good reputation for manufacturing garments for world-renowned brands like Marks & Spencer, GAP, Van Hussein, and a few others. Those buyers also had invested in high-tech equipment and quality-control experts in the Tony Group factories. The Tony Group had been running smoothly. The total export income of Sri Lanka was US$12,050 million in 1996, and the Textiles & Garments Sector accounted for US$6,484 million which was more than 53 percent of the total export income of the country. The Tony Group accounted for approximately US$200 million in the year 1996.

Tony arrived in 30 min with his team of professionals, including the Finance Director (experienced accountant), Financial Controller (chartered accountant), Chief Operating Officer and three Joint Managing Directors. Tony explained how he developed the group. He stressed the national importance of his business, especially in employment generation and foreign exchange earnings.

Participant Observation 8.2 The proposal

According to Tony, his group had 15,000 permanent employees and another 15,000 contract workers. He tabled the details of his factories with their capacity and locations. He blamed the newly appointed Labour Government for not allocating an adequate quota for him to work on full capacity and therefore, he complained that his group was running with a negative cash flow due to under-utilization of assets. He was in the midst of a crucial issue and stated that he had no money to pay the salary bill for the current month. The best employees were leaving the group and he said he had no option other than agreeing to the foreclosure suggested by the banks.

He also pointed out that the new owners of his business would not be able to maintain the same rapport with the international buyers as he did and, the whole episode would end up in a tragedy, pushing 30,000 workers out into the streets. He was pleading to the Chairperson to bail him out, assuring that this credit line would rejuvenate the whole group and the future cash flows would be very healthy with the orders in their hands. He also requested a facility of Letter of Credit (LC) to import the necessary fabrics and accessories.

At the request of the Chairperson, he explained his cash-flow situation and borrowing positions. The total borrowings of the Tony Group were more than US$20 million. This comprised US$5 million from Bank of Lanka, US$5 million from SET Bank, US$2 million from HAT Bank and a syndicated loan of US$8 million from all three banks. He had borrowed from three private banks and the Bank of Lanka, the largest state-owned bank in Sri Lanka, which had structured a syndicated loan as well. The monthly commitment for servicing these loans was US$0.7 million. Due to non-servicing of interest, almost all the banks had classified Tony Group as a defaulter and had reported this to the Credit Information Bureau (CIB), which is the central monitoring authority for defaulters.

When evaluating a client's creditworthiness, the first thing a credit officer has to do is call for a CIB report of the client. Tony had applied for a facility from the Soft Bank a week ago, and the team leader of the Corporate Credit Division had declined it on the basis of the CIB report.

Tony was very convincing and very politely explained how the senior officers of the Soft Bank rejected his proposal and praised the Chairperson for his visionary leadership and patriotic attitudes in advance.

Participant Observation 8.3 The negotiation

The Chairperson showed his grief about the situation and blamed the politicians, regulators and bankers for not identifying the country's needs.

He said, "This is the whole problem with our bankers. They are guided by some stupid rules called banking practices and ruin people like you [Tony]. They will never think beyond that cage of banking practices. Even our state policies do not have provisions to support people like you. Now, you have generated 30,000 jobs. If they lost their jobs and you lose your business, the cost will be much more to the economy in the long run." Then, in a disappointed tone, he said, "I don't know, when these people will learn these things?...."

After listening to overjoyed remarks by Tony who too joined the Chairperson to criticize the prevailing systems, the Chairperson suggested "…tell me Tony how much do you want? And what is the collateral you can offer?".

Tony said, "Sir, all my assets have been taken by the banks, I can give my personal guarantee and the secondary mortgage of the assets which have already been mortgaged to the other banks, and Chairman sir, believe me and I will not let you down."

"Your house?" the Chairperson queried. "That's in my wife's name and I am sorry sir, I can't draw her into this, she will eat me", Tony answered.

Then the Chairperson asked, "Can you give your wife's personal guarantee?".

But Tony politely disagreed saying, "I am very sorry sir, but do not worry I will never let you down and I will pay every cent due to your bank on time, and I am not going to deal with any of those other banks in the future, my one and only bank is your bank for the rest of my life".

The Chairperson smiled and said, "Ok, tell me your requirement, have you got a good accountant? Do you have the cash-flow projections?".

"Yes sir, our cash flow is always positive but not adequate to service the loans because of production hiccups and not performing our factories in their full capacity. If you help, we can have enough orders to fill the factories to perform at full capacity then we will have cash surplus of US$1 million initially and would grow to US$1.5 million in three months' time. To do this I need US$2.5 million advance to pay my salary bill and other statutory dues. And I need a LC facility of US$1.5 million to finance input materials for uninterrupted production." (A monthly cash flow projection was tabled and explained it by the Financial Controller and the Chairperson looked convinced).

"Is that all you need? Tell me right now, you will not be facilitated under the normal banking practices by our credit officers in the bank."

Tony was jubilant and said, "That's what I wanted your Honour. I do not need a cent more than what is necessary", and he invited the Chairperson to chair his board: "Why don't you come and chair our Board meetings as well?".

The Chairperson thanked him for the invitation and said, "I don't want to interfere with your business, you are the best person to manage your business, but you can give us good publicity." "Of course, sir", Tony readily agreed.

The following day the business page of a leading newspaper carried an article entitled, "Soft Bank rescued Tony Group". The news spread fast, and Tony announced that Mr. Perera had helped him. He issued a special circular to the Tony Group employees saying that they all must do banking with the Soft Bank explaining that the Soft Bank was a truly kind bank, while criticizing all other banks for advocating him to sell the factories.

The Chairperson was very happy about the publicity given and the copies of the paper cutting of the news item circulated among the Board members at the next Board meeting. He assured the board that he will bring more and more business and expressed his regrets that he could have done that before, meaning involving himself more in the day-to-day activities of the bank.

Participant Observation 8.4 The decision

The Chairperson approved all the facilities amounting to US$4 million. He was very critical of the bank officers who rejected Tony's proposal purely on the CRIB report and said, "I think we need a good accountant, not a banker, to manage Tony's facilities" and appointed the CFO as the credit officer in charge for the Tony Group. The CFO was instructed to monitor the cash flow position thoroughly and whenever there was a situation which warranted over the limit borrowing, he had to report to the Chairperson.

Tony was jubilant on the quick decision made by the Chairperson and offered him a seat on the main Board of the Tony group with a brand-new Mercedes, which was still in the harbour, imported under the permit granted for exporters. The Chairperson politely refused the offer and appointed his CFO as an observer in the Board of the Tony Group. The Chairperson, pointing the CFO, said "He will be my eyes and ears in your Board".

8.9   Case Studies

Example 8.4 The role of bank lending in sustaining income and wealth inequality.

Narratives: (Although short because they were made for a case study), these narratives are constructed in the form of long novels in a design aimed at narrative research (**Narratives** 8.1, 8.2, 8.3 **and** 8.4).

Narrative 8.1 Background

Superclean Services was a sole proprietorship of Mr. Silva of Dehiwela. He started this janitorial service business in 1991. His first client was the Dehiwela branch of the Soft Bank, which was the first branch of the bank. His service was very satisfactory so that he was contracted with all janitorial services of the bank. As the bank expanded its branch network rapidly Superclean's business boomed, becoming the sole janitorial service provider to the bank.

Mr. Fernando, the manager of Dehiwela branch (BM) handled the Superclean account very systematically, financing its initial capital investment requirements to serve the newly opening branches. The recovery process was streamlined by directing the payments from the Soft Bank to the loan account of the Superclean. Superclean was exceptionally good at meeting the needs of "finishing up" at branch openings, mostly at short notice, sometimes at concurrent branch openings. Further, he did these services all free of charge for the opening day polish-ups, strengthening the relationship with the Bank. The business is small scale and neither Mr. Fernando nor Mr. Silva worried about hiring qualified accountants to manage the accounts and finances of Superclean.

Silva had accumulated a considerable amount of assets in his business, and always had substantial payments due from the bank, against which he had facilities, including an overdraft. His ambition was to cater to big institutions like the parliament, airport, embassies, etc. For that purpose, he invested more in the business, which created a liquidity problem. Fernando was worried about continuing requests for additional facilities and/or overdue loans, and Silva's argument was that there was money overdue from the bank. But he did not have the accounting expertise to prove his claim.

Silva used to send a team to polish the house floors of bank officers, who did him even the slightest favour, at Christmas or at New Year. Such personal favours to the bank officers gradually became as dues and his inability to cope with the requests made by down-the-line staff (in addition to the managers) caused irritation amongst staff circles. The staff made it a point to complain that the Superclean services were not up to standard and branch managers started cutting payments for the slightest lapse or error of Superclean staff. Further, branch managers sought permission from the CEO of the bank to look after their own janitorial needs and started recruiting Superclean employees on a contract basis.

Meanwhile, Mr. Nath, who was in-charge of janitorial services at ABC Ltd., which was an established large company in the same business, approached the Soft Bank through the Personal Assistant (PA) to the CEO. Finally, the PA, who was a classmate of Mr. Nath, became in full charge of the bank's janitorial services and started to divide branch janitorial work between the two contenders, ABC and Superclean. Still, it was expected that Superclean should attend to the opening day's work of new branch openings, which was offered free, and thereafter it was the PA who decided "to give which branch to whom".

Narrative 8.2 Action against Mr. Silva

Fernando, manager of the Dehiwela Branch, was promoted and transferred to head office and the credit officer of the branch, where Silva had his account, became powerful in the branch. The credit officer, who was not on very good terms with Silva over some misunderstanding, started treating Superclean account in the strictest sense of discipline, ignoring delays in receiving bank's payments thereto and thus depriving him from business expansion. The PA's intervention aggravated the issue. Silva was experiencing a very severe liquidity crunch, lost business gradually because the Soft Bank:

Did not issue any performance bonds for Superclean to canvass new clients.

Returned all cheques if without fund and no arrangements made.

Did not extend advances against payments from the bank and did not allow withdrawals from his account until "overdues" are cleared (despite payments to him from the bank being still overdue for months.

It was later revealed that the branch has over-recovered (charged) interest, substantial enough to recover half the term loan. The Superclean was too small to afford qualified or experienced accountants and totally relied on the bank statements.

Soft Bank classified the Superclean account as non-performing and transferred the file to the recoveries department for closely monitored recovery action, failing which, the Soft Bank could serve a Letter of Demand proceeding towards legal action.

Narrative 8.3 Negotiations

After careful examination of the account, realizing that the bank had overcharged interest and pushed the client to the present condition, Manager–Recoveries (M–R) comprehended Silva's plight in all aspects, i.e.:

Overcharging of Interest.

The non-recovery of capital whilst funds actually being available (due from the bank) causing a loss situation to the client.

The deprivation of business expansion was "criminal" negligence in banking sense.

By this time, Silva had lost all his credibility and business integrity and was unable to pursue tender bids without performance bonds. The only business that Superclean had at this time were a few branches in the conflict areas (the North and East) where ABC was reluctant to serve.

The Manager–Recoveries called for explanations from the new branch-manager of the Dehiwela branch on the overcharged interest. A bank officer said that, "This action probably may not have been taken to take action against the errant staff but may be to build up a case and find an avenue to help the victim of errant action of the bank".

Narrative 8.4 Decisions

Decision 8.1

The Manager–Recoveries was a professional experienced banker and was working with Soft Bank from the beginning. He had authority to re-structure loans for the purpose of recovery/settlement and if new facilities were required, he had to recommend them to the CEO for approval. He had no authority to approve new facilities. But realizing the lapse on the bank's part, and identifying the need of resurrecting Superclean, Manager–Recoveries made the following decisions.

Set-off of overcharged interest against capital outstanding.

Re-scheduling the Term Loan; and

Appropriation of money due from the bank for the services rendered to bank branches.

The Manager–Recoveries, going beyond his authority, further approved the following facilities for reasons unknown:

Performance bond facility to secure future contracts.

A small overdraft; and,

Additional facilities as and when he successfully negotiates business on assignment of payments direct to the bank.

Mr. Silva started building up his business outside the bank and was able to secure a few lucrative contracts in outstations. Some friends helped him because he deserved it but with the blemish created by the bank's unreasonable action, he lost a few good contracts. However, he was gradually coming out of his commitments but was rejected again by a higher officer who strictly followed "banking rules".

Decisions 8.2

The Manager–Recoveries lost his job for approving facilities (including Superclean), without authority. The succeeding Chief Manager–Recoveries (CM—R), a retired senior banker assigned on contract for special recovery jobs declined the credit applications of Superclean for new facilities on grounds that Silva should settle a substantial part of the existing commitments for him to recommend a new facility and/or did provide a positive cash flow projection to convince the decision-maker.

Silva is now in a pathetic situation, struggling to save his house which was mortgaged to Soft Bank, and looking for an opportunity to present his case to the Chairperson. However, he did not have the necessary social network to reach the Chairperson as Tony did nor had an appealing project to present at an investor promotion forum as Yousef did. He had a certain amount of wealth, different networks and a customer base although he was not in the affluent class such as Tony or Yousef. He is small and powerless.

8.10 Unobstructive Data

Uninterrupted data include the use of independent, non-reactive sources without the involvement of the researcher and includes documentary evidence, physical evidence, and archival analysis. With the development of electronic environments such as the Internet, what is unique and unpublished or undisclosed becomes increasingly useless and wasted. Web pages, for example, can contain links to many other websites or pages, challenging the notion that a document should be an essential and independent report.

8.10.1 Physical Measures

Webb et al. (2000), Identify four categories of physical measurement data. These include natural developments (natural accretion such as museums), controlled developments (controlled accretion such as the use of the Internet), natural erosion (how many times a product has been used, and natural erosion such as clothing monitoring) and controlled erosion (how long a person has walked. Controlled erosion-related measurements, such as measuring the sole of the foot for measurement.

8.10.2 Documents (Secondary Sources)

Running records: Institutional documents such as annual reports, board meeting reports, action reports and political and judicial reports.

Episodic records: These records are more personal, for example, personal diaries, correspondence, and so on.

Digital archives: With the development of the Internet and the World Wide Web, these traditional methods of document processing have changed significantly.

Visual research: Sleeman (2002) points out, with the growth of electronic environments such as the internet, what is unique and published or unpublished is increasingly blurred. Web pages, for example, can contain links to many other sites or pages, challenging the notion of a document as an integral and independent record. With the growth of the use of visual images in scientific inquiry, Bell and Davison (2013) offer a useful categorization based on a distinction between empirically-based methods and theory-based methods.

Visual media include two-dimensional still images, such as photos, cartoons, maps, charts, logos, and diagrams; Two-dimensional motion pictures, television, video and interactive web pages and other multimedia including 3D and live media such as clothing and architecture. Words and numbers-based approaches have become prevalent in the social sciences, and there is a general awakening of visually inclined theoretical frameworks and methods of visual research. Visual methods can rearrange parts that cannot be reached by other methods. With the development of visual images of the use of scientific inquiry, a useful classification is made based on the difference between empirically-based methods and theory-based methods.

Over two evenings of ground-breaking lectures Dr Lez Henry helps us to meet that ch
powerful lectures, he will lay out an untold history of the pseudoscience that created c
around colour. Helping us understand how they became erroneous sociological const
built into the modern world. If we are to overcome the ideological myth of white superi
inferiority that's deeply embedded in our society and consciousness, then we must st
understand its history. If you think to live untainted or free from the doctrine of colouri
This series will make you think again, this is thesilver bullet.

2019 Lecture Dates:
May 7th in Birmingham
Sept 3rd in Manchester

Empirically-based: Visual research methods have historically been seen as part of the postmodernist tradition, and visual data such as photographs are used as factual evidence to support a realistic narrative. The disadvantage here is that the information provided by the visual data on how to analyse and interpret is limited. It also ignores the cultural and historical context in which the images are produced. There are three approaches to this. These are visual content analysis, semiotics, and critical visual analysis.

Theory-based: These approaches include aesthetics such as art theory, fashion and dresses, semiotics and rhetoric, and critical visual analysis. All of these data collection methods have their inherent advantages and disadvantages. They should be carefully weighed before using any method.

While validity and reliability are important issues, they are not sufficient. In considering the value of a study, plausibility and credibility must also be considered. In writing reports, researchers should provide sufficient evidence to convince an audience. (Hammersley, 1992).

8.11 Summary

- Instead of the term 'data collection', the term 'intelligence gathering' is used to include sources such as memory recollection and experience reconstruction.

- Quantitative data collection methods are structured rather than qualitative data collection methods such as interviews, open questionnaires, participant observations and experience reconstruction.
- Quantitative data collection methods include live surveys, questionnaire surveys, mobile and booth surveys, follow-up studies, webinars, live polls and formats and systematic observations.
- Data collection can be done through primary and secondary sources. The primary sources are the participants, including the researcher, and the secondary data are published documents and archival records.
- In-depth interviews are often conducted in a structured, semi-structured or non-structured manner with a small number of participants. Telephone Interviews Video conferences are also available today as telephone interviews are available for free or at a very low cost.
- Today, Zoom, Skype, Viber, WhatsApp and Messenger are widely used for interviews. Phone conversations are shorter compared to face-to-face interviews and may be at risk of having less depth and breadth.
- Surveys are a method of getting feedback on a topic from a given type of questionnaire. Samples are usually large, but the depth of reaction is often not as complex as that of focus groups or interviews.
- Common scales are scales like Likert.
- Sample selection can be made using different sampling methods. They range from random, purposive to convenient samples.
- Always do a pilot test with several participants and test the questionnaire.
- Details of the questionnaire and instruction set should be clearly provided. The purpose of the research is to clearly state how long it will take to complete, anonymous or confidential, incentives for participation, and timing of feedback submissions. Giving an incentive to survey participants is often effective.
- There are many ways to collect observational data: naturalistic observation (non-participant or non-reactive): ethnography, and participant observation.
- There are four types of observers; announced participants; Observers know that they are being monitored. Unannounced participant (secret participant); observers do not know that they are being observed. Announced observer (open Observer); They know that they are being observed; undercover observer (Secret Observer); Observers do not know that they are being observed.
- Ethnography: The study of conversations, events, customs, and communications within a community or the observation of a group of people within that community. The methodology of self-anthropology should be carefully selected and carefully considered for the implications and evidence required for the project. Among the important things to consider when collecting anthropological data is the selection of a field, setting a time frame, gaining access, obtaining informed consent (invisibility, building support and identity identification) and designing an exit mechanism Should have.
- Data collection by recording in detail the summary of interviews, summaries of interviews and responses recorded by the researcher or with the permission of the

participants, the manner in which the discussions took place, the amount of time spent, the amount of data collected, etc. It can build confidence in the field action procedure and highlight the validity of the data.

- Facilitates data management by facilitating data management and data analysis approach by categorizing interview responses according to the excitation methodology.
- Uninterrupted data is available without researcher intervention and includes documentary evidence, physical evidence, and archival analysis.
- Physical measurement data include measurements related to natural developments (natural accretion such as museums), controlled developments (controlled accretion such as Internet use), natural erosion, and controlled erosion (such as controlled erosion).
- **Running records**: Institutional documents such as annual reports, board meeting reports, action reports and political and judicial reports.
- **Episodic records**: These records are more personal; Personal diaries, correspondence, etc.
- Visual media includes two-dimensional still images such as photos, cartoons, maps, charts, logos and diagrams.
- Visual research methods have historically been seen as part of the postmodernist tradition, using visual data such as photographs as factual evidence to support a realistic narrative. These include visual content analysis, semiotics, and critical visual analysis.
- **Theory-based**: This approach includes art theory, fashion and dresses, semiotics, rhetoric and critical visual analysis.

References/Further Reading

Arksey, H., & Knight, P. (1999). *Interviewing for social scientists*. London: Sage.
Atkinson, P., & Hammersley, M. (1994). Ethnography and participant observation. In N. K. Denzin, & Y. S. Lincoln (Eds.), *The Sage handbook of qualitative research* (pp. 248–261). Thousand Oaks, CA: Sage.
Bailey, C. A. (2007). *A Guide to qualitative field research*. Thousand Oaks, CA: Sage.
Blaxter, L., Hughes, C., & Tight, M. (2008). *How to research*. Berkshire, England: Open University Press.
Blaxter, L., Hughes, C., & Tight, M. (2010). *How to research*. UK: McGraw-Hill Education.
Brinkman, S., & Kavale, S. (2015). *Interviews: Learning the craft of qualitative research interviewing*. Thousand Oaks, CA: Sage.
BRM. https://research-methodology.net/sampling-in-primary-data-collection/.
Czarniawska-Joerges, B. (1992). *Exploring Complex Organizations: A Cultural Perspective*, Newbury Park, CA: Sage.
Czarniawska, B. (2004). *Narratives in finance science research: Introducing qualitative methods*. London: Sage. https://doi.org/10.4135/9781849209502.n5.
De Vaus, D. A. (2002). *Surveys in social research*. London: George Allen & Unwin.
Dolnicar, S. (2013). Asking good research questions. *Journal of Travel Research, 52*(5), 551–574.
Focus group. https://medium.muz.li/how-to-conduct-a-focus-group-d401ab7081eb.

Gillham, B. (2007). *Developing a questionnaire*. London: Continuum.

Gray, E. D. (2018). *Doing Research in the Real World* (4t ed.). London, Sage.

Halcomb, E. J., Gholizadeh, L., DiGiacomo, M., Phillips, J., & Davidson, P. M. (2007). Literature review: considerations in undertaking focus group research with culturally and linguistically diverse groups. *J Clin Nurs., 16*(6), 1000–11. Doi: https://doi.org/10.1111/j.1365-2702.2006.01760.x.

Hammersley, M. (1992). *What's Wrong with Ethnography?* Routledge, London.

Krueger, R. A., & Casey, A. (2009). *Focus groups: A practical guide for applied research*. Thousand Oaks, CA: Sage.

Langer, J. (2001). The mirrored window: focus groups from a moderator's point of view. Ithaca, NY: Paramount Market.

Lee, R. M. (2000). *Unobstructive measures in social research*. Buckingham: Open University Press.

Liamputtong, P. (2011). *Focus Group Methodology: Principles and Practice*. Sage.

Roulet, T. J., Gill, M. J., Stenger, S., & Gill, D. J. (2017). Reconsidering the Value of Covert Research: The Role of Ambiguous Consent in Participant Observation. *Organizational Research Methods, 20*(3), 487–517. https://doi.org/10.1177/1094428117698745

Roulston, K. (2010). *Reflective interviewing: A guide to theory and practice*. London: Sage.

Saliya, C. A. (2010). *The Role of bank lending in sustaining income/wealth inequality in Sri Lanka*. PhD Thesis, Auckland University of Technology, New Zealand. https://openrepository.aut.ac.nz/handle/10292/824

Saliya, C. A. (2021). Enterprising mother: A grounded Theory Emerged from Fijian Working Women. Grounded Theory: A International Journal. Under review.

Saliya, C. A., & Hooper, K. (2020). The Role of Credit Weapon and Income/Wealth Inequality: A Sri Lankan Case Study. *Journal of Applied Economic Sciences, 15*(2), 425–436. http://cesmaa.org/Docs/JAES%20VolumeXV%20Issue2(68)Summer2020.pdf

Seidman, I. (2013). Interviewing as Qualitative Research: A guide for Researchers in Education and Social Sciences. NY: Teachers College Press.

Sleeman, D. (2002). Knowledge technologies: A reuse perspective. *Research and Development in Intelligent Systems* XVIII, 3–6. https://doi.org/10.1007/978-1-4471-0119-2_1

Smith, J., & Hodkinson, P. (2005). Relativism, criteria and politics. In N. Denzin, & Y. Lincoln (Eds.), *Handbook of qualitative research, 3*, 915–932. London: Sage.

Webb, E. J., Campbell, D. T., Schwartz, R. D., & Sechrest, L. (2000). Unobstructive Measures: Nonreactive Research in the Social Sciences. Thousand Oaks, CA: Sage.

Youtube Links

How to use Google forms: https://www.youtube.com/watch?v=BtoOHhA3aPQ.

Sampling techniques: https://www.youtube.com/watch?v=6skCMCdh3FY.

Sampling and Sampling Methods: https://www.youtube.com/watch?v=VoaQcff-uxQ.

Survey monkey: https://www.youtube.com/watch?v=7xdCDJxxoRk.

Types of Data: Categorical vs Numerical Data. https://www.youtube.com/watch?v=DUcXZ0 8IdMo.

Part III
Research Traditions: Quantitative Strategies

Chapter 9
Research Traditions

...articulating the research problem and choosing your inquiring strategy depends on each other, and figuring out a problem depends on the ontological and epistemological stance of the researcher and preferred methodology (Saliya, 2021a).

Understanding various strategies in advance will help you in many ways.

Choosing a topic; to choose your research area and topic as well as articulate the research problem. This is because, especially when following qualitative strategies, interpreting the problem to suit your strategy and choosing the strategy according to the problem are intertwined (Saliya, 2021a). Furthermore, a general understanding of these strategies is important as they depend on your current attitudes, beliefs and skills (paradigm, perspectives, worldview and preferred theories) (Saliya, 2021a). Acknowledgement of the researcher's standpoint will be extremely useful during discussions with the peers/supervisor and the reader to assess the interpretations of results.

Comprehending literature; the foundation for choosing a research strategy should be laid while reviewing the literature related to your topic. Having a general idea of these strategies is also very helpful in reading and understanding certain advance scholarly articles when you do the literature review.

Deciding on the scope & dept; on the other hand, understanding various strategies within the schools of quantitative, qualitative or mixed methods will have a decisive impact on the scope and depth of your investigation. If you want to discover a general perception of people on something, you will need a survey. That means you will need a larger sample. On the other hand, if you care more about how people think or feel about something, you may want them to tell their own stories or discuss the topic with you in detail (interviews). You do not need a large sample (N = number) like a survey. But you must look deeper (D = deep). These two factors are mutually exclusive, meaning that the greater the number (N) the greater the compromise of depth (D), because time and resources are limited). This does not mean that interviews or case studies cannot be done with a large sample without compromising dept, but

C. A. Saliya, *Doing Social Research and Publishing Results*,
https://doi.org/10.1007/978-981-19-3780-4_9

Table 9.1 Research examples by sample size (N) and exploration depth (D)

Focus	Examples	Sample (N)	Dept (D)	Type of strategy
In-depth reading of texts emphasizing physical structure. Concept exploration	Filming/editing a movie, searching for word-of-mouth for a lecture	Small	Deep	Textual analysis, Case studies, In-dept interviews, Narrative research
Brief reading of many texts of the same type	Examining common features and differences between weaves	Medium	Medium	Content analysis, Grounded theory
A brief reading of many different types of texts	Ideological discussions, log entries	Large	Medium	Discourse analysis Action research
Obtaining specific information on a topic	Getting different opinions from random or specific groups	Medium	Medium	Focus groups Phenomenology
Getting personal perspectives on a topic	Submitting a questionnaire per person in a group	Large	Narrow	Surveys (Quantitative research)
Monitoring and investigating communication within a community	In-depth and long-term study of a community for influence and evidence	Small	Deep	Ethnography Case studies
Taking and testing a variety of sources	Examination of common features/differences by statistical methods	Large	Narrow	Metasynthesis
Determining differences by manipulating variables	In one group in a controlled environment	Small	Deep	Experiments

it may be a very resource consuming, difficult exercise as it may seem. Table 9.1 shows some examples of research types in terms of sample size (N) and depth of exploration (D).

This introductory and Chap. 10 discuss data types, variables, verification of normality tests, overdispersions and corrections (reliability tests), extraction and reduction of variables (data reduction) etc. Chapter 11 discusses the basic statistical principles and concepts (p-value, F-statistics, t-statistics, Chi-square, statistical power, effect size, confidence levels, confidence intervals, significant levels, RMSEA, CFI and TLI of SEM etc.), and methods of verifying the goodness of fit or model fit. Chapter 12 discusses statistical testing techniques, model building and path diagrams, and Chap. 13 outlines Structural Equation Modeling (SEM), a very complex statistical technique. Chapter 14 discusses the principles of Econometrics.

The real-world examples, which are recent publications discussed in this book, would provide you with the practical experience needed to submit a research paper appropriately to an international journal. I am confident that the analyses provided here as examples are mainly from a critical point of view, therefore, will be more useful for further research.

9.1 Validity/Conformability and Reliability/Credibility

Acceptance of research results, regardless of the strategy used, is based on the validity/consistency and reliability of the research data and methods used. Validity and reliability are the concepts of the quantitative tradition, and accordingly, the qualitative concepts are confirmability and credibility. A detailed study of these concepts should be done in accordance with the chosen research method, using specific sources.

> O'Leary (2004) suggests an additional indicator that can be used to assess credibility in change-oriented research: "usefulness". If a research objective is to expose an inequitable/unjust situation, then a measure of success or credibility will be how useful the research outcome is in proposing a remedy.

The reliability of qualitative data depends primarily on the severity or intensity of the methods used by the researcher. 'Usefulness' can be used as an additional indicator to assess the reliability of research that seeks social change. If exposing an unequal/unfair situation is a research objective, then how effective the research results are in proposing a remedy is a measure of success or reliability.

9.1.1 Validity/Confirmability

Internal validity is achieved if the changes in the dependent variables can be unambiguously traced to the influence of the independent variable, i. e. if there are no better alternative explanations beyond the study hypothesis (Bortz and Doring, 2006, p. 53).

This means the validity of methods, measuring instruments/equipment and the measuring process. That is, internal validity/construct validity, external validity or research design validity, measurement/criterion validity and content validity.

Internal validity is the ability to analyse research results without uncertainty clearly. Internal validity can be obtained if changes in the dependent variables on the effect of the independent variable can be found with certainty, viz., ensuring that there are no better alternative explanations than research hypotheses.

Measurement validity is how the measurement represents the actual structure–what we are trying to measure is subject to measurement. Face validity is how well we measure what we want to measure.

External validity is to what extent the results would be applicable beyond the actual research conditions and individuals involved.

Content validity is the inclusion of all the important dimensions of the variable. Using the 10-choice scale to capture data using the Likert scale has more content validity than the 5-option optional scale, as a 10-choice scale can provide dimensions greater than the 5-choice scale.

> The polygraph–or lie detector–is supposed to measure whether a person is lying or telling the truth. But it measures heart rate, perspiration, and other physiological manifestation of nervousness. The concept of a lie is translated into nervousness, which in turn is translated into the physiological manifestations of nervousness. It looks that it has a face validity-people telling lies are often nervous, but it lacks face validity because being nervous and telling a lie is not the same thing and do not always coincide (Remler and Van Ryzin, 2011).

Polygraph Expert Shows How to Beat a Lie Detector Test

Source https://www.youtube.com/watch?v=N3fHkCFxgQQ

9.1.2   Reliability/Credibility

Reliability is the consistency of a measure. Just because a measurement is reliable does not always mean it is a valid measurement. For example, a weighing scale can always give the same weight, but if the scale is faulty, it will show the wrong weight.

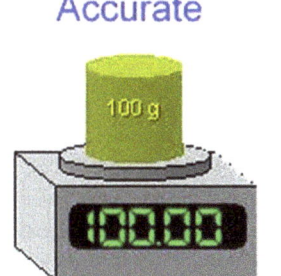

9.1.3 Cherry-Picking

…avoid what we call 'cherry picking'. Cherry-picking means limiting your sample to the sources that will support your argument, which can be avoided by focusing on questions rather than answers (Gournelos et al., 2019, p. 12).

For example, if you are studying actors, your sample should not be limited to the ones you like or the most beautiful/handsome actors.

9.2 Summary

- All researchers should follow reflexive awareness and make an informed choice.
- It is not logical for academic researchers to be confined to quantitative or qualitative schools.
- The research plan systematically aligns the pros and cons and the appropriate methods according to the context.
- The purpose of the quest for knowledge can be an 'expansion of understanding', a 'liberation process' or a 'relentless revelation'.
- Research methods can vary from basic, applied evaluative to critical/radical ethnography.
- There are three main categories: quantitative, qualitative and mixed strategies.
- Defining the problem according to the strategy and choosing the strategy according to the problem are intertwined.
- The suitable strategy depends on the attitudes, beliefs and skills of the researcher.
- Depth (D) is required when a sample is small; large (N) samples are not practical for depth exploration.
- Avoid cherry-picking.
- When selecting the sample, attention should be paid to the questions that are addressed rather than the expected responses/answers.

- Before focusing on data collection, one should have an overview of the most commonly used strategies as research issues, research planning, and philosophical theories in the background should be considered.
- Acceptance of research results is based on the validity/consistency and reliability/acceptability of the research data and methods used.
- Validity and reliability are the concepts of the quantitative tradition, and accordingly, the qualitative concepts are confirmability and credibility.
- A valid measurement is not always valid just because a measurement is reliable.
- Validity applies to all internal, external, content and measurement tools equipment.

References/Further Reading

Creswell, J. W. (2003). *Research design: Quantitative, qualitative and mixed methods approach* (2nd ed.). Sage.

Denzin, N. K., & Lincoln, Y. S. (Eds.). (2005). *The Sage handbook of qualitative research*. Sage.

Gournelos, T., Hammonds, J. R., & Wilson, M. A. (2019). *Doing academic research: A practical guide to research methods and analysis*. Sage.

Gray, D. E. (2018). *Doing research in the real world*. Sage.

O'Leary, Z. (2004). *The essential guide to doing research*. Sage.

O'Leary, Z. (2005). *Researching real-world problems*. Sage.

Remler, D. K., & Van Ryzin, G. G. (2011). *Research methods in practice: Strategies for description and causation*. Sage Publications, Inc.

Saliya, C. A. (2016). Doing research in business management; How to choose your philosophy and methodology? *SSRN Journal of Social Science*. https://doi.org/10.2139/ssrn.2767924

Saliya, C. A. (2017). Doing qualitative case study research in business management. *International Journal of Case Studies Journal, 6*(12), 96–111. http://www.casestudiesjournal.com/Volume%206%20Issue%2012%20Paper%2010.pdf.

Saliya, C. A. (2021a). Conducting case study research: A concise practical guidance for management students. *International Journal of KIU, 2*(1), 1–13. https://doi.org/10.37966/ijkiu20210127.

Wolcott, H. F. (1994). *Transforming qualitative data: Description, analysis, and Interpretation*. Sage.

Youtube Links

Deduction, Induction, Abduction-Georgia Tech-KBAI: Part 5. https://www.youtube.com/watch?v=-nn3XMoPC7s.

How to Argue - Induction & Abduction: Crash Course Philosophy #3. https://www.youtube.com/watch?v=-wrCpLJ1XAw.

The differences between quantitative and qualitative research. https://www.scribbr.com/methodology/qualitative-quantitative-research/.

The Ultimate Guide to Qualitative vs. Quantitative Research. https://www.simplilearn.com/tutorials/data-analytics-tutorial/qualitative-vs-quantitative-research.

Chapter 10
Data and Variables

This chapter describes some important principles of statistical techniques used within the quantitative strategy. Quantitative statistics is a technically specialized field. There are even more subtle micro-specializations in that specialization. There are statistical software related to these specializations. Commonly used software are MS Excel, SPSS, AMOS, Eviews and Mplus. The first edition of the Statistical Package for Social Sciences (SPSS) was released in 1968. You can get a significant understanding of this by using the Youtube links https://www.youtube.com/watch?v=Bku 1p481z80&t=205s; https://www.youtube.com/watch?v=hKA2VQ60bxg.

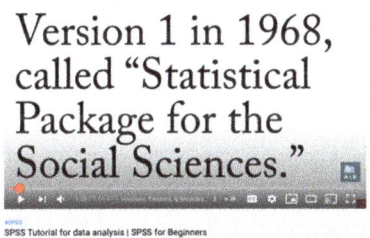

Follow this link to Laerd https://statistics.laerd.com/features-overview.php, a step-by-step website that explains the selection, analysis and presentation of statistical testing techniques using SPSS. The research papers discussed in this book are examples of data analysis using this software.

Proficiency in this software and techniques should be mastered under the guidance of a mentor and/or the use of hundreds of YouTube tutorials available on the Internet. There are many good guides available on the internet to learn software from scratch for a beginner. Once you understand the basic concepts described here, it will not be difficult to understand such Youtube video clips. I hope that by installing SPSS on your computer and practising the exercises in these Youtube video clips, again and again, it will not be difficult to acquire the required special skills.

The quantitative analysis strategy you choose depends on many factors, including (a) the nature of the data; Types of variables (b) Objectives of the research problem and analysis such as comparison of mean and ratios, correlation estimation and forecasting, etc. (c) Number of samples/groups, number of datasets, compatibility/independence between datasets, etc.

The statistical techniques discussed here are the most used statistical techniques needed to select the most appropriate strategy for your research. It is not within the scope of this book to describe how to employ that technique to your data set properly. Therefore, specialization is necessary by studying and practising relevant sources presented here. Part V shows the practical application of some of these techniques as well as how they are appropriately presented to suit the aims of international academic journals.

This chapter is organized as follows.

- Types of data and input data in SPSS.
- Normality tests, overdispersions and corrections.
- Reliability and validity tests.
- Data reduction, PCA, EFA and cluster analysis etc.

10.1   Types of Data

Data classification varies according to the nature of the data and the techniques used. The classification may also vary depending on the amount of data.

10.1.1   Data Classification for Statistical Techniques

All data are broadly divided into numerical data and non-numerical nominal data for statistical purposes. Quantitative data can again be divided into normal continuous data and non-normal categorical data. There are three types of categorical data that do not show this average distribution. They are binary data, ordinal data and count or discrete data. Table 10.1 describes this classification. Table 10.2 provides examples for this data classification.

10.1.2   Parametric and Non-parametric Statistics

Parametric statistics assumes that sample data has a probability distribution. Conversely, non-parametric data makes no assumptions about a parametric distribution. Fortunately, the most frequently used parametric analyses have non-parametric counterparts. This can be useful when the assumptions of a parametric test are violated because you can choose the non-parametric alternative as a backup analysis.

Table 10.1 Data classification for the purpose of statistical tests

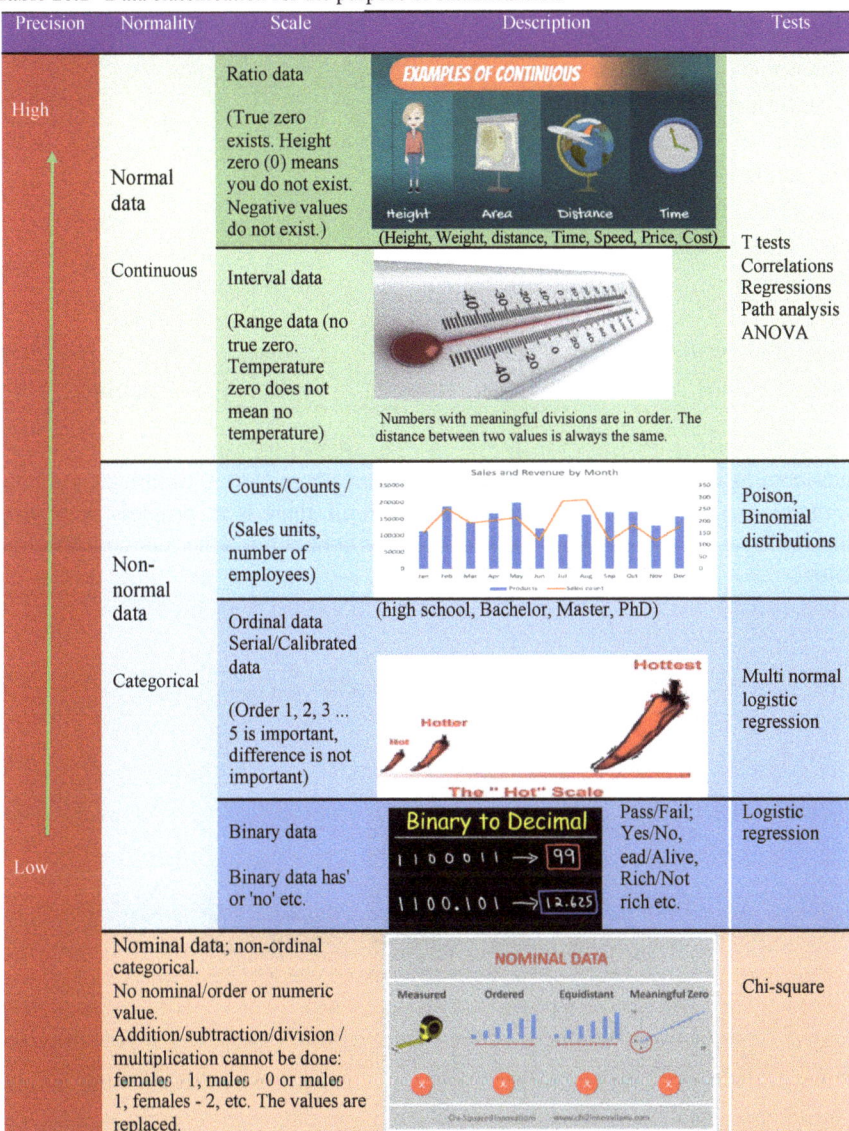

Precision	Normality	Scale	Description	Tests
High ↑ ... Low	Normal data Continuous	Ratio data (True zero exists. Height zero (0) means you do not exist. Negative values do not exist.) Interval data (Range data (no true zero. Temperature zero does not mean no temperature)	**EXAMPLES OF CONTINUOUS** Height / Area / Distance / Time (Height, Weight, distance, Time, Speed, Price, Cost) Numbers with meaningful divisions are in order. The distance between two values is always the same.	T tests Correlations Regressions Path analysis ANOVA
	Non-normal data Categorical	Counts/Counts / (Sales units, number of employees)	Sales and Revenue by Month	Poison, Binomial distributions
		Ordinal data Serial/Calibrated data (Order 1, 2, 3 ... 5 is important, difference is not important)	(high school, Bachelor, Master, PhD) Hottest Hotter Hot The "Hot" Scale	Multi normal logistic regression
		Binary data Binary data has' or 'no' etc.	**Binary to Decimal** 1 1 0 0 0 1 1 → 99 1 1 0 0 . 1 0 1 → 12.625	Pass/Fail; Yes/No, ead/Alive, Rich/Not rich etc.	Logistic regression
	Nominal data; non-ordinal categorical. No nominal/order or numeric value. Addition/subtraction/division / multiplication cannot be done: females 1, males 0 or males 1, females - 2, etc. The values are replaced.		**NOMINAL DATA** Measured / Ordered / Equidistant / Meaningful Zero	Chi-square

This boundary is subjective and depends on the topic and purpose. For example, 10,000 data from 1 to 10,000 with a unit spacing can be easily identified as continuous (with parametric assumptions such as normally distributed) and can be considered for parametric tests even with a small sample of 1 to 150, but is not always recommended. Continuous data are technically an infinite number of steps and form a continuum. The time it takes to find something on a website is continuous because it can be

Table 10.2 Examples for data classification

Typology of variables and data		
Quantitative variables for parametric analysis		
Continuous data	Counts/Discrete data	
Product price, Cost per unit	Units sold, points scored	
P/E ratio, Dividend yield	Number employees	
Age, Weight, Height	Number of meetings/visits	
Categorical variables for non-parametric analysis		
Ordinal data	Binary	Nominal data
Quality; High, Average, Low	Visited/Not visited	Gender (male/female)
Better, Same, Worse	Pass/Fail	Colours
Disagree, Neutral, Agree	Agreed/Disagreed	Areas; Provinces, Districts

measured in as little as 11.427533 s. Time forms a gap from 0 to infinity. As long as there is no (negative) data beyond the zero limit, there is no problem in treating the predictive values, which include decimals, as continuous. Check out the Youtube below.

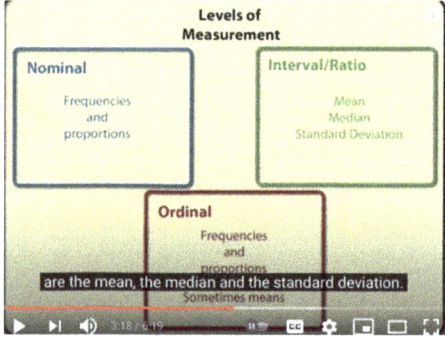

Source Types of Data: Nominal, Ordinal, Interval/Ratio https://www.youtube.com/watch?v=hZxnzfnt5v8

Counts are actually the frequency at which a variable event occurs. Non-integer values can be predicted by treating the count's data as a continuous variable, but this may not be a problem for a large dataset. When there is a very large amount of data, many of the problems with counts variables can no longer be problems.

For example, although the scores recorded in cricket matches are counts data, it can be considered as an intermediate continuous variable. But the number of times New Zealand has won one-day international cricket in the last ten years is not a continuous variable, but only a counts variable.

A **continuous** variable can be converted to a categorical variable using limits or making groups. For example, age can be categorized into continuous 'age groups, such as the elderly, the mature, the middle-aged, the young, and so on, for example, 'Old' group can be over 80 years of age, 'mature' will be 80 years or less and over 60, 'middle aged' can be between 60 and under and over 40, 'young' will be 40 and younger up to 20, and 'less than 20' can be grouped as 'children' etc. Similarly, The factor of height (continuous) can be converted into ordinal data as tall, average and short.

It is easy to summarize categorical variables. Therefore, quantitative variables are often translated into categorical variables for descriptive purposes. However, converting a continuous variable into categorical data reduces the amount of information. Therefore, data classification is often useful for summarizing results but is not generally useful for statistical analysis.

A response can be ranked by using a survey question that expects an order of response (strongly agrees, to some extent agrees, cannot say, disagrees, strongly disagrees, etc.). Such scales are called Likert scales (introduced by Rensis Likert), and such survey questions are called Likert scale questionnaires. Likert scales are either ordinal or interval, and many statistical analysts believe that they are infinite scales because, when well-constructed, there is an equal distance between each value. Therefore, general statistical techniques such as ANOVA or regression are used if a Likert scale is used as a dependent variable in an analysis.

Nominal variables have unrelated values such as high/low, good/bad, high/low, e.g. *Colour*, sexuality, marital status.

Nominal variables have unrelated values such as high/low, good/bad, high/low, e.g. Colour, sexuality, marital status.

10.1.3 Variables

When the data is arranged as a series, it is called a variable. In statistical analysis, when data are processed as a series, they are identified as variables. Quantitative variables are assumed to have a normal probability distribution. Analyzing these variables gives the maximum benefits/advantages of complex parametric analysis. Techniques such as Poison distribution are used for variables that are assumed to have no normal distribution. These categorical variables can be converted into quantitative data when a very large amount of data is used. Then the assumption of a normal distribution can be applied, and all the statistical techniques can be used to get maximum results. Table 10.2 provides examples for identifying data as quantitative and categorical variables.

The definitions and classifications of data types and variables presented in this section are not unique. They can vary from case to case depending on their size, i.e., non-continuous data can be treated as a continuous variable as they increase in size. You should never argue too much about the nature of a particular variable in your analysis! It is not a very effective exercise. Data/variables can be categorized according to their respective statistical techniques in analysis.

10.1.4   Data Feed to SPSS

The data classification in the use of SPSS is as follows. 'Scale' is the term used by SPSS for quantitative variables (both continuous and counts data). SPSS uses the term 'Ordinal' for both graded and binary data and the term 'Nominal' for nominal data for categorical variables.

The following is an example of a data model of SPSS software. The data set consists of three samples of postgraduate students. These three samples are students in three classes. There are 30 students in Class 1 (Day class), 60 students in Class 2 (Night class) and 180 students in Class 3 (Online class).

| Day class (1) | Night class (2) | Online class (3) |
| 30 students | 60 students | 180 students |

Six variables are identified of these samples;

1.   Gender
2.   Vaccinated or not
3.   Employed or not
4.   Age
5.   IQ value
6.   English marks

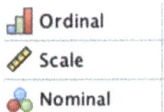

Ordinal

Scale

Nominal

The **variable view** and **data view** of this data set in the SPSS software are shown
in the two screenshots below. The bottom-line states that it is *Data View* or *Variable
View*. Click on the cell in the 'measure' column of the variable view and select the
Ordinal, Scale and Nominal data type from the drop-down menu. The first row of
the SPSS screenshot of the *data view* mode shows the three data types, while the
Variable view of SPSS screenshot below shows the types of data in the last column,
whether they are Normal, Ordinal or Scale.

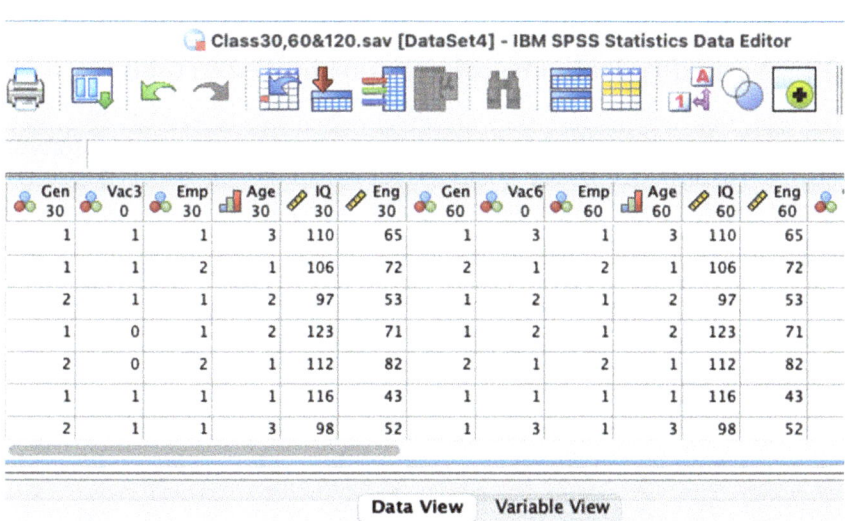

10.1.5   SPSS Statistical Techniques

The following site, Laerd, provides a step-by-step guide to the use of SPSS for statistical techniques, from data input to results interpretation. Since it is a website that is accessible to all for free, what is discussed here is the familiarisation of user interfaces of the software, with examples. https://statistics.laerd.com/spss-tutorials/chi-square-goodness-of-fit-test-in-spss-statistics.php.

10.2   Normality Tests, Overdispersion and Corrections

The hardest part is the fieldwork and collecting data; the computer and the software would do the analysis part but would not answer all our questions. Of course, a sense of the data must be obtained first, and the reliability of the data must be confirmed before proceeding to a full analysis. Mistakes are unavoidable in engaging in analytical work. These omissions are like traps. Identifying these traps, preventing errors, and avoiding mistakes are also important parts of any quality control system. The following methods are commonly used to verify this.

- Removing outliers
- Normal distribution of frequency: Measuring Skewness and Peakedness
- Identifying overdispersion.

10.2.1   Outliers

First, outliers, suspicious or abnormal data, i.e. data that differ significantly or do not match with other members of the sample, should be identified and removed. This can be done using Boxplot or Scatterplot. When summarizing data and examining the data, charts can show particular outliers and patterns. The data shown by the green dots in the first image and the red dot in the second image in Fig. 10.1 are abnormal outliers.

> …producing an appropriate chart will identify outliers and skewed data. A big difference between the mean and median indicates skewed data or influential outliers. (Marshal, 2021)

Data with average continuous distribution are summarized using median and standard deviation, and if the data is unclear or impractical, it is best to summarize using the median (middle value) and quarterly (upper quarter–lower quarter).

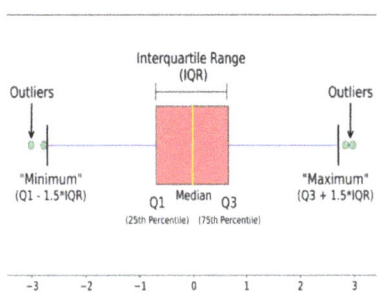
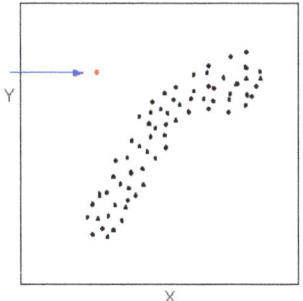

Fig. 10.1   Outliers. *Sources* https://idatassist.com/do-you-know-what-outliers-in-your-data-really-mean/

10.2.2   Normal Distribution and Normality Tests

The main principle to be verified when examining parametric variables is whether the frequency distribution of the collected data is shown symmetrically with a normal distribution (Fig. 10.2). There are many normality tests: the Shapiro–Wilk test, the Kolmogorov–Smirnov test, the Lilliefors test, the Cramer–von Mises test, the Anderson–Darling test, the D'Agostino–Pearson test, the Jarque–Bera test, chi-squared test, a skewness test and Kurtosis test If the data is not of normal distribution, that data is not considered valid for parametric statistical analysis (t-tests, ANOVA and regression, etc.). By watching this video https://youtube/iMak-EW4HtM you can get some idea of the general distribution. General distribution tests include the Shapiro–Wilk test, the Kolmogorov test, the Lilfors test, the Anderson–Darling test, and the Augustine test. These tests can be easily performed by computer software such as SPSS. Skewness, Kurtosis and Jarque–Bera tests are described below.

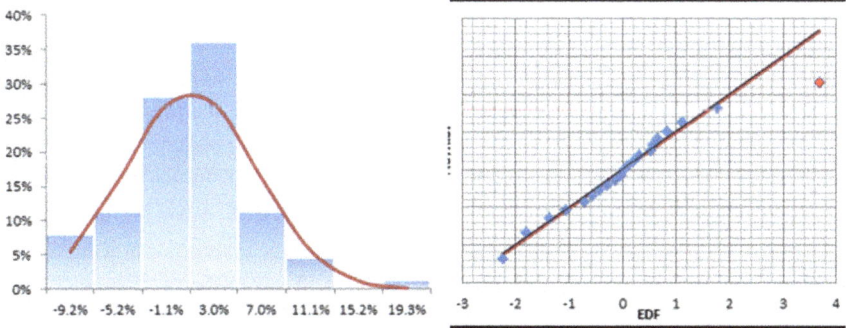

Fig. 10.2   Close normal distribution. *Source* https://www.flickr.com/photos/spiderxl/7824940578

Is my data normally distributed?	In this example, the exam scores should be approximately normally distributed for **both** males and females.
This screencast covers:	
8.4: How to check whether your data have a normal distribution using the chi-squared goodness-of-fit test	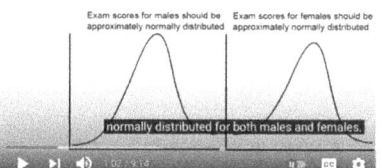
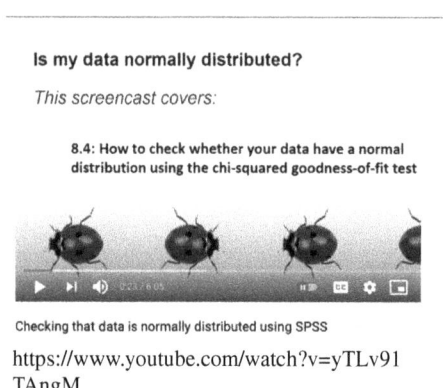	
Checking that data is normally distributed using SPSS	Normality test using SPSS: How to check whether data are norm
https://www.youtube.com/watch?v=yTLv91 TAngM	https://www.youtube.com/watch?v=IiedOy glLn0

10.2.3   Chi-Square Test for Normal Distribution

What is done here is to compare the value of this Chi-square with the critical value of the Chi-square table in terms of the degree of freedom and the expected level of significance. If the Chi-square value is larger than the table value, it can be concluded that the data does not show a normal distribution. (https://www.statisticshowto.com/chi-square-test-normality/) (https://statistics.laerd.com/spss-tutorials/chi-square-goodness-of-fit-test-in-spss-statistics.php).

How to perform this test using SPSS will be discussed in the next chapter under the Chi-square test.

10.2.4   Plotting a Histogram or QQ

Plotting a histogram or QQ plot of the variable of interest will give an indication of the shape of the distribution. Histograms should peak in the middle and be approximately symmetrical about the mean. If data is normally distributed, the points in QQ plots will be close to the line. The shape of the data distribution should be maximal in the centre of a histogram (Fig. 10.2 left graph) and roughly symmetrical around the centre and will be marked near the line in the QQ graph (Fig. 10.2 right graph).

Properties of a normal distribution.

- Mean, mode, and mode are all the same.
- The curve is symmetrical at the centre (i.e. around the centre).
- Half of the values extend symmetrically to the left and half to the right in the middle.
- The total area under the curve is considered to be 1, then the left half is 0.5, and the right half is 0.5.

10.2.5 Skewness and Peakedness

The criteria used to calculate the symmetry (Bell shape) of the frequency distribution are skewness and Peakedness (Fig. 10.3). By preparing an appropriate graph, data leading to inaccuracies and distorted chambers can be identified. The large difference between the mean and the mean indicates the absurd and distorted chambers.

Skewness: As shown in Figs. 10.3 and 10.4, the symmetry of the frequency distribution can be distorted by weighing the frequency distribution chamber to the left (negatively) or to the right (positively).

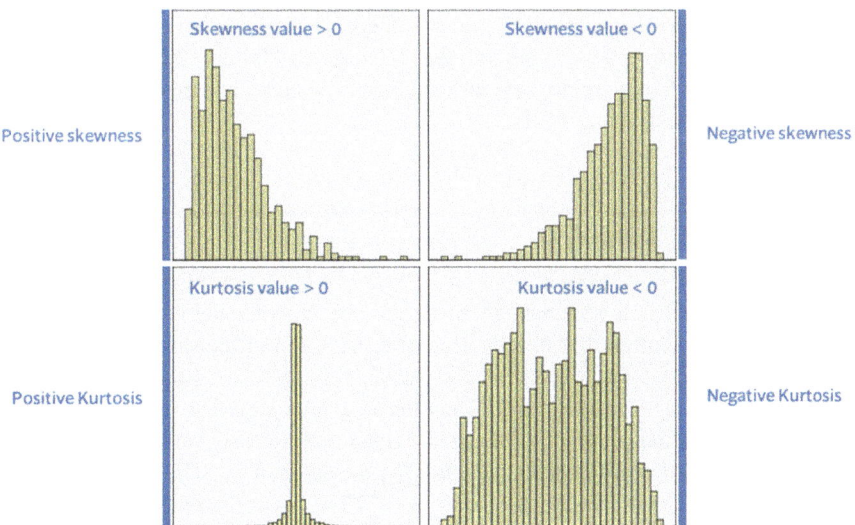

Fig. 10.3 Skewness and peakedness. *Source* https://statistics.laerd.com/premium/spss/tfn/testing-for-normality-in-spss.php

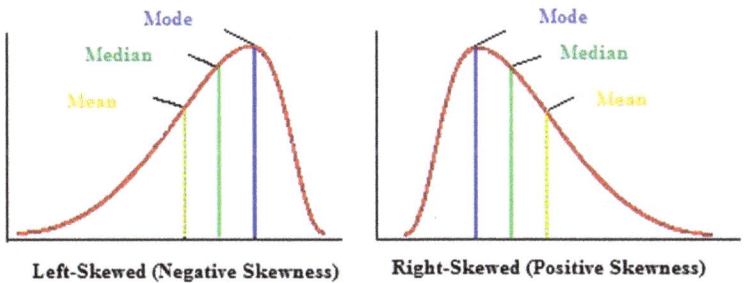

Fig. 10.4 The Skewness of distribution. *Source* https://www.statisticshowto.com/pearson-mode-skewness/

Pearson mode and Pearson median are the two most commonly used criteria. The Pearson mode compact is used when the sample data exhibits a strong mode. If the data includes multiple modes or weak modes, Pearson's neutral module is used. If the value of the curvature is less than −1 or greater than +1, the distribution is weighted to the left (Negative < −1) or to the right (Positive > + 1), respectively.

The distribution is moderately drawn if the value is between −1 and −0.5 or between 0.5 and 1. If the chamber is between −0.5 and 0.5, the distribution is defined as approximately symmetric.

Peakedness: The second graph in Fig. 10.4 measures the average (Bell shape) average (not too high, not too flat) of the frequency distribution by the kurtosis value. Higher heights are called positive Kurtosis, and flat heights are called negative Kurtosis. The Kurtosis values for asymmetry are considered acceptable between –2 and + 2 to confirm the distribution of one variable (George & Mallery, 2010) and up to 3 (Saliya, 2021b) when there are several variables. The next chapter will discuss how to perform these tests using SPSS.

10.2.6   Jarque Bera Test

The Jark-Bera test is a goodness of fit test to see whether there is a bell-shaped symmetrical distribution that fits the normal distribution of the sample data, taking into account both factors; Skewness and Peakedness. All Jark-Bera test values are positive. If it is significantly larger than zero, it indicates that it has no normal distribution. Here the presence of these skewness and kurtosis values is decisive in the selected level of significance represented by the value of 'p'. The valid significant level in general statistics is considered to be 5%. That is, confidence level is greater than $100\% - 5\% = 95\%$. SPSS does not have a Jark-Bera test facility. Once the chamber and the coefficient are found, those values can be substituted, and Jark-Berra can be calculated using the following equation. This equation is easy to compute using MS Excel when there are several variables.

The test statistic **JB** is defined as:

$$JB = (n/6)^* \left(S^2 + \left(C^2/4\right)\right)$$

where:

- **n**: the number of observations in the sample
- **S**: the sample skewness
- **C**: the sample kurtosis.

Example 10.1 Stock Market Development and Nexus of Market Liquidity.

Test your understanding: The following Table 10.3 produced by SPSS shows Skewness, Kurtosis value and Jark-Berra values. In the blanks provided to the right of those columns, you can mark your judgments as Yes (Validity) or No (Invalid).

The interpretation of this result is described in Box 10.1.

Table 10.3 Descriptive and normality statistics produced by SPSS

Variable	Descriptive and normality statistics								
	Distribution				Normality				
	Mean	Std Dev	Skewness	?	Kurtosis	?	Jarque-Bera	p-value	?
SMD	1.097	3.703	1.429		5.432		14.088	0.001	
MCAP	13.603	6.784	0.432		2.649		0.871	0.647	
LQDT	0.266	0.393	2.616		9.948		75.638	0.000	
INF	3.960	3.304	0.290		4.714		3.273	0.195	
FDI	5.927	5.620	0.899		3.334		3.346	0.188	
BNKD	61.623	22.405	−0.301		1.527		2.531	0.282	
GDP	5.610	2.510	0.866		2.732		3.073	0.215	
INV	19.740	3.226	1.309		4.238		8.385	0.015	
SAV	0.112	0.089	0.373		1.952		1.655	0.437	
COC	0.149	0.314	−0.041		2.184		0.673	0.714	
GVTEF	−0.362	0.288	−0.514		2.221		1.664	0.435	
PS_AV	0.391	0.394	−0.146		1.730		1.699	0.428	
RGLQ	−0.366	0.260	0.001		2.667		0.111	0.946	
ROL	−0.277	0.394	0.080		1.801		1.464	0.481	
DMCY	−0.266	0.452	−0.762		2.031		3.258	0.196	

(Answer: Skewness NO for SMD and LQDT; Kurtosis NO for SMD, LQDT, INF, FDI and INV; Jarque–Bera No for all)
Source Saliya (2020a)

Box 10.1 Normality tests interpretations

Means and standard deviations of study variables are presented in Table 10.3. Skewness of all the variables were below 1.5 (except for LQDT which had skewness of 3.043 and Kurtosis value of 9.948) providing evidence for the normal distribution. Kurtosis statistics shown for the variables INF, FDI and INV are also higher than the critical value of 3 (Westfall, 2015) and all other variables fall between 1.5 and 3. According to Westfall (2015), kurtosis says very little about the peak or centre of a distribution; its only unambiguous interpretation is in terms of tail extremity. The Jarque–Bera test is a goodness-of-fit test of departure from normality, based on both the skewness and kurtosis (Jarque, 2011). Higher p-values of Jarque–Bera statistics confirm 12 variables as not normally distributed.

10.2.7   Overdispersion

Overdispersion is a problem encountered in the analysis of count data that can lead to invalid inference if unaddressed. The decision about whether data are over-dispersed is often reached by checking whether the ratio of the Pearson chi-square statistic to its degrees of freedom is greater than one.

10.3   Reliability Test

Reliability can be divided into two main parts: test the reliability of the consistency of a set of variables used to measure a variable (Chronbach's alpha) and the reliability of the data.

10.3.1   Reliability Test of Consistency

Cronbach's alpha is a measure of internal consistency (how closely related the items are as a group). If the questions relate to the same issue, participants will be expected to get similar scores on each question. Cronbach's alpha ranges from 0 to 1, and scores are expected to be between 0.7 and 0.9.

Chronbach's Alpha–α coefficient: Commonly used to measure the reliability or internal consistency of a sample/data set. Chronbach's alpha is used to test the reliability of multiple query surveys using the Likert scale. Likert scale questions measure latent or uncontrollable variables; It is very difficult to measure a person's conscience, attitudes or perceptions. Chronbach's Alpha will tell you how closely relevant the responses given to such questions in a survey area and how closely they relate. If Cronbach's alpha value is greater than 0.7, the data set is shown to be reliable (Table 10.4).

[**SPSS**: Analyse > Scale > Reliability Analysis *Make sure* 'Alpha' is selected as the Model in the dialogue box that appears > OK].

Table 10.4   Cronbach's alpha interpretation

Cronbach's alpha	$\alpha \geq 0.9$	$0.8 \leq \alpha < 0.9$	$0.7 \leq \alpha < 0.8$	$\alpha < 0.7$
Internal consistency	Very high consistency (the items are so similar that some may not be needed)	Good	Acceptable	Poor internal consistency

Test: Cronbach's Reliability Test

Results:

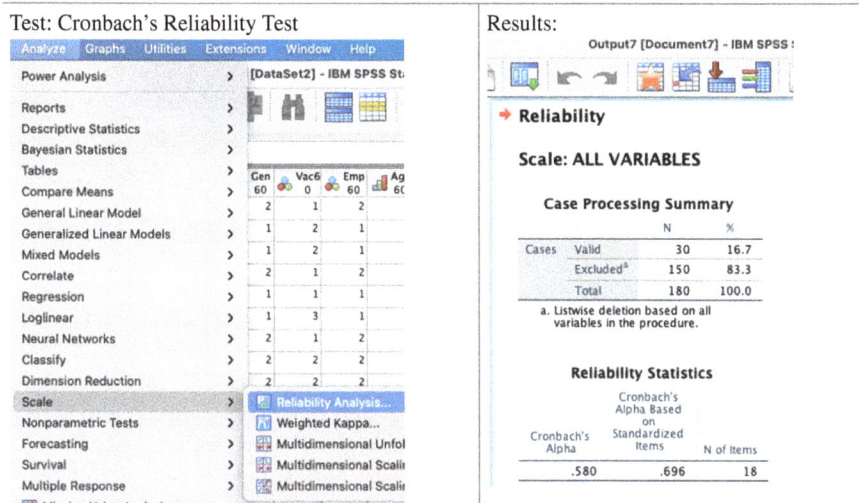

Presenting this result in APA style is as follows…

A questionnaire was employed to measure different, underlying constructs consisted of six questions. The scale did not have an acceptable level of internal consistency, as determined by a Cronbach's alpha of 0.580.

This Cronbach's test, when it is less than 0.7 (Cronbach's alpha = 0.580, standardized = 0.696), does not show that the data of our three hypothetical samples (18 variables of 6 in each of the three classes) are so reliable with respect to internal consistency. Table 10.4: Refer to Chronbach's Alpha Interpretation.

The alpha value ranges from 0 to 1. The expected value for reliability is between 0.7 and 0.9. If the alpha value is greater than 0.9, it indicates that the similarity between the data is unnecessarily high, and if it is less than 0.7, it has very poor internal consistency. The following example illustrates the reliability of the data using this Cronbach's alpha in Example 10.2. The following Table 10.5: Reliability Statistics shows how that information is presented in the research paper. How this is described in the paper is discussed below (Box 10.2: Method and procedures). Cronbach's Alpha value is 0.784, so it is acceptable.

Table 10.5 Reliability statistics

Cronbach's Alpha	No. of Items
0.784	18

Source Saliya (2021b) (Example 10.2)

Example 10.2 Driving Forces of Individual Investors in Stock Market Participation (Saliya, 2021b).

Box 10.2 describes how this process worked for research published under the title Driving Forces of Individual Investors in Stock Market Participation.

Box 10.2 Method and procedures and reliability test (Cronbach's Alpha) interpretation

We used the questionnaire to gather knowledge and behaviour of the people in relation to the SPX in several respects using 18 items. Five items concerned demographic factors: Age, Assets, Level of Education, Income level and Origin (Indigenous, Fiji-Indian and Other). Thirteen items had Likert scale answers: Strongly agree, Agree, Uncertain/NA, Disagree and Strongly disagree. There were 162 participants with university education. Two research assistants were involved in collecting data. The survey was carried out with employed university students majoring in Banking and Finance.

We initially formulated 18 items, and their construct reliability was assessed using Cronbach's alpha (range > 0.7) (Hair et al., 1998, 2012), and this resulted in 0.784 (as shown in Table 10.5: Reliability test).

10.3.2 Assessing Agreement Between Raters/Instruments

Cohen's Kappa: Assessing consensus among appraisers for categorical variables. Kappa checks only if the general data fits exactly; weighted kappa is more appropriate. For sequential or ordinal data, cuts are more appropriate if specific adjustments are required (Table 10.6).

[**SPSS**: Analyse > Descriptive Statistics > Crosstabs > Statistics > 'Kappa' > OK].

Table 10.6 Cohen's kappa interpretation

Cohen's kappa	<0	0–0.2	0.21–0.4	0.41–0.6	0.61–0.8	0.81–1
Strength of agreement	Poor (Agreement worse than by chance)	Slight	Fair	Moderate	Good	Very good

10.3.3   Intraclass Correlation Coefficient-(ICC)

ICC is a measure of consensus of continuous measurements for two or more assessors. The screenshot below shows how to find ICC using SPSS. The inter-class correlation coefficient is also defined using Table 10.6, as in Cohen's Kappa.

[**SPSS**: Analyse > Scale > Reliability Analysis > Statistics > *Check the box* 'Intraclass correlation coefficient' > OK].

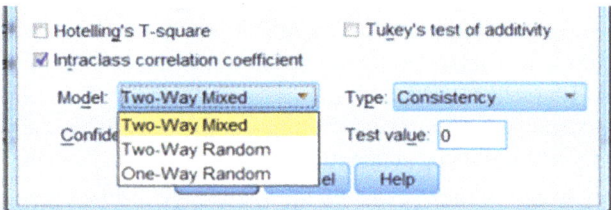

10.4   Data Reduction

Usually, when there are more than two variables, the variable classification and/or data reduction is done by multivariate techniques. The table below contains some of the most commonly used techniques (Table 10.7).

10.4.1   Principal Component Analysis (PCA)

The principal component analysis aims to reduce the number of inter-correlated variables to a smaller set, which almost explains the overall variability. It produces new variables, which are linear combinations of the original variables called Principal Components (PC's) or factors. These new variables can be used in further analysis,

Table 10.7 Multivariate techniques for data reduction

Purpose	Data reduction		Identify groups of similar subjects
Dependent variable (type)	2 + (Scale/binary although ordinal often used)	2 + (categorical)	2 + (Any)
Analitical technique	Principal components analysis	Factor analysis	Correspondence analysis	Cluster analysis

e.g. regression. Principal components analysis is not usually used to identify under-lying latent variables, but if the interpretation of which variables contribute most to each PC is of interest.

[**SPSS**: Analyse > Dimension Reduction > Factor > OK].

Tables 10.8 and 10.9 shows the results of this data reduction technique.

10.4.2   Communalities Table

The classification table provided by the SPSS software contains the ratio of each variable in the 'Extraction' column. Variation is explained by the main components (PCs) / factors extracted. The strongest variables are the highest coefficients. Opti-mally, the coefficient should be greater than 0.5. Example 10.2, Tables 10.8 and 10.9 below illustrate the process of extracting the main components.

Table 10.8 Total variance explained

Component	Initial Eigenvalues			Extraction Sums of Squared Loadings		
	Total	% of Variance	Cumulative %	Total	% of Variance	Cumulative %
1	3.290	19.353	19.353	3.290	19.353	19.353
2	1.945	11.439	30.792	1.945	11.439	30.792
3	1.703	10.018	40.809	1.703	10.018	40.809
4	1.519	8.937	49.747	1.519	8.937	49.747
5	1.196	7.033	56.779	1.196	7.033	56.779
6	1.103	6.488	63.268	1.103	6.488	**63.268**
7	0.939	5.521	68.789			
8	0.876	5.155	73.944			
9	0.845	4.971	78.915			
10	0.715	4.204	83.119			
11	0.671	3.945	87.064			
12	0.496	2.919	89.982			
13	0.458	2.692	92.675			
14	0.424	2.496	95.170			
15	0.329	1.937	97.107			
16	0.264	1.555	98.662			
17	0.227	1.338	100.000			

Extraction Method: Principal Component Analysis

Table 10.9 Component matrix

Variable	Component					
	Power	Knowledge	Social	Attitude	5	6
Income	0.529	−0.482	0.5	−0.103	0.079	0.109
Age	0.568	−0.116	0.48	−0.332	−0.118	−0.099
Assets	0.612	−0.544	0.276	−0.126	0.068	−0.095
Origin	−0.18	0.401	0.547	0.037	−0.131	0.217
SPX	−0.299	0.33	0.223	0.237	**0.285**	−0.094
SPX Co.1	0.391	0.451	0.085	−0.328	−0.143	0.253
Trading2	0.477	0.389	−0.003	−0.24	0.47	−0.031
Risky3	0.296	0.185	0.473	0.456	0.293	−0.275
Peers4	0.112	0.557	0.226	0.075	−0.386	−0.357
Averse5	0.566	0.366	−0.176	−0.098	0.035	−0.429
Self Cf6	−0.005	0.146	0.079	0.619	0.303	0.192
Karma7	0.445	−0.096	−0.273	0.308	−0.383	−0.136
Riskyii8	0.629	0.205	−0.258	0.154	0.212	0.196
No Greed9	0.493	−0.109	−0.041	0.51	−0.319	0.391
Karmab10	0.519	0.173	−0.184	−0.092	0.198	0.432
Lazy11	0.484	0.017	−0.099	0.335	−0.296	−0.155
SPX Return12	0.128	0.448	−0.174	−0.289	−0.217	0.189
Useless13	0.303	−0.16	−0.556	−0.021	0.257	−0.263

Extraction Method: Principal Component Analysis

KMO–Kaiser–Meyer–Olkin (Measure of Sampling Adequacy): This measure varies from 0 to 1, but the closer to 1, the better. It should be above 0.6 to use PCA. By default, SPSS only retains PCs with eigenvalues above 1 (Kaiser Criterion), but this method sometimes selects more PC's than needed. The Scree plot is an alternative method for choosing the optimal number of PCs. It plots the eigenvalues for each PC. Table 10.8 shows 17 key components with eigenvalues, and the interpretation in APA style is provided in Box 10.3 Total variance explained interpretation.

Scree plot Is an alternative method for selecting the optimal number of key components (PCs). Figure 10.5 shows a graph line of eigenvalues and a blue line indicating the limit on the number of PCs. Figure 10.5: Scree Plot Chart and Critical Limit provided by SPSS Software.

[**SPSS**: Analyse > Dimension Reduction > Factor > OK].

Fig. 10.5 Scree plot chart and critical limit

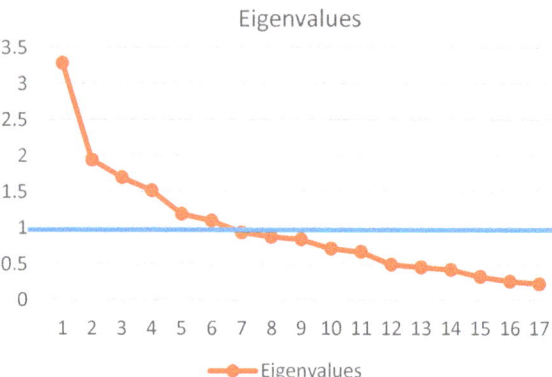

Eigenvalues

Box 10.3 Total variance explained interpretation (APA)

Then we ran a Principal Component Analysis (PCA) which is a variable-reduction technique in SPSS, and the results produced six principal components covering 63.27% of the cumulative variance of the variance in the original variables (see Table 10.8: Total Variance Explained). Based on the behavioural theories, knowledge of the Fijian context, together with the level of contributions of each factor and correlations and component matrices among all items, we identified four substantively important and empirically powerful dimensions and constituent items (Table 10.9: Component Matrix).

10.4.3 Component Matrix

Component matrix: Shows the component loadings (correlations between individual variables and the PC's). Although the values can range between −1 and +1, we chose to suppress coefficients between –0.3 and +0.3 and concentrate on those with higher loadings. Accordingly, components 5 and 6 are excluded and components 1–4 are selected, and label names are suggested. They are Socio-Economic, Knowledge, Cultural and Psychological.

The APA style presentation of the above Table 10.9 is shown in Box 10.4 as follows;

Box 10.4 Principal Component Analysis interpretation

For further analysis, based on CFA results, we computed composite measures for each factor by adding constituent items. For example, the composite measure for attitude was created by adding indicators F1 and F2. As shown in Fig. 10.5: this was a process combining theories, local knowledge and empirical findings (data). We tested: (a) first-order Confirmatory Factor Analysis and (b) a second-order CFA in the Structural Equation Modeling (SEM) framework (Bollen, 1989). After confirming the model fit, we ran the regression between the hypothesized dimensions and the actual investments of the respondents in the SPX. Each variable is represented by one item, as indicated in Table 10.9. The responses to F1 and F2 were used as indicators of Attitude, F3–F9 (seven items) were used as indicators of Perceived Power, while F10 and F11 and F12 and F13 were used as indicators of Social Norms and Knowledge dimensions respectively.

As shown in Table 10.9 of the above Example 10.2, the coefficient of a factor outside the norm is 0.226, but the factor is chosen because the authors consider it important. Factors such as attitudes towards Karma, Knowledge and Attitude towards the stock market, Self-confidence, Greed for money, Reluctance to take risks, and Laziness are represented by other questions in the questionnaire. Using this model, it is possible to study the attraction of any other stock exchange elsewhere.

10.4.4   Correlation Matrix

The correlation model is created to assess the correlation between variables aimed at determining the contribution of each variable to the elimination of factors with similar effects. Here the factors are identified as variables. If $r > 0.9$ is for two variables, then they are too closely related and represent only one factor, so only one is required. The correlation coefficients between the 13 variables in Example 10.3 are represented by Table 10.10. Note the 'correlation is significant at the 0.01 level (2-tailed)' below this table, which will be discussed in the next Chap. 11.

If $r > 0.9$ for two variables in the correlation matrix, they are too related, and only one is needed. If one variable consistently has correlations under 0.1, it is not related enough to the other variables and is likely to form a Principal Component of its own. The fact that the correlation coefficients are not close to 0.9 indicates that the factors represented by these variables are different from each other, which means that it is valid to treat those variables separately. As shown in Table 10.10 correlation template, all variables make a significant contribution.

[**SPSS**: Analyse > Correlate > Bivariate > OK].

Table 10.10 Descriptive statistics and correlations among all study variables

	SMP	IIN	AM	RT	SE	SC	OT	RA	CN
IIN	0.330[a]								
AM: Ach. Motivation	0.144	0.123							
RA: Risk-taking	0.106	0.213[a]	0.406[a]						
SE: Self-Efficacy	0.106	0.207[a]	0.360[a]	0.109					
SC: Self-Control	0.1	0.019	0.263[a]	0.259[a]	0.287[a]				
OT Optimism	0.111	0.004	0.468[a]	0.259[a]	0.331[a]	0.143			
RA: Risk Appetite	0.193[b]	0.116	−0.250[a]	−0.181	−0.06	−0.098	−0.086		
CN: Contention	0.333[a]	0.104	0.12	0.316[a]	0.341[a]	0.201[b]	0.148	0.095	
MA: Mkt. Awareness	−0.091	0.162[b]	0.241[a]	0.264[a]	0.141	0.029	0.246[a]	0.045	0.0
OS: Op. Skills	0.023	0.074	0.255[a]	0.270[a]	−0.076	0.045	0.363[a]	−0.026	−0.0
ED: Knowledge & Ed	0.154	0.153	0.194[b]	0.292[a]	0.075	0.178[b]	0.135	−0.195[b]	−0.17
OG: Origin	0.092	−0.031	−0.043	−0.07	0.105	0.222[a]	0.013	−0.055	−0.0
PM: Perception SPX	0.198[b]	0.109	0.209[a]	0.028	−0.012	0.205[a]	0.147	0.103	−0.1
AGE	0.147	0.251[a]	0.054	0.258[a]	0.073	0.184[b]	0.123	−0.105	−0.0
Mean	1.23	3.16	3.49	2.99	3.09	3.13	3.16	3.25	3.2
Std. Deviation	0.634	1.99	1.269	1.16	1.2	1.137	1.263	1.182	1.2
Skewness	2.476	−0.16	−0.533	0.269	−0.082	−0.041	−0.197	−0.127	−0.3

[a] Correlation is significant at the 0.01 level (2-tailed)
[b] Correlation is significant at the 0.05 level (2-tailed)
Source Saliya (2020b)

Example 10.3 Role of Enterprising personality in motivating stock investors: Case of Fiji, forthcoming.

APA style interpretation is given in Box 10.5.

> **Box 10.5 Correlation matrix interpretation**
> The correlation matrix was produced to assess the correlations between variables to determine the similarities of contributions of each variable in order to eliminate variables with similar influences. All variables show distinct contributions, as shown in Table 10.10: Correlation matrix.

10.4.5 Exploratory Factor Analysis

There are two types of Factor Analysis. Exploratory Factor Analysis (EFA) aims to group together and summarise variables that are correlated and can therefore identify possible underlying latent variables which cannot be measured directly, whereas Confirmatory Factor Analysis (CFA) tests theories about latent factors. Confirmatory Factor Analysis is performed using additional SPSS software.

The exploratory factors and the main components are very similar. Factor analysis is most commonly used to measure latent variables behind directly measurable variables, such as Likert-style questionnaires.

10.4.6 Cluster Analysis

Cluster analysis is a multifaceted technology used to group people/variables. There are several types of cluster analysis.

Hierarchical clustering: Hierarchical clustering in SPSS (each subject starts from its clusters and clusters are merged until all subjects are the same cluster).

[**SPSS**: Analyse > Classify > Hierarchical Cluster > OK].

K-means clustering does not require an inequality model for all pairs, combined into a cluster with a mean close to its value for each subject.

[**SPSS**: Analyse > Classify > K-Means Cluster > OK].

All of this variable classification and/or data reduction multivariate techniques presented in Tables 10.8 and 10.9 minimize, simplify and intensify the number of variables.

10.5 Summary

- To gain expertise in SPSS, AMOS, Eviews and Mplus software, all you must do is install the software and study the YouTube tutorials available on the Internet, preferably under expert guidance.
- Once you understand the basic concepts described here, it will not be difficult to understand such Youtube video clips given below under further readings; Youtube videos.
- The choice of quantitative analysis strategy depends on many factors: the nature of the data; types of variables; research problem; objectives of the analysis; sizes and number of samples, number of datasets, etc.
- Here is only an outline of the most widely used statistical techniques. The most commonly used methods are illustrated in the relevant examples.
- There are four types of data in terms of scale precision or accuracy: Ratio (highest-precision), Interval (Ordinary), Ordinal (Nominal-lowest precision) and Nominal (lowest-precision).
- Numerical data can be classified as normal continuous and non-normal categorical (Binary, Counts and Ordinal). Non-numeric data is called nominal data.
- This classification is easy to study (a) Normal continuous (b) Categorical binary (c) Categorical counts, (d) Categorical ordinal and (e) Nominal for the use of statistical techniques.
- SPSS classification is Scale (Continuous and Counts), Ordinal and Nominal.
- When data is arranged as a series, it is called a variable.
- The definitions of data types and variables are not unique; they can vary from case to case, depending on their size.
- Care should be taken in using an appropriate strategy to avoid errors and omissions: removal of abnormal/external data; quality assurance; extraction and reduction; ensuring reliability; including verification of goodness of fit or model fit, etc.
- The Shapiro–Wilk test, the Kolmogorov–Smirnov test, and the Skewness, Kurtosis, and Jarque-Berra values are widely used to confirm the average distribution of categorical data.
- Chronbach's Alfa and Cohen's Kappa are widely used to ensure authenticity.
- Factor Component Analysis (PCA) reduces the number of interrelated variables to a small set.
- If the correlation coefficient for the two variables is $r > 0.9$, then they are too close and represent only one factor.
- Component matrix: Displays component loadings. Although the correlation coefficient values may be in the range between -1 and $+1$, only components with high loadings are selected, small coefficients between -0.3 and $+0.3$ are disregarded.
- Exploratory Factor Analysis (EFAs) summarizes variables and identify latent variables in the background that cannot be directly measured.
- Confirmatory Factor Analysis (CFA) tests theories about latent factors.
- Clusters are a multivariate technology used to group people/variables.

References/Further Reading

Bollen K. A. (1989). *Structural equations with latent variables.* Wiley George and Mallery.

George, D., & Mallery, M. (2010). *SPSS for windows step by step: A simple guide and reference, 17.0 update* (10a ed.). Pearson.

Ghasemi, A., & Saleh Zahedias, S. (2012). Normality tests for statistical analysis: A guide for non-statisticians. *International Journal of Endocrinology and Metabolism. 2012 Spring; 10*(2), 486–489. https://doi.org/10.5812/ijem.3505. (Published online 2012 Apr 20).

Hair, J. F., Sarstedt, M., Ringle, C. M., & Mena, J. A. (2012). An Assessment of the Use of Partial Least Squares Structural Equation Modeling in Marketing Research. *Journal of the Academy of Marketing Science, 40*, 414–433. https://doi.org/10.1007/s11747-011-0261-6.

Marshal, E. (2021). The Statistics Tutor's Quick Guide to Commonly Used Statistical Tests. University of Sheffield. https://www.statstutor.ac.uk/resources/uploaded/tutorsquickguidetostatistics.pdf.

Patten, M. L. (2014). *Understanding research methods: An overview of the essentials.* Pyrczak Publishing.

Saliya, C. A. (2020a). Dynamics of credit decision-making: a taxonomy and a typological matrix. *Review of Behavioral Finance.* https://doi.org/10.1108/RBF-07-2019-0092.

Saliya, C. A. (2020b). Stock Market Development and Nexus of Market Liquidity. *International Journal of Finance and Economics.* https://doi.org/10.1002/ijfe.2376.

Saliya, C. A. (2021b). Driving Forces of Individual Investors in Stock Market Participation. *Review of Economics and Finance, 19*, 73–79. https://refpress.org/ref-vol19-a8/.

Yapa, B. W., & Simb, C. H. (2011). Comparisons of various types of normality tests. *Journal of Statistical Computation and Simulation. 81*(12). https://www.tandfonline.com/doi/pdf/, https://doi.org/10.1080/00949655.2010.520163.

Youtube Links

Calculating and Interpreting Cronbach's Alpha Using SPSS https://www.youtube.com/watch?v=Kz8OdR6lV44.

Checking that data is normally distributed using SPSS https://www.youtube.com/watch?v=yTLv91TAngM.

Flow chart for choosing tests https://www.youtube.com/watch?v=tfiDu--7Gmg.

Jarque-Bera test in MS Excel https://www.youtube.com/watch?v=3or8IwMUgjs.

Learn SPSS in 15 minutes https://www.youtube.com/watch?v=TZPyOJ8tFcI&t=738s.

Normal Distribution Definition and Properties https://www.youtube.com/watch?v=iMak-EW4HtM.

Normality test using SPSS: How to check whether data are normally distributed https://www.youtube.com/watch?v=IiedOyglLn0.

SPSS Tutorial for data analysis | SPSS for Beginners: https://www.youtube.com/watch?v=Bku1p481z80&t=29s

SPSS for Beginners https://www.youtube.com/watch?v=hKA2VQ60bxg.

SPSS step by step Tutorials https://www.spss-tutorials.com/basics/.

SPSS step by step application https://statistics.laerd.com/features-overview.php.

Types of Data: Categorical vs Numerical Data: https://www.youtube.com/watch?v=DUcXZ08IdMo.

Types of Data: Nominal, Ordinal, Interval/Ratio https://www.youtube.com/watch?v=hZxnzfnt5v8.

Chapter 11
Relevant Statistical Concepts

This chapter discusses some of the basic concepts of statistical data analysis. Although these concepts are somewhat difficult to understand without context, it is useful to gain some sense of them. These basic concepts include hypothesis testing, p value, t tests, critical value, F test, Z score, degree of freedom (df), significant levels, confidence levels and intervals, normal distribution, Poison distribution and Chi-Square (χ^2–distribution) as well as the goodness of fit tests. It also discusses how to use **SPSS** to verify the validity of a data set using the three samples (classes) of postgraduate students mentioned in the previous chapter.

The use of complex formulas, lengthy calculations, and various tables is gradually replaced by user-friendly statistical software and path analysis tools and techniques. Statistical assessments, estimates and predictions have become much simpler and easier due to the ability to obtain the 'magical value' of p using computer software.

11.1 Hypothesis Testing

The nature of hypothesis testing was already discussed, assuming it as a court case in Chap. 6. It is the responsibility of the complainant (researcher) to provide adequate evidence. The accused is assumed 'innocent' until proven guilty based on sufficient evidence. That is, the hypothesis (Null hypothesis-H0: Innocent) cannot be refuted (must be accepted; Retain the null: H0). Hypothetical testing is the testing of the validity and adequacy of evidence. If the evidence presented to substantiate the allegation is sufficient, the hypothesis of 'innocence' is refuted (Reject the null: H0).

The hypothesis testing assumes that the sample data do not provide evidence of an abnormality. That is, the sample represents the normal condition; That it is no different from a normal situation; That the difference is zero ($\bar{x} - \mu = 0$); That there is no effect, That the evidence is not statistically significant. The researcher hopes to reject the null. Therefore, the researcher expects a low p (Sig.) value from the statistical test or statistics within critical values.

© The Author(s), under exclusive license to Springer Nature Singapore Pte Ltd. 2022   171
C. A. Saliya, *Doing Social Research and Publishing Results*,
https://doi.org/10.1007/978-981-19-3780-4_11

The reliability of evidence is expected to be at 95% probability in the application of statistical techniques. That is, the value represents the remaining 5% (0.05) of insufficient evidence. The researcher (plaintiff) expects the value to be less than 5%; then, the H0 will be refuted on the ground that the adequacy of the evidence is more than 95% means the research has discovered something abnormal (new knowledge).

11.1.1 Type I and Type II Error

When the jury does not really know whether the accused is guilty or not, they can make a mistake. The reality is that the person may actually be guilty (null false; innocence is false) or innocent (null true; innocence is true), but the jury concludes contrarily, then these errors occur. The possible conclusions of the jury are analysed and presented in Fig. 11.1.

A type I error is similar to the guilt of an innocent person, and in a hypothetical test, the probability of this error can be assumed at 10%, 5%, or 1% levels of significance; the sensitivity level is chosen. In contrast to this case, if the accused is indeed guilty, and there is no enough evidence to prove beyond a reasonable doubt (less than 5% suspicion or more than 95% confidence), the jury will have to acquit the accused. This is called a Type II error. Therefore, both errors can't occur simultaneously.

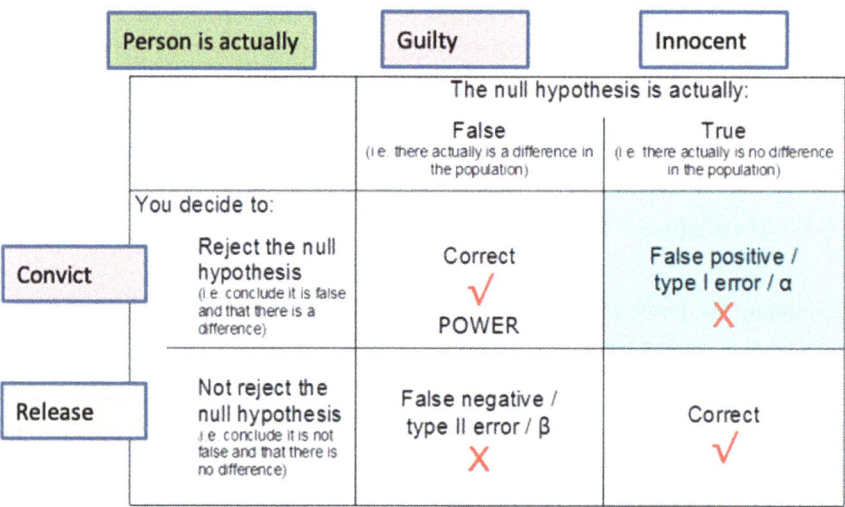

Fig. 11.1 Conclusions that the jury can reach, *Source* Marshal and Boggis (2014)

11.2 *p* **Value**

'*p*' (lower case italics) denotes a probability value. The *p*-value is the most common measure of significance. It provides crucial information but certainly does not tell the whole story. *p* evaluates only the probability of a false acceptance (α; probability of *false acceptance* of the null hypothesis), not the probability of a *false rejection* of the null hypothesis. The significant level of general statistics, or the level of significance, is considered to be 5% (0.05). That is, the reliability is 95% ($1 - \alpha$ = 100% – 5% = 95%). Although most authors consider $p < 0.05$ as statistically significant, a more reliable level of 99% can be expected in more sensitive research. Then $\alpha = 1\%$ ($p < 0.01$: less than one percent probability of error). You should note that the reject decisions of the null hypothesis are a statistical assessment of the adequacy of evidence at a given probabilistic significant level.

This significance is usually represented by small asterisks (*). Ex: $p < 0.05$ denoted with one asterisk (0.023^*), $p < 0.01$ with two asterisks (0.002^{**}), $p < 0.001$ with three asterisks ($<0.000^{***}$). According to the traditions of the American Psychological Association (APA), another standard is to present only three decimal places, without '0' (zero) in front.

The APA style is to present terms such as *p* value, z test, F test and t test etc., without hyphenated, but if presented as an adjective, it is the z-test score, t-test value (hyphenated). If values can be greater than 1, it is necessary to put '0' in front when presenting decimal places, but no '0' is necessary when presenting decimals of values that do not exceed 1, such as probability values.

> Nouns (*p* value, z test, t test) are not hyphenated, but as an adjective, they are t-test results, z-test scores.
>
> Place a zero before the decimal point if the statistic can be greater than one (0.26 lb). If the number cannot be greater than one, leave out the decimal point ($p = 0.015$).

11.2.1 Critical Value

Critical values for a test of hypothesis depend upon a test statistic, which is specific to the type of test, and the significance level, α, which defines the sensitivity of the test. A value of $\alpha = 0.05$ implies that the null hypothesis is rejected 5% of the time when it is in fact true. The choice of α is somewhat arbitrary, although in practice values of 0.1, 0.05, and 0.01 are common. Critical values are essentially cut-off values that define regions where the test statistic is unlikely to lie; for example, a region where the critical value is exceeded with probability α if the null hypothesis is true. The *null hypothesis is rejected if the test statistic lies within this region which is often referred to as the rejection region(s).* NIST/SEMATECH (2012, p. 7.3.1.3.1).

When the test statistics (probability α) exceeds the critical value (z > 1.96 or > 2.58 etc.), the *p* value is lower than the sensitivity level that the researcher expects to bear (<0.1, < 0.05 and <0.01). It then rejects the null hypothesis. If the test statistics stays within the critical value z < 1.96 or < 2.58 etc.), the *p* value is higher than the significant level (>0.1, > 0.05 or > 0.01, so, Retain the null).

It is important to understand the relationship between these two concepts, as some statistical software reports p values instead of critical values. For example, the two tail critical value of z is 1.96. If the test statistics are less than 1.96, the p value is greater than 0.05 (Retain the null), and if the test statistics are greater than 1.96 (in the rejection zone), the p value is less than 0.05 (Reject the null).

11.2.2   Significance Level: α and Confidence Levels and Intervals

The significant level or level of significance of an event is the probability that the event will happen by accident. If the level of significance is slightly lower, that is, if the probability of occurring by accident is slightly lower, it indicates that the event is important. Suppose μ is the expected focal point for the study question. And if Q is the tolerable level of variation above or below the expected value, then the reliable return interval would be between $\mu - Q$ (the lower confidence level) and $\mu + Q$ (the upper confidence level).

If confidence interval includes zero, that means that there is a good chance of finding no difference between groups, no correlation etc. (https://www.scribbr.com/). However, a high p value means that the results could have occurred under the null hypothesis of no relationship between variables or difference between groups.

Both levels must be either negative or positive so that there is no zero between the levels. Attitude and Knowledge variables, as shown in Table 11.1, indicate that the *p* values (0.116 and 0.625) are not (important) at the 5% significance level (>0.05).

Table 11.1 Regression results: the four dimensions on IB

	Standardized coefficients	t	p	95.0% Confidence Interval for IB	
	Beta			Lower bound	Upper bound
(Constant)		7.283	0.000	1.697	2.96
Attitude	0.135	1.580	0.116	−0.012	0.109
Power	−0.246**	−2.595	0.010	−0.062	−0.01
Social Norms	0.159*	2.004	0.047	0.001	0.138
Knowledge	0.041	0.489	0.625	−0.044	0.072

Dependent Variable: IB: Investment Behavior

Note * p < 0.05 and ** p < 0.01, Significant at the 5 and 1 percent levels respectively

Those variables' lower and upper confidence levels cross the zero, i.e., a lower level with a negative value and a higher level with a positive value. However, p vales (0.010 < 0.05 and 0.047 < 0.05) show that the factors Power and Social Norms are important, and both of their upper levels are either negative (–0.62 & –0.1) or positive (+0.001 & +0.138) do not cross zero. Therefore, these results suggest that the variables Attitude and Knowledge are not statistically significant, and the factors Power and Social Norms are statistically significant.

However, the decision to reject always carries a significant level. In fact, there is some probability of error. This is also known as a Type I error. This is the probability of not refuting the H0. Type I error is the false rejection of the null hypothesis, and type II error is the false acceptance of the null hypothesis.

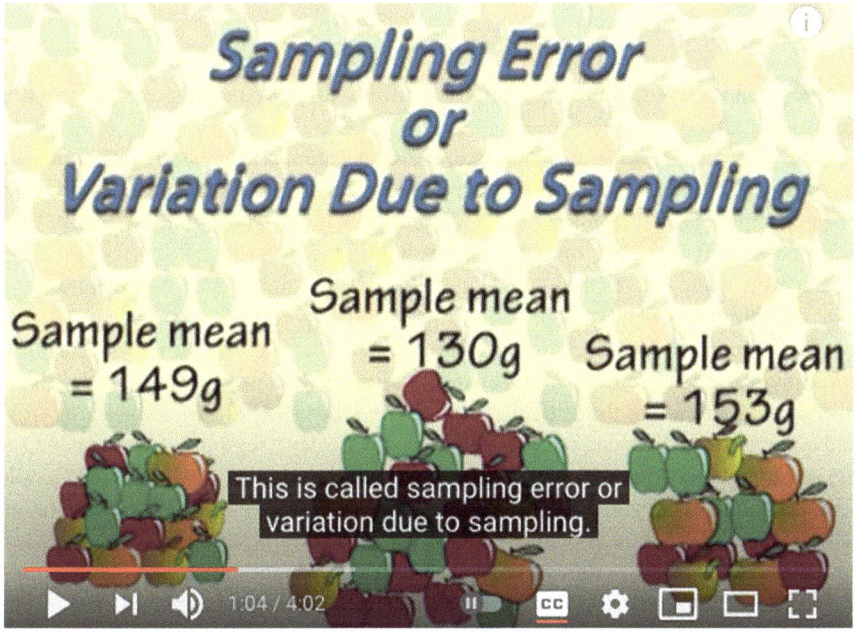

Source Understanding Confidence Intervals: Statistics Help: https://www.youtube.com/watch?v=tFWsuO9f74o

11.3 Z Test

The Z test is used to assess whether the mean of a sample corresponds to the mean of the population at a given confidence level (when the variance is known and the sample size is large). Since this test is evaluated using p statistics using statistical computer software and presented with a z value, there is no need to bother to search

Table 11.2 Significance levels (alpha) and corresponding critical values

α	z_α	α	z_α	α	z_α	α	z_α	α	z_α
0.50	0.00	0.050	1.64	0.030	1.88	0.020	2.05	0.010	2.33
0.45	0.13	0.048	1.66	0.029	1.90	0.019	2.07	0.009	2.37
0.40	0.25	0.046	1.68	0.028	1.91	0.018	2.10	0.008	2.41
0.35	0.39	0.044	1.71	0.027	1.93	0.017	2.12	0.007	2.46
0.30	0.52	0.042	1.73	0.026	1.94	0.016	2.14	0.006	2.51
0.25	0.67	0.040	1.75	0.025	1.96	0.015	2.17	0.005	2.58
0.20	0.84	0.038	1.77	0.024	1.98	0.014	2.20	0.004	2.65
0.15	1.04	0.036	1.80	0.023	2.00	0.013	2.23	0.003	2.75
0.10	1.28	0.034	1.83	0.022	2.01	0.012	2.26	0.002	2.88
0.05	1.64	0.032	1.85	0.021	2.03	0.011	2.29	0.001	3.09

for these values using tables etc. But it is useful to have an understanding of what is going on in this process.

The reliability level is the significance level balance. That is, the significance (alpha) is 5%, while the reliability level $(1 - \alpha)$ is 95%. 100% is the total distribution of the sample. That is, the 95% reliability level is the field around the centre, with 2.5% on the left and 2.5% on the right, excluding the bell-shaped distribution. Risk The critical value for $z\alpha$ at alpha 5% is 1.96, with an alpha value corresponding to 0.025 (Table 11.2).

The value Q, which is accepted as the tolerable deviation from the expected critical value mentioned above, is obtained by multiplying the z value calculated by the above formula by 1.96 (5% of the critical $z\alpha$ found using the table is 1.96). Subtracting from and adding that to the mean of the population (μ) creates a region ($\mu - Q$ and $\mu + Q$ as the lower and higher confidence limits). This means that if the median of the sample remains within that region, it can be assumed that the sample represents the population with 95% confidence or less than 5% risk. When doing these tests using statistical software such as SPSS, you will be given a p value that takes all of this into account.

The Fig. 11.2 shows the interstellar region in blue ($X = \mu$), which is true of this hypothetical hypothesis. The two yellow tails at the two corners are the rejection zones. For each significance level, α, the Z-test has a critical value. For example, the significance level $\alpha = 0.05$ has a critical value of 1.96. If the Z-test statistic is greater than this critical value, this may provide evidence for rejecting the null hypothesis.

For each significance level, α, the Z-test has a critical value. For example, the significance level $\alpha = 0.05$ has a critical value of 1.96. If the Z-test statistic is greater than this critical value, this may provide evidence for rejecting the null hypothesis.

The alpha 5% distribution is 2.5% on one side (a tail, tail). Since Z tables are designed with only one side in mind, you should find a value of 97.5% (0.975) in the table. That is, the value of both tails in the bell-shaped distribution is 5% or 2.5% of the value of one tail. Then you should find the value of 0.975 in the table. Figure 11.2: Z Critical value table. The horizontal corresponding value to the left of that value is

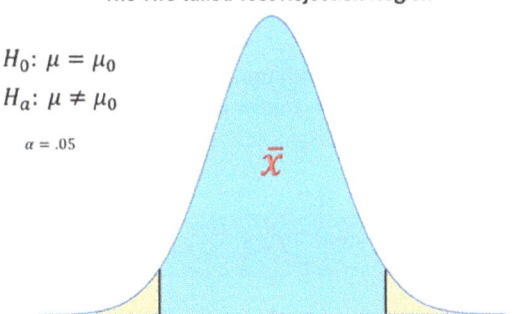

The two-tailed test.

$H_0: \mu = \mu_0$

$H_a: \mu \neq \mu_0$

$\alpha = .05$

Statistics 101: Single Sample Hypothesis Z-test Concepts

Fig. 11.2 The two-tailed test. *Source* Statistics 101: Single Sample Hypothesis Z-test Concepts: https://www.youtube.com/watch?v=HoqzIR8xj4s and https://www.youtube.com/watch?v=tsPv-ffN-0M

1.9, the corresponding vertical value is 0.06, then the critical $z\alpha$ 5% value is 1.96. Table 11.2 is a table created by combining these columns and rows. The $z\alpha$ value corresponding to its level of $\alpha = 0.025$ (97.5%) can be found directly in 1.96. Watch the videos shown in Fig. 11.2 and Table 11.3 for more explanations.

Table 11.3 Z Critical value table

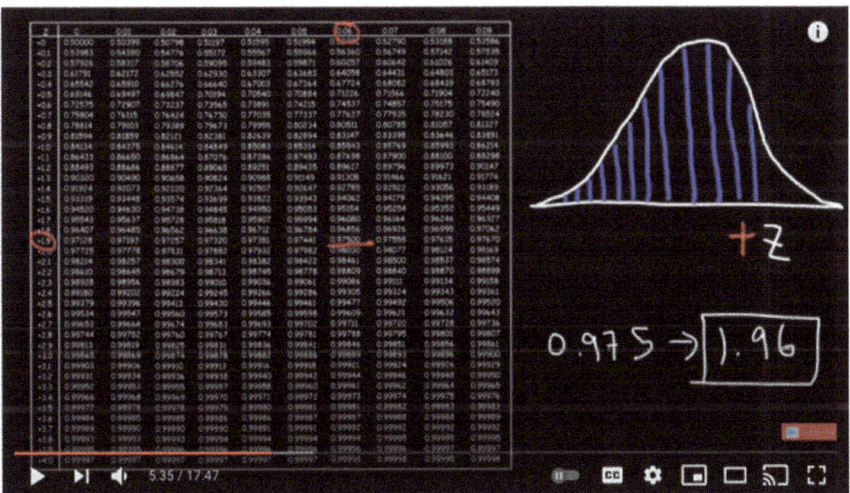

P-Value Method For Hypothesis Testing

Source (https://www.youtube.com/watch?v=8Aw45HN5lnA)

11.3.1 Z Test by SPSS Syntax

The Z test cannot be performed with the SPSS menus. You have to use Syntax for that. The syntax for the Z test is as follows, and you can get it from this link: http://www.how2stats.net/2014/03/one-sample-z-test.html.

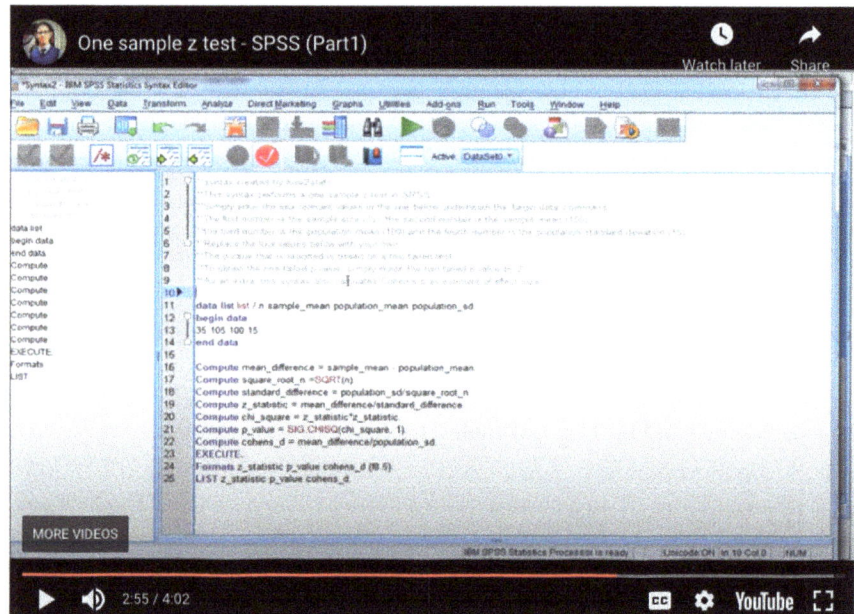

11.3.2 Z Test Versus t Test

The equations for the z-value for continuous data are as follows. To solve these equations you need to know the standard deviation.

$$Z = (X-\mu) / SE \qquad X = \text{Sample mean,}$$
$$Z = (X-\mu) / (\sigma/\sqrt{n}) \qquad \mu = \text{Population mean}$$
$$SE = \text{Standard Error} = (\sigma/\sqrt{n})$$
$$\sigma = \text{Standard Deviation}$$

Use the following examples to study how z test and t test tests apply. These three examples illustrate the calculation of z-text scores and t-test scores for standard deviation, both known and unknown, using samples of 5, 9, and 40 students in a school with 75 students. These three examples illustrate how the z test is used when the conditions that must be met for the t test are not met. You can find out more by looking at the related Youtube link (examples given here using the link: https://www.youtube.com/watch?v=YsalXF5POtY).

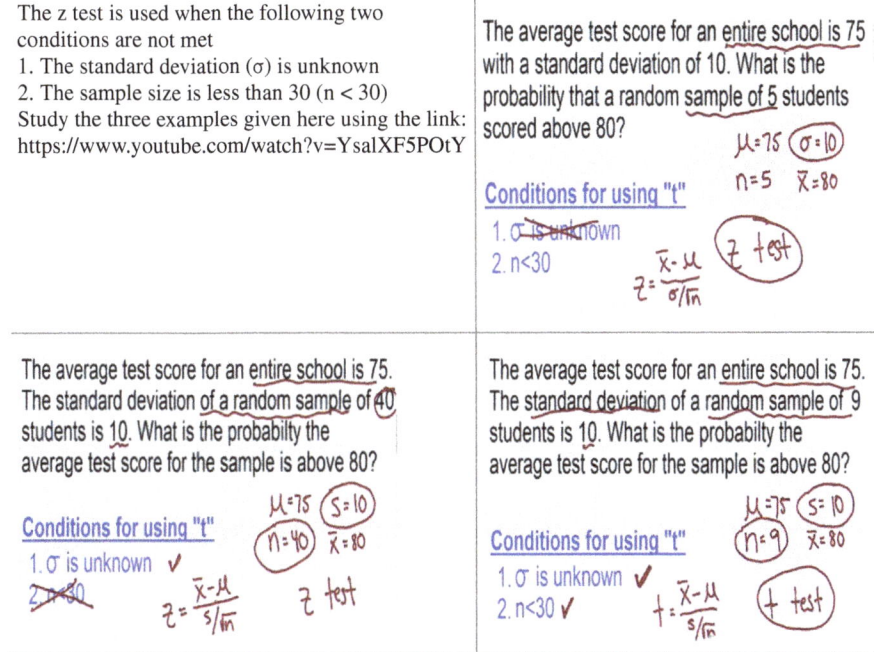

11.3.3 Calculating a Z Statistic in a Test About a Proportion

The population of Sri Lanka was published in the 2012 Census, as shown in Fig. 11.3. Using a student sample from a private class in Fig. 11.3, we examine whether the observed proportions are significantly different from these actual proportions.

The equation for calculating z for ratios is as follows. To solve these equations, the standard error (SE) is approximated and as shown in the following equation. (For ease of calculation, 'Other' religious groups were removed, and Hindu and Islamic denominations were rounded up. The calculation of the Z value can be learned from the following video: https://www.youtube.com/watch?v=8Aw45HN5lnA).

$$Z = \frac{(X - \mu)}{\left[\sqrt{(\mu x (1 - \mu)/n)}\right]}$$

Religious group	Percentage	Population
Buddhism	70.11%	14,272,056
Hinduism	12.58%	2,561,299
Islam	9.66%	1,967,523
Christianity	7.62%	1,552,161
Other	0.03%	6,400
Total	100%	20,359,439

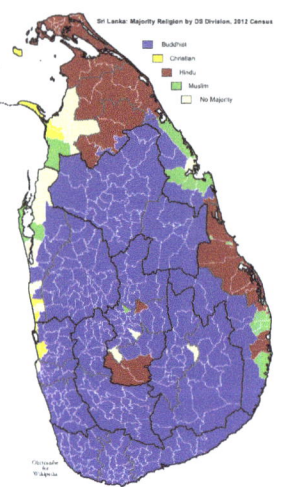

Fig. 11.3 Population of Sri Lanka 2012, *Source* Census of Population and Housing of Sri Lanka, 2012

Test: H_0: There is no difference between the proportions of the student sample and the actual census.

Results: Tables 11.4 and 11.5 show the z-test scores comparing the sample data and the census data.

Interpretation: This study was conducted to determine if the religious composition of the sample class was significantly different from the actual proportion reported in the census. A z test was used to compare the sample of 200 students (x) with the census (μ). The null hypothesis (H0) is that "the religious composition of the class does not differ significantly from the true religious composition of Sri Lanka." The ratios of each religion in the class are between $z = -1.12$ and 0.74, which are

Table 11.4 The z-test scores

Religion	Sample observation	Census proportions (%)	z-test Score	5% significance	H_0: No difference
Hindu	24 (12%)	12.6	−0.25570	<1.96>−1.96	Do not reject H_0
Catholic	11 (5.5%)	7.6	−1.12071	<1.96>−1.96	Do not reject H_0
Islam	20 (10%)	9.7	0.14335	<1.96>−1.96	Do not reject H_0
Buddhist	145 (72.5%)	70.1	0.74136	<1.96>−1.96	Do not reject H_0
Total	200 (100%)	100			

Table 11.5 The calculation of the z-test scores

Religion	x	Census proportion μ	$(x - \mu)$	$1 - \mu$	$\mu \times (1 - \mu)$	$\mu \times (1 - \mu)/n$	$\left[\sqrt{(\mu \times (1 - \mu)/n)}\right]$	$Z = \frac{(x - \mu)}{[\sqrt{(\mu \times (1 - \mu)/n)}]}$
Hindu	12.00%	12.60%	(0.006)	0.87	0.11	0.00055	0.02347	−0.25570
Catholic	5.50%	7.60%	(0.021)	0.92	0.07	0.00035	0.01874	−1.12071
Islam	10.00%	9.70%	0.003	0.90	0.09	0.00044	0.02093	0.14335
Buddhist	72.50%	70.10%	0.024	0.30	0.21	0.00105	0.03237	0.74136
Total	1	100%						

within the critical region <1.96 and >−1.96, the sample class's religious composition reflects the true representation of the religious composition in Sri Lanka in 2012.

We will test these results in the Chi-square discussion in the next chapter. There we do not examine each religion separately; instead, we check whether the distribution of religions in the sample differs significantly from the census proportions.

The presentation in APA style is as follows. This presentation is only for Hindus.

A difference between the Hindu proportion of 12.6% of the census of Sri Lanka in 2012 and the Hindu proportion of 12% in the sample of 200 students would, $z = -0.256 < 1.96$, is not statistically significant.

11.4   T Tests

One sample test: Compares the mean of a single group with a given mean. For example, when the average sales volume is given (last year or month) check the increase or decrease in sales for this month.

When the population parameters (mean and standard deviation) are not known, a t test is used to compare the means of two samples.

$$t = (\mathbf{x_1} - \mathbf{x_2})/((\sigma/\sqrt{\mathbf{n_1}}) + (\sigma/\sqrt{\mathbf{n_2}}))$$

Where 1 and 2 represent samples 1 and 2 and where $\mathbf{x}$ and σ are the parameters of samples 1 and 2.

Paired *t* test: A paired t test compares the means for the difference between two variations in the same sample or population. For example, checking progress by comparing trainees' performance before and after training in a training program. The SPSS paired t test will be conducted in the next chapter to compare the English marks of the Day class and the Night class one year before and after.

Independent *t* test: **Two sample *t* test** or **Student's *t* test** is used to check if there is a statistically significant difference between the means of two unrelated groups. For example–comparing boys and girls. In the example of the above three classes, the SPSS Independent T-Test will be conducted in the next chapter to test whether there are any similarities between the English scores of Day and Night classes.

If the sample size of each group is the same, it is considered a 'balanced' design. If the sample size is not the same in each group, the assumption that the design is balanced is violated, and the validity of the test is adversely affected. Although this is difficult to achieve in practice, a 'balanced' plan is more appropriate. This requires data processing; learn it from this link: https://statistics.laerd.com/premium/spss/istt/independent-t-test-in-spss-6.php#independent-variable.

The Youtube video below shows you how to do this test with an Excel spreadsheet. The conditions to be met for this test are shown in the screenshot below as assumptions.

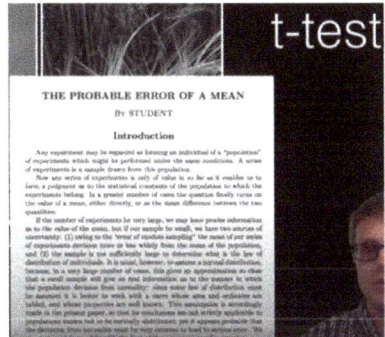 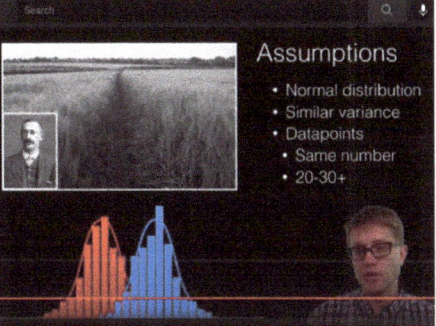

Source t test: https://www.youtube.com/watch?v=pTmLQvMM-1M

11.5   Chi-Square (χ^2–Distribution)

There are two types of Chi-square tests.

(a)   Chi-square goodness of fit:

  - The chi-square goodness of fit test checks whether the sample data has a normal (standard) distribution. This test is widely used in simple applications as well as in more complex situations. We discussed earlier that comparison with the critical value could determine whether a classification variable has a normal distribution. The following is a technical method used by SPSS using the *p* value.
  - Second, chi-square goodness of fit also tests the suitability of the model (goodness of fit: adaptability) when using techniques such as complex reflex models and structural equation modelling (SEM). It is also discussed later with an example.

(b)   A chi-square test for independence:

  - To compare two variables in a random table and see if there is a correlation between them, we examine whether the distribution of the categorical variables is different from each other. (How this test is performed by ANOVA using the F test when there are more than two variables will be discussed later).

11.6   Obtaining Statistics of Normal Distribution Using SPSS

The SPSS screenshots show how to examine the distribution of IQ values (intelligence values) between the two samples Day class (30 students) and Online class (180 students) (See SPSS step by step tutorials for details: https://www.spss-tutorials.com/basics/; SPSS step by step application: https://statistics.laerd.com/features-overview.php).

11.6.1   Outliers

The SPSS Output screenshot below shows that 14 out of 30 data in the Day class (Intellegence1) are outliers; the intelligence of four people is unusually high, and the intelligence of ten is unusually low.

[**SPSS**: Analysis > Descriptive Statistics > Explore > Click and select variables into Dependent List > Click Plots > Then check Normality plots with tests/Histogram/Boxplots Boxes > Click OK].

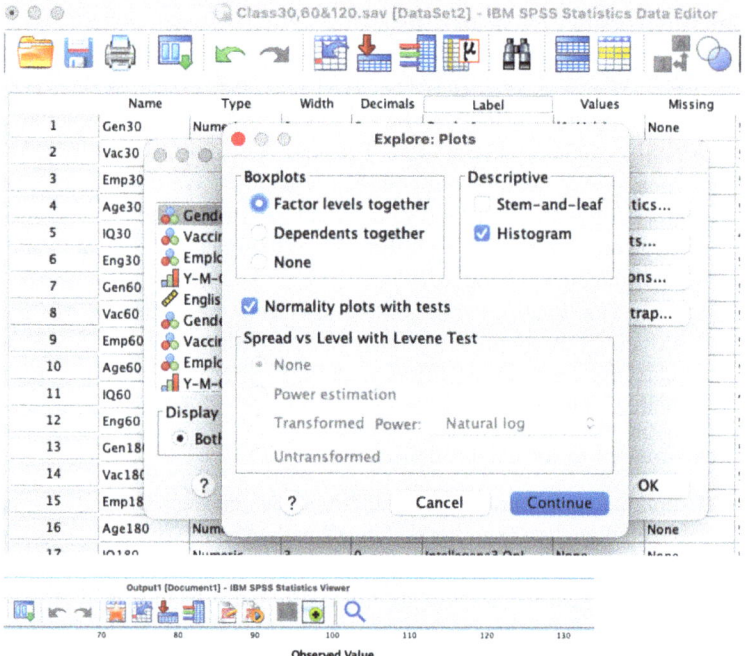

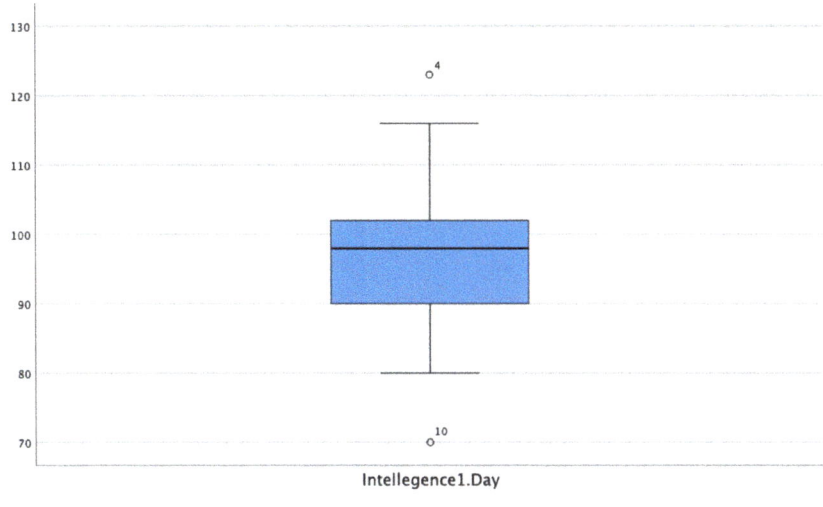

11.6.2 Histogram or QQ

Graphing a Histogram or QQ Using SPSS Day class results can be seen below.

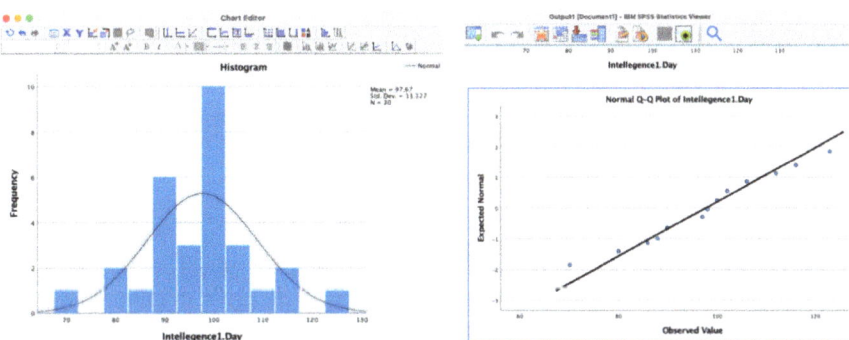

The presentation in the APA style is as follows.

Intelligence scores were approximately normally distributed for Day class, as assessed by visual inspection of its histogram. Intelligence scores were normally distributed for the Day class, as assessed by visual inspection of Normal Q-Q Plots.

11.6.3   Kolmogorov–Smirnov and Shapiro–Wilk Tests

The results of the Kolmogorov–Smirnov and Shapiro–Wilk index tests using SPSS are as follows.

The null hypothesis of the Shapiro–Wilk test and the Kolmogorov–Smirnov test (H0) is that data distribution is similar to normal distribution. Rejecting the null (H0), $p < 0.05$, means that data distribution is not the same as a normal distribution.

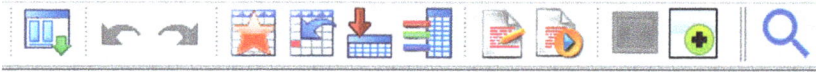

Output1 [Document1] - IBM SPSS Statistics Viewer

Tests of Normality

	Kolmogorov-Smirnov[a]			Shapiro-Wilk		
	Statistic	df	Sig.	Statistic	df	Sig.
Intellegence1	.143	30	.119	.973	30	.619
Intellegene3	.203	30	.003	.902	30	.010

a. Lilliefors Significance Correction

Sig. Is the value of p. The Kolmogorov–Smirnov and Shapiro–Wilk indices have *p* values of 0.119 and 0.619 and 0.003 and 0.010 in Class 1 (Day) and 3 (Online), respectively. That is, both Kolmogorov–Smirnov and Shapiro–Wilk indicators show that Day class intelligence values have a normal distribution and Online class intelligence values do not have a normal distribution, similar to the above histograms and Q-Q Plots.

The APS style presentation is as follows.

> Intelligence scores were normally distributed for the Day class ($p > 0.05$); however, Intelligence scores were not normally distributed for the Online class ($p < 0.05$) as assessed by Shapiro–Wilk's test and Kolmogorov–Smirnov normality test. **ALTERNATIVELY, for example**, The assumption of normality for Intelligence scores was satisfied for all group combinations of Day, Night and Online classes, as assessed by Shapiro–Wilk's test (p > 0.05) etc.

11.6.4   Skewness, Peakedness and Kurtosis Values

The following is how Kurtosis values are tested for skewness and Peakedness using SPSS.

[**SPSS:** Analysis > Descriptive Statistics > Descriptives > Options > Then check Kurtosis and Skewness Boxes Normality Continue > Click OK].

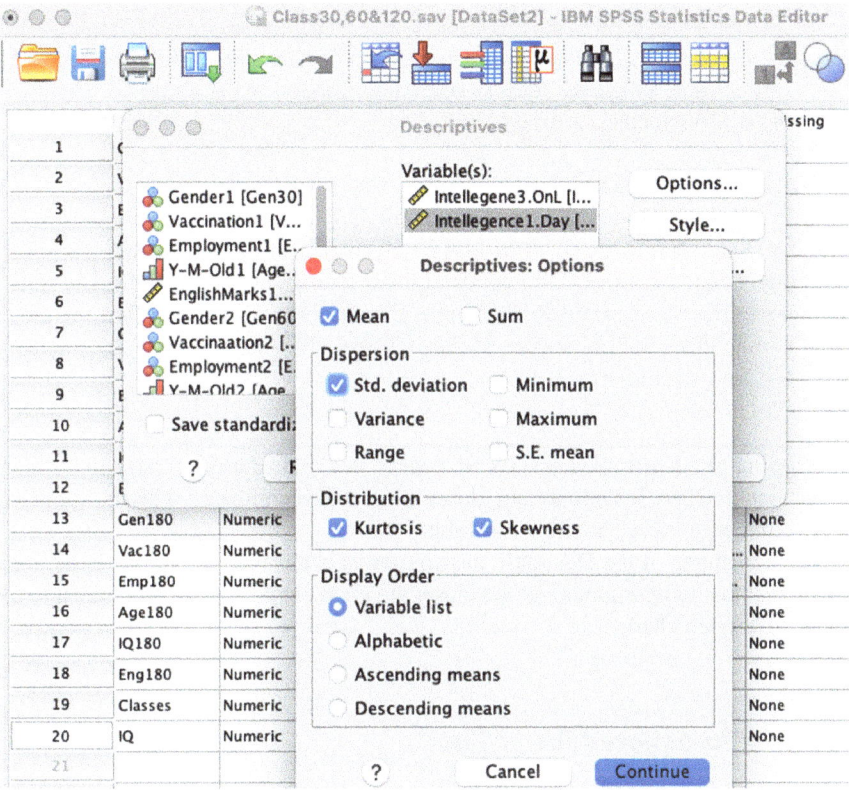

Kurtosis values for Skewness and Peakedness are as follows.

Descriptive Statistics

	N	Mean	Std. Deviation	Skewness		Kurtosis	
	Statistic	Statistic	Statistic	Statistic	Std. Error	Statistic	Std. Error
Intellegene3.OnL	180	106.59	8.169	.650	.181	-.736	.360
Intelligence1.Day	30	97.67	11.327	-.086	.427	.582	.833
Valid N (listwise)	30						

The presentation to the APA style is as follows.

Intellegence scores were normally distributed for the Day class with a skewness of –0.086 (standard error = 0.427) and kurtosis of 0.582 (standard error = 0.833). Intellegence scores were not normally distributed for the Online class with a skewness of 0.650 (standard error = 0.181) and kurtosis of –0.736 (standard error = 0.360).

ALTERNATIVELY, for example, The assumption of normality for Intelligence scores was satisfied for the Day class with a z score of –0.2 (–0.086/0.427) for skewness and 0.658 (0.582/0.833) for kurtosis (at the significance level of 0.05, which equates to a z-score of $\pm$ 1.96), and the assumption of normality for Intelligence scores was not satisfied for the Online class with a z-score of 3.5 (0.650/0.181) for skewness and –2.04 (–0.736/0.360) for kurtosis.

Note that the Kurtosis values for Skewness and Peakedness in Class 1 (Day) are lower than in Class 3 (Online) and closer to zero. It is compared with the statistical details of these classes, as shown in Table 11.6.

The bell shape of the Day class distribution is better than the online class, indicating that the compartment and the closet are closer to zero. It is also shown in the above histogram chart.

11.6.5 Chi-Square Fitness Test

The students in the example samples to observe the ratios of the observed age groups (to check if they are normally distributed between the three age groups) (Chi-square goodness of fit test).

The Null Hypothesis (H0): There is no significant difference between the three age groups in the samples.

Alternative hypothesis (H1): There is a significant difference in composition between the three age groups in the samples.

If $p > 0.05$, the composition of the youth, middle, and adult age groups are not statistically significant. It can be concluded that there is no significant difference, so retain the null.

There are several ways to do this. The following are the chi-square statistics obtained from a single sample by non-parametric tests. The same result can be obtained through the non-parameter Legacy Dialog feature. The screenshot below shows the results of the chi-square fitness test to see if the data has a normal distribution.

Table 11.6 Comparison of average distribution values with statistical details of classes

Class	N	Mean	SD	Kolmogorov–Smirnov	p value	Shapioro–Wilk	p value	Skewness	Kurtosis
1 Day	30	97.67	11.327	0.143	0.119	0.973	0.619	0.086	0.582
3 OnL	180	106.59	8.169	0.203	0.003	0.902	0.010	0.650	−0.736

[**SPSS**: Analyse > Nonparametric Tests > Legacy Dialog > Chi-Square > OK].

or [**SPSS**: Analyse > Nonparametric Tests > One sample > Select fields > OK].

Both of these pathways have the same results. According to the first method, the three classes have to be tested separately. Presenting the results of the second method is simpler and easier to understand. Here all three classes can be tested simultaneously. It also shows the hypothesis and the conclusion of whether it should be rejected or accepted.

The first method for Day class

Chi-square fitness test results to see if the data has a normal distribution.

Test: **Results:** Retain the null ($p = 0.082$, > 0.05)
Step 1

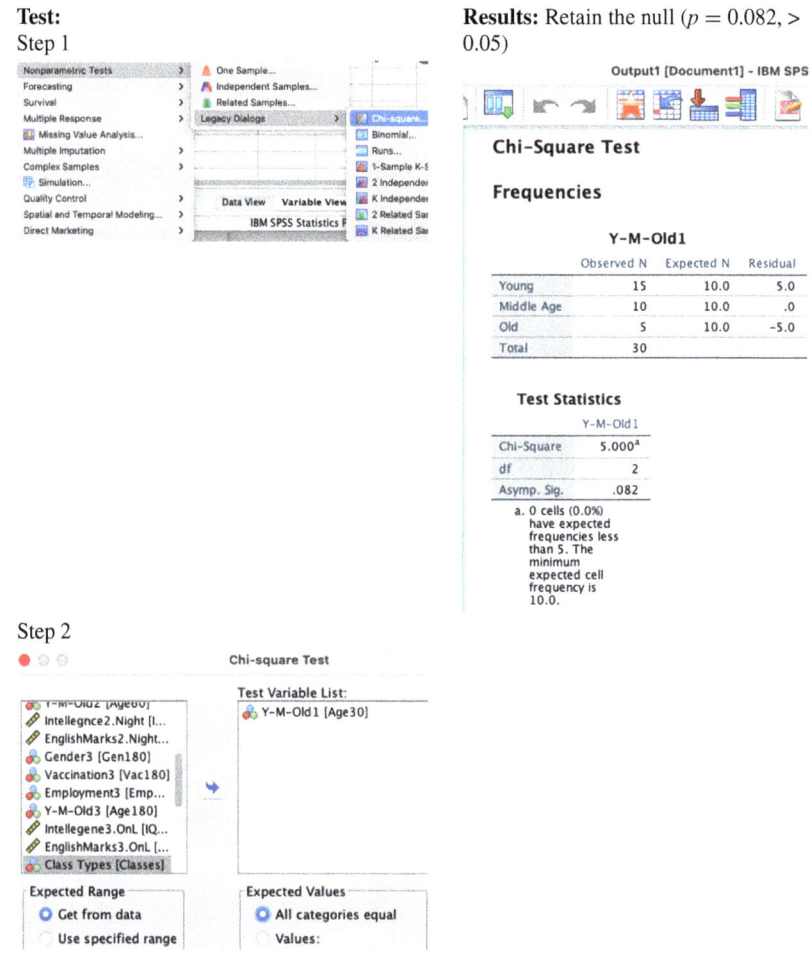

Step 2

Of the 30 students, 15 were young, 10 were in middle age, and 5 were old. A chi-square goodness-of-fit test was conducted to determine whether an equal number of participants from each of the three groups were in the Day class. The minimum expected frequency was 10. The chi-square goodness-of-fit test indicated that the number of young, middle age and old students in the Day class was not statistically significantly different ($\chi^2(2) = 5, p = 0.082$), with just half of the students were young.

ALTERNATIVELY: The minimum expected frequency was 10. The chi-square goodness-of-fit test indicated that the Day class students equally represented the three age groups ($\chi^2(2) = 5$, p = 0.082).

This should be done separately for the three classes.

The hypothetical hypothesis of the kai-type test for goodness of fit (H0) is that the sample distribution is normal, with no significant difference. The above test shows that students are equally divided between age groups. A P sensitivity exceeding the expected sensitivity limit of 0.05 indicates that there is no significant evidence that the Day class age group is evenly distributed.

The second method is to have all three classes simultaneously:

→ Nonparametric Tests

Hypothesis Test Summary

	Null Hypothesis	Test	Sig. [a,b]	Decision
1	The categories of Y–M–Old1 occur with equal probabilities.	One–Sample Chi–Square Test	.082	Retain the null hypothesis.
2	The categories of Y–M–Old2 occur with equal probabilities.	One–Sample Chi–Square Test	.007	Reject the null hypothesis.
3	The categories of Age1Y–M–Old3 occur with equal probabilities.	One–Sample Chi–Square Test	<.001	Reject the null hypothesis.

a. The significance level is .050

b. Asymptotic significance is displayed.

One–Sample Chi–Square Test

APA: The chi-square goodness-of-fit test indicated that the number of young, middle age and old students in the Day class was not statistically significantly different ($p = 0.082, > 0.05$) while they are significantly different for the Night class ($p = 0.007, < 0.05$) and Online class ($p < 0.001$).

The screenshot above provides the results of a chi-square goodness-of-fit test to see whether the composition of the young, middle-aged and adult age groups is evenly distributed or different. From this table, we can see that the statistics in the Day class (Class No. 1) are not statistically significant: p = 0.082 > 0.05; therefore, we cannot refute the null hypothesis (Retain the null) that suggest that young, middle and adult age group composition is not statistically significantly different (similar 1/3, 1/3, 1/3).

However, the Night Class (Class 2) and Online Class (Class 3) test results are different. As shown in the screenshot above, we can see that the test statistics of the two classes Night and Online are statistically significant: they are $p = 0.007 < 0.05$ and p < 0.001, respectively, so we have to refute the null hypothesis, and it can be concluded that the age group composition is statistically, significantly different. Since the p value of the online class is less than 0.01, it is more than a 99% level of probability that the composition of the online class is not evenly distributed among the age groups. Here, too, it should be noted that although the compositions of both classes have the same ratios as a percentage, it is suggested that if the sample is small, the composition is not evenly distributed, and if the sample is large, this difference is significant. That is, the larger the sample, the greater the effectiveness of the results.

The presentation to the APA style is as follows.

A chi-square goodness-of-fit test was conducted on Age groups composition. There were no statistically significant differences in the number of students of different age groups in the Day class ($p > 0.05$) while there were statistically significant differences in the number of students of different age groups in the Night class ($p < 0.05$) and Online class ($p < 0.05$).

11.6.6 Chi-Square Test for Independence; Distribution of Religions

Let us examine this test with a kai-type test for independence. It does not examine religions separately but only examines whether the distribution of religions in the sample as a whole differs significantly from the average distribution. To do this, we first need to calculate the expected average religious distribution of the sample value of 200, as shown in Table 11.7.

Calculate the expected religious distribution in Table 11.7: 200 students.

[**SPSS**: Analyse > Nonparametric Tests > Legacy Dialog > Chi-Square > OK].

Test: Step 1

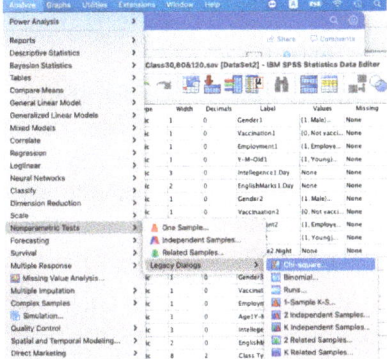

Test: Step 2

The average distribution values should be included in the same order as the religions in the sample as follows

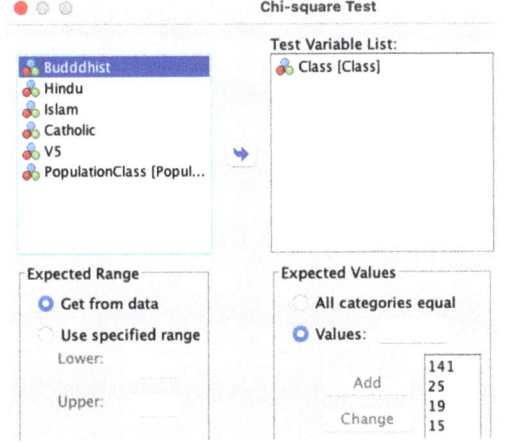

Results:

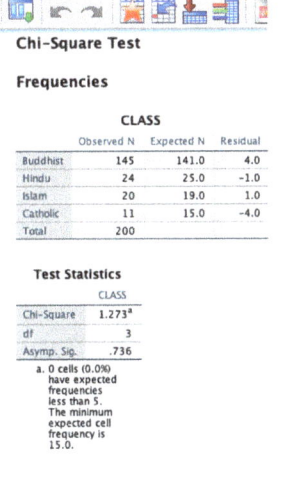

Chi-Square Test

Frequencies

CLASS

	Observed N	Expected N	Residual
Buddhist	145	141.0	4.0
Hindu	24	25.0	-1.0
Islam	20	19.0	1.0
Catholic	11	15.0	-4.0
Total	200		

Test Statistics

	CLASS
Chi-Square	1.273ᵃ
df	3
Asymp. Sig.	.736

a. 0 cells (0.0%) have expected frequencies less than 5. The minimum expected cell frequency is 15.0.

Interpretation: Because the $p = 0.736 > 0.05$, it does not reject the null hypothesis; that is, the religious distribution of the sample does not change significantly. The z test using proportions provided the same result separately for each religion.

APA presentation is as follows;

The minimum expected frequency was 15. A chi-square goodness-of-fit test was conducted to determine whether the religious groups of the study had the same proportion of religious groups as those in the general population. The minimum expected frequency was 15. The chi-square goodness-of-fit test indicated that the three-body composition types were similarly distributed in the participants recruited to the study as the general population ($\chi^2(2) = 1.273$, $p = 0.736$).

Pearson's chi-square test assess the extent to which change is likely to occur by chance.

Second, Pearson's chi-square testing assess the extent to which the observed difference between samples is likely to occur randomly.

The purpose of the following chi-square test is to examine whether there is a difference in employment between men and women in the example above. The null hypothesis (H0) for this test is that there is no relationship between gender and employment. The alternative hypothesis is that there is a relationship between gender and employment (e.g., women's employment is higher than men's, or vice versa).

The following photo screenshot shows how to follow this method.

[**SPSS**: Analyse > Descriptive > Crosstabs > Statistics > Chi-Square > OK].

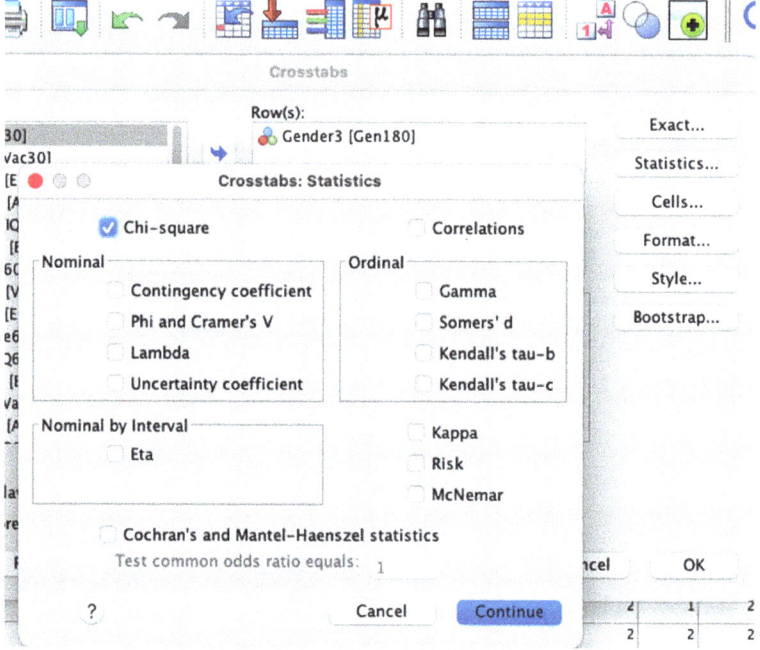

The screenshots below show the results of a chi-square test that examines the relationship (independence) between gender and employment: the left side of the screenshot represents the Day class, and the right side represents the online class results.

The relationship between gender and employment.

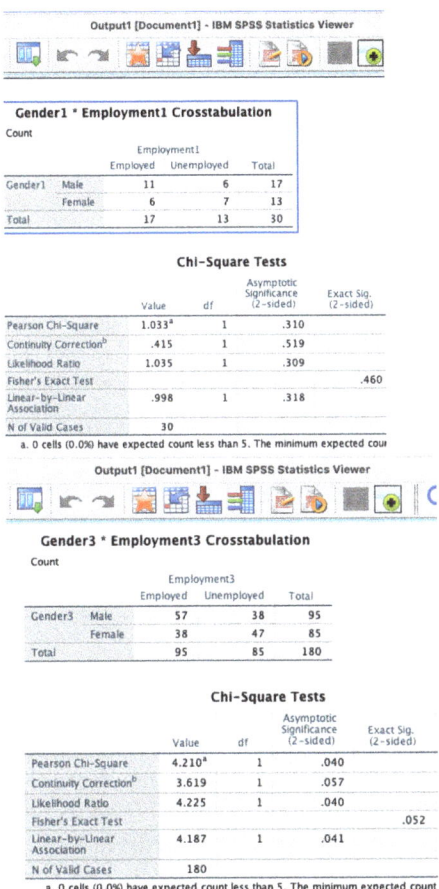

Gender1 * Employment1 Crosstabulation

Count

		Employment1		Total
		Employed	Unemployed	
Gender1	Male	11	6	17
	Female	6	7	13
Total		17	13	30

Chi-Square Tests

	Value	df	Asymptotic Significance (2-sided)	Exact Sig. (2-sided)
Pearson Chi-Square	1.033ᵃ	1	.310	
Continuity Correctionᵇ	.415	1	.519	
Likelihood Ratio	1.035	1	.309	
Fisher's Exact Test				.460
Linear-by-Linear Association	.998	1	.318	
N of Valid Cases	30			

a. 0 cells (0.0%) have expected count less than 5. The minimum expected cour

Output1 [Document1] - IBM SPSS Statistics Viewer

Gender3 * Employment3 Crosstabulation

Count

		Employment3		Total
		Employed	Unemployed	
Gender3	Male	57	38	95
	Female	38	47	85
Total		95	85	180

Chi-Square Tests

	Value	df	Asymptotic Significance (2-sided)	Exact Sig. (2-sided)
Pearson Chi-Square	4.210ᵃ	1	.040	
Continuity Correctionᵇ	3.619	1	.057	
Likelihood Ratio	4.225	1	.040	
Fisher's Exact Test				.052
Linear-by-Linear Association	4.187	1	.041	
N of Valid Cases	180			

a. 0 cells (0.0%) have expected count less than 5. The minimum expected count

A chi-square test of independence was conducted between gender and Employment for two classes: Day and Night. All expected cell frequencies were greater than five. There was not a statistically significant association between gender and Employment, for the Day class $\chi^2(1) = 1.033$, $p = 0.310$ but there was a statistically significant association between gender and Employment, for the Night class $\chi^2(1) = 4.210$, $p = 0.040$.

From this table, we can see that the test statistics in Class 1 are not statistically significant: because in Class 1, $\chi^2 = 1.033$, $p = 0.310 > 0.05$, therefore, "do not reject" the null hypothesis: there is the relationship between gender and employment. And by retaining the null, it can be concluded that there is no relationship between gender and employment. That is, femininity and non-employment are independent of each other, and femininity does not affect unemployment.

But the results of Class 3 are different. As shown in the second photo screen, we can see that the test statistics are statistically important: $\chi^2 = 4.210$, $p = 0.040$, < 0.05 so that we can reject the null. It can be concluded that there is a significant association (statistically significant association) between gender and employment. That is, femininity and non-employment show an association, and that femininity affects employment.

It should also be noted that the unemployment rate in both classes as a percentage varies considerably (female unemployment in Class 1 is $7/13 = 53.8\%$ of total unemployment and male unemployment is $6/13 = 46.15\%$). If the sample is small, it is not statistically significant. This reflects the sample power, which is related to the size of the sample. In the hypothetical test technique, the larger the sample strength/size, the higher the probability of reaching the critical value, and the lower the p value. When the critical value is exceeded ($p < 0.05$), the null hypothesis is rejected (Reject the null).

NOTE: In contrast to the definition of kai-types chi-square in complex model-fit tests such as SEM, exceeding the phase value suggests that the model is not suitable.

11.7   Degree of Freedom (d.f.)

Statistically, the degree of freedom: the moving number/space (df–DOF) is the number of free/optional numbers that can vary in statistics in the final calculation. Since you have two samples in the F test (1 and 2), you have two moving numbers: one for the core and the other for the core. Various d.f. The critical F values relative to the values at the significance level of 5% (Alpha $\alpha = 0.05$) are given in Table 11.7.

Table 11.7   Relative to d.f. F critical values (Level of significance, $\alpha = 0.05$)

$\alpha =$	0.050						F-table										
							$dF_1(v_1)$										
$dF_2(v_2)$	1	2	3	4	5	6	7	8	9	10	11	12	15	25	40	60	120
1	161.4	199.5	215.7	224.6	230.2	234.0	236.8	238.9	240.5	241.9	243.0	243.9	245.9	249.3	251.1	252.2	253.3
2	18.5	19.0	19.2	19.2	19.3	19.3	19.4	19.4	19.4	19.4	19.4	19.4	19.4	19.5	19.5	19.5	19.5
3	10.13	9.55	9.28	9.12	9.01	8.94	8.89	8.85	8.81	8.79	8.76	8.74	8.70	8.63	8.59	8.57	8.55
4	7.71	6.94	6.59	6.39	6.26	6.16	6.09	6.04	6.00	5.96	5.94	5.91	5.86	5.77	5.72	5.69	5.66
5	6.61	5.79	5.41	5.19	5.05	4.95	4.88	4.82	4.77	4.74	4.70	4.68	4.62	4.52	4.46	4.43	4.40
6	5.99	5.14	4.76	4.53	4.39	4.28	4.21	4.15	4.10	4.06	4.03	4.00	3.94	3.83	3.77	3.74	3.70
7	5.59	4.74	4.35	4.12	3.97	3.87	3.79	3.73	3.68	3.64	3.60	3.57	3.51	3.40	3.34	3.30	3.27
8	5.32	4.46	4.07	3.84	3.69	3.58	3.50	3.44	3.39	3.35	3.31	3.28	3.22	3.11	3.04	3.01	2.97
9	5.12	4.26	3.86	3.63	3.48	3.37	3.29	3.23	3.18	3.14	3.10	3.07	3.01	2.89	2.83	2.79	2.75
10	4.96	4.10	3.71	3.48	3.33	3.22	3.14	3.07	3.02	2.98	2.94	2.91	2.85	2.73	2.66	2.62	2.58
11	4.84	3.98	3.59	3.36	3.20	3.09	3.01	2.95	2.90	2.85	2.82	2.79	2.72	2.60	2.53	2.49	2.45
12	4.75	3.89	3.49	3.26	3.11	3.00	2.91	2.85	2.80	2.75	2.72	2.69	2.62	2.50	2.43	2.38	2.34
15	4.54	3.68	3.29	3.06	2.90	2.79	2.71	2.64	2.59	2.54	2.51	2.48	2.40	2.28	2.20	2.16	2.11
25	4.24	3.39	2.99	2.76	2.60	2.49	2.40	2.34	2.28	2.24	2.20	2.16	2.09	1.96	1.87	1.82	1.77
40	4.08	3.23	2.84	2.61	2.45	2.34	2.25	2.18	2.12	2.08	2.04	2.00	1.92	1.78	1.69	1.64	1.58
60	4.00	3.15	2.76	2.53	2.37	2.25	2.17	2.10	2.04	1.99	1.95	1.92	1.84	1.69	1.59	1.53	1.47
120	3.92	3.07	2.68	2.45	2.29	2.18	2.09	2.02	1.96	1.91	1.87	1.83	1.75	1.60	1.50	1.43	1.35

This d.f. Or the DOF concept is also used for statistical tests, such as the Chi-Square described below. These judgments are made using p value in these experiments using computer software.

11.7.1 F-Test

The technique used to study the independence of behaviour of more than two variables is called 'Analysis of Variance' (ANOVA). In an ANOVA test, the F value is calculated and compared with the critical value. Fisher's test is a comparison of the distribution of two sets of data to test whether they belong to the same population. That is, whether the absolute precisions are equal or unequal is checked using the ratio of the two variables (S2–variances) [F = S21/S22].

Sources Analysis of Variance (ANOVA) https://www.youtube.com/watch?v=ITf4vH hyGpc

Traditionally, the large S2 value is taken as the numerator and the small S2 value as the denominator. The F-test result is always a positive number (since the square multiplier-X^2-is always positive). If the result is not much different from one (1), then S21 and S22 are not much different, and in practice, the calculated F, compared to the corresponding F value in the F-table (F must be less than the critical value), is valid. As mentioned above, there is no need to confuse these tables when calculating F by computer software, and whether the F-test result is important or not, depending on the significance level (10%, 5%, 1% or 0.1%) selected by the p value. A judgment can be reached.

This is further discussed in Chap. 14 in Example 12.2–Table 14.2 ARDL bounds test F-statistics for cointegration and the critical values.

Let us test using SPSS to see whether there is a significant difference between the values of the above three samples. A sample of **30 students** from all three classes was used for this test. This test requires data processing (it requires processing into a single column). It can be seen in this link: One-Way ANOVA Test in SPSS: https://www.youtube.com/watch?v=OEOeXpxSjf8.

[**SPSS**: Analyze > Compare Means > One-Way ANOVA > OK].

Test:

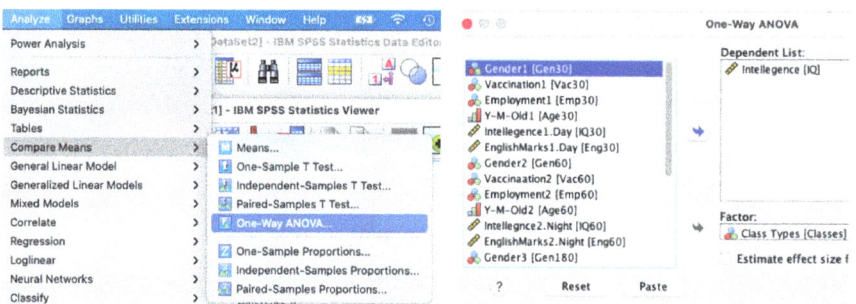

The null hypothesis (H0) for this test is that there is no significant difference between the intelligence values of the students in the three classes. The alternative hypothesis (H1) is a significant difference between the students' intelligence in the three classes (e.g., the intelligence in the Night class is generally higher or lower than the students in the other classes).

Results:

ANOVA

Intelligence

	Sum of Squares	df	Mean Square	F	Sig.
Between Groups	1397.067	2	698.533	6.554	.002
Within Groups	9272.933	87	106.585		
Total	10670.000	89			

Interpretation: Here $p < 0.05$ (p value 0.002) means that H0 should be rejected (Reject the null), that is, there is a significant difference in intelligence between classes.

APA format:

> A one-way ANOVA was conducted to determine if the Intelligence values were different among groups. A sample of 30 students was taken from each class: Day, Night and Online (n = 30 each totalling to N = 90). the differences between Intelligence levels between classes was statistically significant, F (2, 89) = 6.554, p = 0.002.

Goodness of fit or model fit

As discussed above, chi-square statistics are widely used to test the suitability of complex regression models (such as Structural Equation Modeling SEM), especially when the sample is not very large. The degree of freedom (d.f. = DOF) Is required to assess the suitability (goodness of fit) of a sample observed by chi-square distribution. The measure of suitability is that the value obtained by dividing the chi-square value by the DOF should be less than the critical value 2.5.

Table 11.8: Chi-square test and DOF show the model fit of the four factors hypothesized in Example 10.2 that may influence investment in the stock market. This study confirms whether these four factors motivate an investor to invest in the stock market in the same way as four distinct, identifiable factors. Accordingly, it was revealed that factors such as 'Attitudes', 'Power', 'Social norms' and 'Knowledge' affect 'Investment intent' from different dimensions. To use this technology (SEM) you need to use AMOS software with SPSS.

Table 11.8 Chi-square test and DOF

CFA order	Chi-square	Sig-(p)	Chi-Sq/degree of freedom	Yes/No
First order	107.77 for 52 d.f.	0.03	107.77/52 = (a)?	(c)?
Second order	124.12 for 55 d.f.	0.043	124.12/55 = (b)?	(d)?

Answers: a. 2.07, b. 2.25, c. Yes, d. Yes

Box 11.1 Chi-square results in interpretation

The first-order CFA model reflected an acceptable model fit (chi-square/df = 2.07, CFI = 0.85, RMSEA 0.08, CI: 0.051, 0.087). Overall, the CFA model provided evidence for a hypothesized first-order factor structure. Accordingly, Attitude, Power, Social Norms and Knowledge are four distinct dimensions of IW.

The second-order CFA model reflected an acceptable model fit (chi-square/d.f. = 2.26, CFI = 0.81, RMSEA 0.08, CI: 0.051, 0.087). Overall, the CFA model provided evidence for a hypothesized factor structure. Accordingly, Attitude, Power, Social Norms and Knowledge are four distinct dimensions of investment intention.

Here the interpretation of the results of the chi-square test is in the exact opposite direction. That is, as the size of the sample increases, its power increases, and it can be suggested that the kai-square value exceeds the phase value, and the model is not suitable. In such cases, the suitability is verified using techniques such as RMSEA, CFI and TLI discussed below.

11.7.2 Structural Equation Modeling–(SEM) RMSEA, CFI and TLI

Does the hypothesized model fit the data well? This is a critical question in almost every application of structural equation modeling (SEM). The model chi-square statistic and several fit indices are commonly reported to address this question. Structural equation modeling relies on, application of the Root Mean Square Error of Approximation (RMSEA), comparative fit index (CFI), and Tucker–Lewis index (TLI) with continuous data. For ordered categorical data, unweighted least squares (ULS) and diagonally weighted least squares (DWLS) too.

The suitability of structural equation modeling for continuous data depends on the following factors; That is, (RMSEA < 0.08); Comparative Fitness Index (CFI > 0.08) and Tucker Lewis Index (TLI > 0.08). For categorical, ordinal data, Unweighted Leas Square (ULS) and Diagonal Weighted Least Square (DWLS) are used.

Table 11.9 SEM model fit measures

CFA Order	Chi-Square	CFI	RMSEA	CI
First order	107.77 for 52 d.f.	0.85	0.08	0.051, 0.087
Second order	124.12 for 55 d.f.	0.81	0.08	0.051, 0.087

Source Saliya (2021b)

Eligibility criteria for this Table 11.9: SEM test should be presented along with the test results. It is discussed with the results of Fig. 13.1: Model pathway analysis in Chap. 13-Structural Equation Modeling Example.

11.7.3 Growth Curve Model-GCM, RMSEA, CFI and TLI

The multilevel latent growth curve model (MLGCM) and the multilevel structural equation modeling framework (MLSEM) have been advocated as a means of investigating individual and cluster trajectories.

The MLGCM, overridden by the multilevel structural equation modeling framework (MLSEM), is increasingly being used as a single and cluster trajectory analysis medium. Growth curve models (GCMs) are statistical methods that allow estimating patterns of change over time, commonly known as time trajectories, time tracks, growth curves, or latent trajectories. Those patterns are static (static/flat- showing no change over time); Gradually increasing or decreasing over time; Linear or curved form, etc. The above model fitness indicators are also used for this technique.

11.8 Other Assumptions and Concepts

Homogeneity of variances: Levene's test is used to test the equality of variances when comparing the means of independent groups, e.g. Independent t-tests and ANOVA. The violation of this assumption is more serious than a violation of the assumption of normality, but both t-tests and ANOVA are fairly robust to deviations from this assumption.

How this test works automatically with the Independent Sample T test can be seen in the next sample t test using SPSS in the next chapter.

Tests for sphericity: a measure of whether variances of the differences between all repeated measures are all equal. If the assumption is not met, the F-statistic is positively biased, leading to an increased risk of a type 1 error. If the p value < 0.05, there are significant differences between the variables. That is, the spherical condition is not met.

The relationship between sample size, sample error, intensity, and power can be summarized as follows.

Sample error is the error caused by observing a sample instead of the total aggregation. That is, the difference between the sample frequency and the unknown parameter values when using sample statistics to estimate a parameter in a compilation. The relationship between size, sample error, intensity, and power can be summarized as follows.

11.8.1   Effect Size

An effect size is a measure of the strength or magnitude of the effect of an independent variable on a dependent variable which helps to assess whether a statistically significant result is meaningful.

Effect Size = [Mean of the experimental group] − [Mean of control group]/Standard Deviation

Although not always meaningful, Cohen provides the following guidance for intensity; 0.2–0.3 can be 'small' intensity, 0.5 to 'medium' intensity and 0.8 to infinity, 'large' intensity.

Partial eta-squared is a measure of variance. It represents the proportion of variance in the dependent variable that is explained by the independent variable. It also represents the effect size statistic. The effects sizes given in Cohen (1988) for the interpretation of the absolute effect sizes are: $\eta 2 = 0.010$ is a small association. $\eta 2 = 0.059$ is a medium association. $\eta 2 = 0.138$ or larger is a large association (Lakens, 2013).

11.8.2   Statistical Power

Statistical power is the ability of a statistical test to identify relationships between variables. If the null hypothesis (H0) is false, the probability of failing to reject it (Type II error: Error of the second model) β is called the force $(1 - \beta)$. Thus, if the probability of a type II error (β) for any test is 0.2, then the test strength is 0.8 (80%). Table 11.10: The relationship between sample size, sample error, intensity and power can be summarized as follows.

Table 11.10 Relationships between sample size, sample error, intensity, and power

Sample size	Statistical power	Sampling error	Error types
Sample size high	Power increases	Sampling error low	The probability of Type II error low
Sample size low but effect size high	Power increases	Sampling error high	The probability of Type I error low

11.9 Summary

- The traditional significance level of statistics is 5% (0.05). That is, confidence is greater than 95% ($1 - \alpha = 100\% - 5\% = 95\%$). A value of '$p$' greater than 0.05 means 'not significant'.
- Small asterisks usually indicate this level of significance. Ex: $p < 0.05$ by one asterisk (0.045^*), $p < 0.01$ by two asterisks (0.002^{**}), $p < 0.001$ by three asterisks (0.000^{***}), etc. (APA).
- Exceeding the test statistical critical value means that the p value is less than 0.05. That is, to reject the null. Staying in the test static zone means that the p value is large ($p > 0.05$, $\alpha = 0.05$), and the null hypothesis cannot be refuted (Retain the null).
- The larger the sample, the greater the potential for exceeding the critical value that determines the validity of a test, and the greater the power of the sample.
- It is important to present results according to the traditions of the American Psychological Association (APA).
- If Q is the magnitude of the change that can be tolerated, then the confidence interval is the difference between $\mu - Q$ and $+ Q$.
- In an ANOVA test, F (Fisher's test) is a comparison of the distribution of two sets of data to see if they belong to the same set.
- Degree of freedom (d.f. = df = DOF) is the number of free or possible alternate numbers that can vary in statistics in the final calculation.
- Z test is a test performed to assess whether the mean of a sample corresponds to the mean of the population at a certain level of significance.
- The Z-test score is the standard deviation from the mean.
- T-tests are tests performed between the mean of two samples or for comparison.
- Chi-square goodness of fit determines whether the sample data is suitable for the test.
- A chi-square test for independence compares two variables in a random table and see if there is a relationship between them, checking that the distribution of the classification variables is different from each other.
- Stability of Structural Equation Formatting (SEM) for continuous or continuous data RMSEA (root mean square error of approximation < 0.08), CFI; (comparative fit index > 0.95) and TLI; (Tucker–Lewis index > 0.95).
- Other validity tests include Homogeneity of variances: Levene's test, and Tests for sphericity.

References/Further Reading

American Psychological Association. (2019). *Publication Manual of the American Psychological Association*, (7th ed).
Analysis of Variance (ANOVA). https://www.youtube.com/watch?v=ITf4vHhyGpc.

Chi-Square Goodness-of-Fit Test in SPSS Statistics. https://statistics.laerd.com/spss-tutorials/chi-square-goodness-of-fit-test-in-spss-statistics.php.

Chi-square test in SPSS + interpretation. https://www.youtube.com/watch?v=wfIfEWMJY3s.

Chi-Square Goodness-of-Fit Test in MS Excel. https://www.real-statistics.com/tests-normality-and-symmetry/statistical-tests-normality-symmetry/chi-square-test-for-normality/.

Chi-Square Test-Reporting a One-Sample. https://www.spss-tutorials.com/spss-one-sample-chi-square-test/.

Confidence Intervals: Statistics Help. https://www.youtube.com/watch?v=tFWsuO9f74o.

Cohen, J (1988). Statistical Power Analysis for the Behavioral Science. *Hillsdale, NJ*. Lawrence.

Glen, S. (2020). *Reporting statistics APA style*. From StatisticsHowTo.com: Elementary Statistics for the rest of us! https://www.statisticshowto.com/probability-and-statistics/reporting-statistics-apa-style/.

How To... Perform a One-Way ANOVA Test in SPSS. https://www.youtube.com/watch?v=OEO eXpxSjf8.

Jarque-Bera test in MS Excel. https://www.youtube.com/watch?v=3or8IwMUgjs.

Lakens, D. (2013). Calculating and reporting effect sizes to facilitate cumulative science: a practical primer for t-tests and ANOVAs. *Frontiers of Psychology*, 4(863), 26 Nov. 2013. https://doi.org/10.3389/fpsyg.2013.00863.

NIST/SEMATECH. (2012). *e-Handbook of statistical methods*. https://doi.org/10.18434/M32189.

One-Way ANOVA Test in SPSS. https://www.youtube.com/watch?v=OEOeXpxSjf8.

P-Value Method For Hypothesis Testing. https://www.youtube.com/watch?v=8Aw45HN5lnA.

P-values and significance tests | AP Statistics | Khan Academy. https://www.youtube.com/watch?v=KS6KEWaoOOE.

Saliya, C. A. (2021b). Driving Forces of Individual Investors in Stock Market Participation. *Review of Economics and Finance, 19*, 73–79. https://refpress.org/ref-vol19-a8/.

Single Sample Hypothesis Z-test. https://www.youtube.com/watch?v=tsPv-ffN-0M.

Single Sample Hypothesis Z-test. https://www.youtube.com/watch?v=HoqzIR8xj4s.

Statistics 101: Single Sample Hypothesis Z-test Concepts. https://www.youtube.com/watch?v=Hoq zIR8xj4s and https://www.youtube.com/watch?v=tsPv-ffN-0M.

Student's t Test. https://www.youtube.com/watch?v=pTmLQvMM-1M.

Syntax for Z test link. http://www.how2stats.net/2014/03/one-sample-z-test.html.

T test Vs Z test. https://www.youtube.com/watch?v=YsalXF5POtY.

t tests and p values. https://www.youtube.com/watch?v=5ABpqVSx33I.

Two sample t- and z-tests (Paired) for matched data in SPSS. https://www.youtube.com/watch?v=vzQGQ62tScQ.

Understanding Confidence Intervals: Statistics Help. https://www.youtube.com/watch?v=tFWsuO 9f74o.

What Is A P-Value? https://www.youtube.com/watch?v=ukcFrzt6cHk.

Which test. https://www.youtube.com/watch?v=rulIUAN0U3w and https://www.youtube.com/watch?v=UaptUhOushw.

Which test2. https://www.youtube.com/watch?v=QrYgXZf-Ay8.

Which test3. https://www.youtube.com/watch?v=I10q6fjPxJ0.

Z Critical value. https://www.youtube.com/watch?v=8Aw45HN5lnA.

Z Syntax. http://www.how2stats.net/2014/03/one-sample-z-test.html.

Z and t test examples given here using the link. https://www.youtube.com/watch?v=YsalXF5POtY.

Z statistics Calculating a z statistic in a test about a proportion. https://www.khanacademy.org/math/ap-statistics/tests-significance-ap/one-sample-z-test-proportion/v/calculating-a-z-statistic-in-a-significance-test.

Z test. http://www.how2stats.net/2014/03/one-sample-z-test.html and https://www.youtube.com/watch?v=8Aw45HN5lnA.

Z using SPSS Syntax. https://www.youtube.com/watch?v=HC-oKjNwiXc and http://www.how2stats.net/2014/03/one-sample-z-test.html.

Chapter 12
Statistical Testing Methods

This chapter discusses statistical testing methods, model building and path diagrams. This book cannot present an in-depth analysis of complex statistics, which requires the use of specialized skills. The boundaries of quantitative and qualitative data analysis are blurring. Translating qualitative data such as attitudes and behaviour patterns into numerical variables and building correlations between those variables has become fascinating research. This chapter is based on the assumption that you have statistical software, especially SPSS, AMOS, and computer capabilities.

If your organization has subscribed to the statistics software as a corporate package, request through the Dean to install it on your computer. Alternatively, SPSS and AMOS software can be downloaded via the https://spss.en.softonic.com link. Available for six months for a small fee at this https://www.hearne.software/SPSS-Selection-v28 link.

According to the data types discussed in the previous chapter, the best data for analysis are those with a continuous distribution normal distribution (normal distribution). So if you do not have a clear idea of what type of data you want to collect, you should try to collect the interval and ratio data as much as possible. That is, to prepare questionnaires with the type of questions that facilitate any value.

Two main types of analysis can be done using statistics. These are descriptive analysis and statistical technical analysis.

12.1 Descriptive Analysis

This means presenting data from charts and tables (pie charts, bar charts, Z-scores, central tendency, etc.). Table 12.1 shows a summary of the different types of detailed data presentations.

© The Author(s), under exclusive license to Springer Nature Singapore Pte Ltd. 2022   205
C. A. Saliya, *Doing Social Research and Publishing Results*,
https://doi.org/10.1007/978-981-19-3780-4_12

Table 12.1 Types of descriptive statistics

Chart	Variable type	Purpose	Statistics
Pie chart or bar chart	One Categorical	Shows frequencies/proportions/percentages	Class percentages
Stacked/multiple bar	Two categorical	Compares proportions within groups	Percentages within groups
Histogram	One scale	Shows distribution of results	Mean and Standard Deviation
Scatter graph	Two scale	Shows relationship between two variables and helps detect outliers	Correlation Coefficient
Boxplot	One scale/one categorical	Compares spread of values	Median and IQR
Line chart	Scale by time	Changes over time	Means by time point
		Comparison of groups	
Means plot	One scale/2 categorical	Looks at combined effect of two categorical variables on the mean of one scale variable	Means

Source Marshal (2021)

The use of SPSS software for this can be practised by following the link: https://www.youtube.com/watch?v=99fGYHGyO5U. By comparing several variables in one graph, information can be presented comparatively very effectively. It may be useful to refer to the detailed statistics and graphs in Example 12.1 (*Source* Saliya and Pandey, 2021).

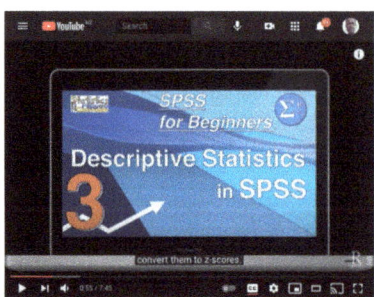

03 Descriptive Statistics and z Scores in SPSS – SPSS for Beginners

12.1.1   Descriptive Statistics and Charts

Examples are given in Figs. 12.1 and 12.2 and Fig. 12.3.

Many attractive graphs can be created using Excel. For example, the six series of data in Fig. 12.1 are comparatively represented using two axes.

Fig. 12.1 Per capita GDP and MCAP

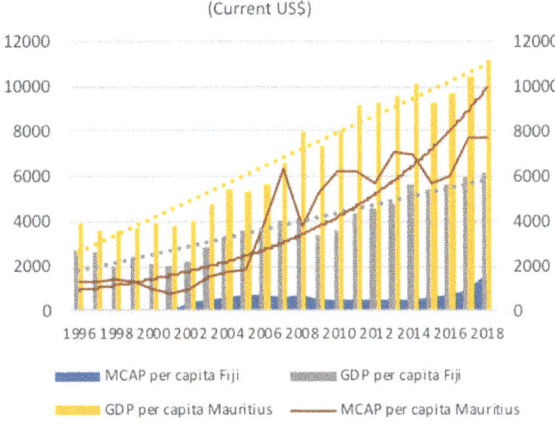

Fig. 12.2 % Contribution of perspectives

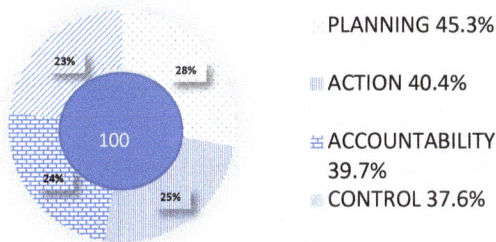

Fig. 12.3 % Contributions of Critical Factors (Criteria). *Source* Saliya (2020a), Saliya and Panday (2021) (Examples 12.1 and 12.2)

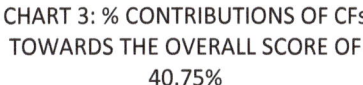

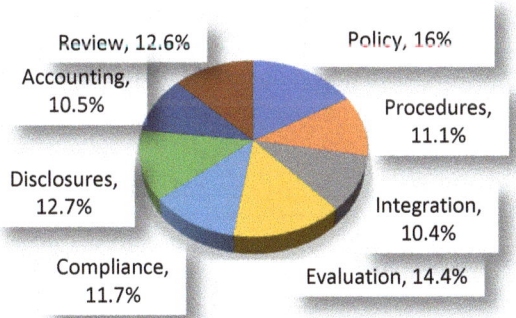

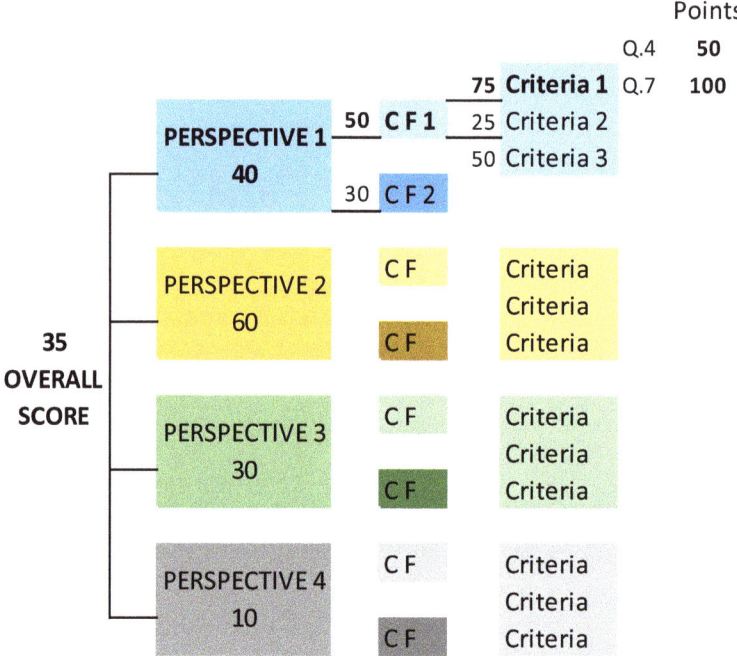

Fig. 12.4   Scoring system. *Source* Saliya and Pandey (2021)

Example 12.1 Financial Battle Against Climate Change—Assessing Effectiveness using a Scorecard.

Box 12.1 Method

The following example (Fig. 12.4 scoring system) demonstrates how the scores are derived for perspectives, from a single response (mean average is taken for 212 responses). Suppose a criterion is assessed by two questions (for example, Q.4 & Q.7) which carry equal weightage and are ascribed 50 points (the answers 'Neutral'), and '100' (Yes). Then the score of the criteria would be 75 [(50 + 100)/2 = 75%)]. If the other two criteria of this CF carry scores of 25 and 50 each, the score of this CF would be 50 [(75 + 25 + 50)/3]. If the score of the other CF is 30 (30%), then the perspective score would be the simple average of 50% and 30% (i.e., 40%). Suppose the other three perspectives carry 60%, 30% and 10%, then the total score would thus be 35% [(60 + 30 + 10 + 40)/4].

12.2   Statistical Technical Analysis

There are two main approaches to choosing an appropriate statistical method. They are based on the nature of the data or variables and the research problem or objectives. The selection of a statistical method according to the research problem or purpose can be summarized as shown in Fig. 12.5, and the selection of a statistical method according to the nature of the data or variables can be summarized as shown in Table 12.2.

12.2.1   Ordinal Data

Some researchers routinely use parametric tests to analyse ordinal data. As a general rule of thumb, ordinal variables with seven or more categories can be analysed with parametric tests if the data is approximately normally distributed. Underlying latent variables are measured using questionnaires with a set of questions. In such a case, the scores are added together, summed, or averaged, converted into scalable variables, and parametric tests are performed.

12.2.2   What Tests?

Alternative to the complex statistical techniques presented in Fig. 12.5 and Table 12.2, the following website (https://maths.shu.ac.uk/mathshelp/WhichTest by Professor Ellen Marshall. PHP) provides a useful guide to choose the basic statistical technique that best suits your data. All you have to do is answer four questions. Four of the following questions will direct you to the relevant technique and provide links to the resources needed to study it.

Regression or ANOVA? Use regression if you have the only scale or binary independent variables. Categorical variables can be recoded to dummy binary variables, but if there are a lot of categories, ANOVA is preferable. Several such experiments arc illustrated in the following paragraphs.

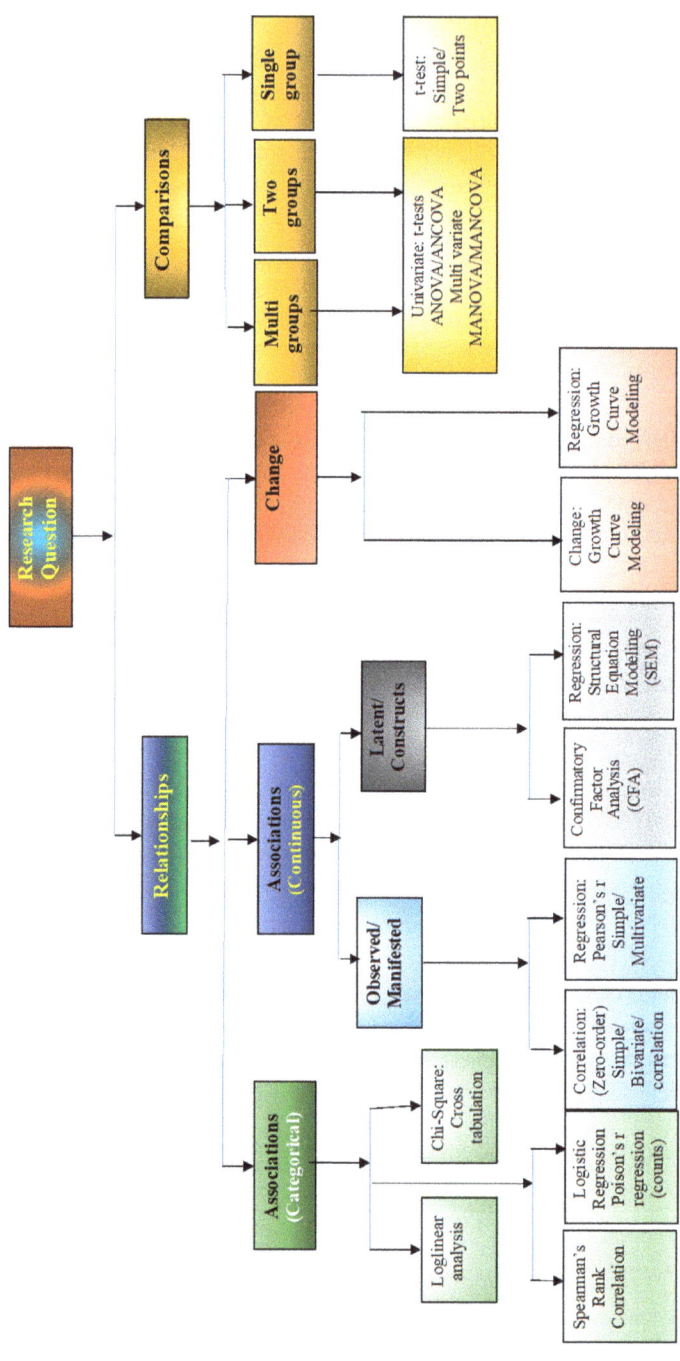

Fig. 12.5 Choosing statistical technique according to the research problem/purpose

Table 12.2 Choosing statistical technique according to the data types or variables

Data distribution	Comparison				Association	Regression	Association of latent variables
	2 Data sets		• Data sets				
	Paired	Unpaired	Paired	Unpaired			
Normal distribution (Mean)	Paired t—test	Unpaired t—test	Repeated measurement ANOVA	One way ANOVA	Pearson correlation	Regression M-level regression	Structural equation modeling (SEM)
Non-normal distribution (Median)	Wilcoxon signed rank	Wilcoxon rank sum Mann-Wallis U	Friedman	Kruskal Wallis	Spearman's rank correlation	Logistics, Poision	Growth curve modeling (GCM)
Nominal (Mode)	z-test for proportion		Chi-Square test (X^2)				

12.3   Commonly Used Statistical Techniques

You can get a basic understanding of data types and related statistical tests by watching the Youtube clip shown in the screenshot below. Youtube clip: https://www.youtube.com/watch?v=I10q6fjPxJ0.

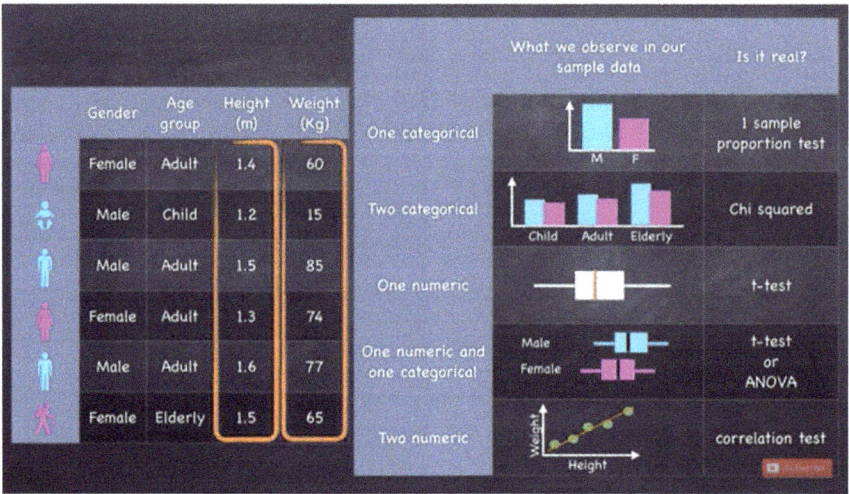

Statistics made easy ! ! ! Learn about the t-test, the chi square test, the p value and more

Source https://www.youtube.com/watch?v=I10q6fjPxJ0

Here are some examples of interpretations of experiments.

12.3.1   Independent Samples t-test and Homogeneity of Variances: Levene's Test

Following the same procedure as in the previous chapter, you can use the 'Comparison means' menu in **SPSS** under the Independent Samples test (Student's test) as shown below. Let us examine whether there are any similarities between the English marks between the two classes. For a balanced test, 30 students from the Night Class are required to take a sample. Then the data processing should be done as described under t test above (https://statistics.laerd.com/premium/spss/istt/independent-t-test-in-spss-6.php#independent-variable).

[**SPSS:** Analyse > Compare Means > Independent Samples T-test > select variables into relevant panes > Define variables > Click OK].

Test:

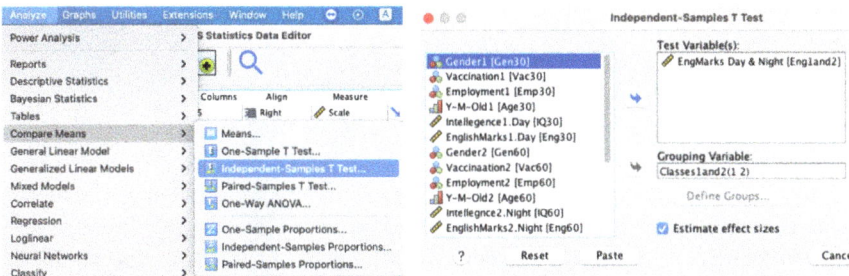

Results:

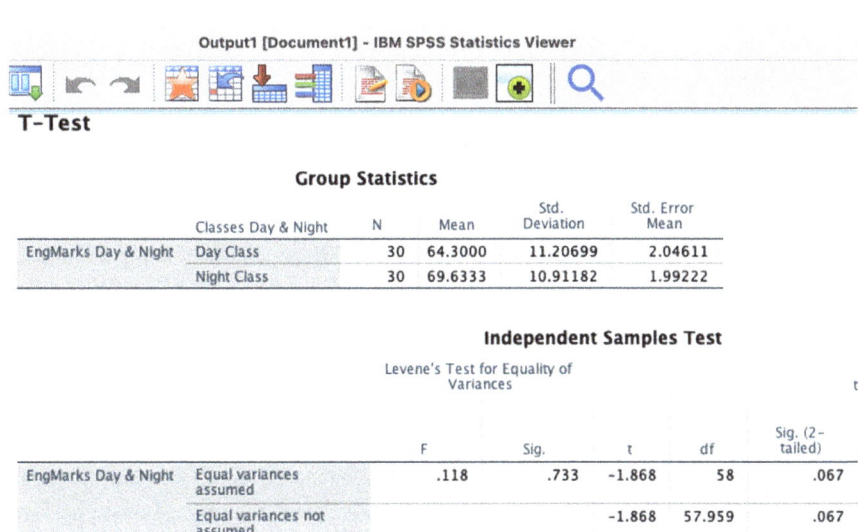

T–Test

Group Statistics

	Classes Day & Night	N	Mean	Std. Deviation	Std. Error Mean
EngMarks Day & Night	Day Class	30	64.3000	11.20699	2.04611
	Night Class	30	69.6333	10.91182	1.99222

Independent Samples Test

		Levene's Test for Equality of Variances		t			
		F	Sig.	t	df	Sig. (2–tailed)	
EngMarks Day & Night	Equal variances assumed	.118	.733	−1.868	58	.067	
	Equal variances not assumed			−1.868	57.959	.067	

Interpretations: Levene's test column should be checked for the similarity of the variables to see if the summation variables are the same (H0: $\sigma 12 - \sigma 22$). In this example, the value of significant is 0.733 (p = 0.733). If the computation variables of both groups are the same, this test will return a p value greater than 0.05 (p > 0.05, Retain the null, H0: $\sigma 12 = \sigma 22$). It shows that the assumption of the homogeneity of the variables is fulfilled however, if the test gives a value less than 0.05 (p < 0.05), the assumption of the homogeneity of the variables is violated, in this example, the assumption of homogeneity of the variables is fulfilled because the coefficient variables in the scores of both groups are p = 0.733 (i.e., p > 0.05).

The H0: hypothesis that there is no difference because the p value of the T test is greater than 0.05 is not refuted (Retain the null). That suggests that there is no difference in the distribution of English scores between the Day Class and the Night class.

There were 30 Day students and 30 Night students. A Welch t-test was run to determine if there were differences in English marks between Day class and Night class due to the assumption of homogeneity of variances being violated, as assessed by Levene's test for equality of variances ($p = 0.733$). There were a few outliers in the data, as assessed by inspection of a boxplot, and English marks for each class were normally distributed, as assessed by Shapiro–Wilk's test ($p > 0.05$). The English marks of the two classes showed that the difference is statistically not significant t $= 1.868$, $p = 0.067$ (> 0.05).

12.3.2   Statistical Techniques for Data that Do not Show the Nature of a Normal Distribution

If the data clearly do not show a normal spreading nature, consider using a nonparametric alternative, such as the Wilcoxon signed ranks tests or the Mann-Whitney U test. In the example above, the difference between the Day class and the Night class is that non-parametric data is used for these age groups.

Mann-Whitney (U test): (non-parametric equivalent to the independent *t*-test).

The Man-Whitney test rates all data and then compares the sum of the ratings. There are two types of Man-Whitney U tests to determine if the groups are the same for each group. If the score distribution for both teams is the same, the neutrals can be compared. If not, the mean series can be compared. The following is how to perform this test using SPSS.

[**SPSS**: Analyse > Nonparametric Tests > Independent Samples > Check Mann-Whitney U test Box > Median test] OR.

[**SPSS**: Analyse > Nonparametric tests > Legacy dialogs > Two Independent Samples > Mann-Whitney U test > Median test].

SPSS: Analyse → Nonparametric Tests → Independent Samples

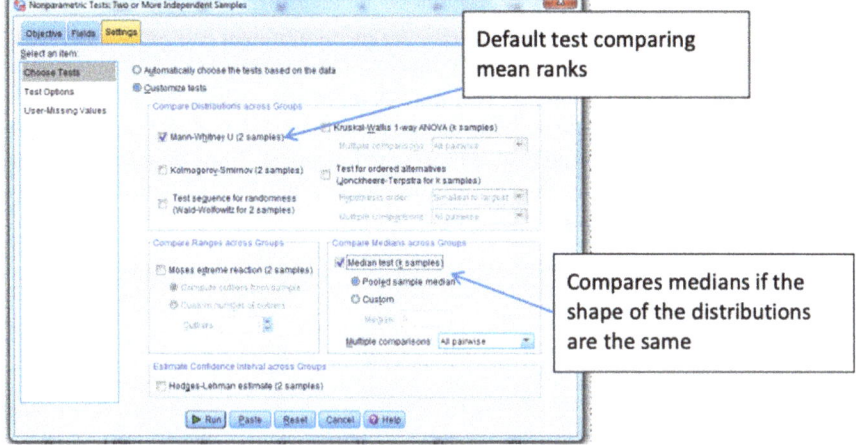

Results:

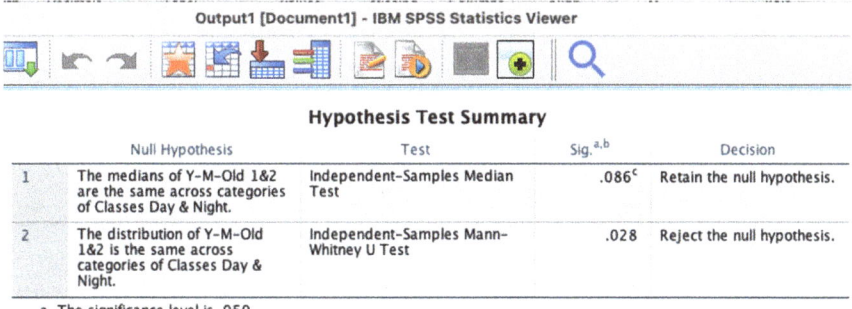

Hypothesis Test Summary

	Null Hypothesis	Test	Sig.[a,b]	Decision
1	The medians of Y–M–Old 1&2 are the same across categories of Classes Day & Night.	Independent-Samples Median Test	.086[c]	Retain the null hypothesis.
2	The distribution of Y–M–Old 1&2 is the same across categories of Classes Day & Night.	Independent-Samples Mann–Whitney U Test	.028	Reject the null hypothesis.

a. The significance level is .050.

b. Asymptotic significance is displayed.

c. Yates's Continuity Corrected Asymptotic Sig.

Interpretation: Although the mediators of the two age groups are similar (p > 0.05, Retain the null), its distribution also shows a statistically significant difference (<0.05) between the two classes (Reject the null).

A Mann-Whitney U test was run to determine if there were differences in Age groups between Day and Night classes. Medians of the Age groups for Day and Night classes were similar, as assessed by visual inspection. Median Age groups was not statistically significantly different, $p = 0.086$, > 0.05. Distributions of the Age groups for Day and Night classes were statistically significantly different, p = 0.028, < 0.05.

Wilcoxon signed-rank test: (non-parametric equivalent to the paired t-test).

The Wilcoxon-marked series test compares the measurements of two interrelated or compatible samples two or three times in one sample and assesses whether the median grades of their sum are different. It is a test that assesses the difference in the pair and is a non-parametric option. Absolute changes are graded, and then the marks of real changes are used to add negative and positive grades. The above example is not really applicable as this should be used for paired samples. To do this, let's compare the day class employment in the above example with the employment status of the same 30 students a year later.

[**SPSS**: Analyse > Nonparametric tests > Legacy dialogs > 2 related samples > select variables > Click OK].

Results:

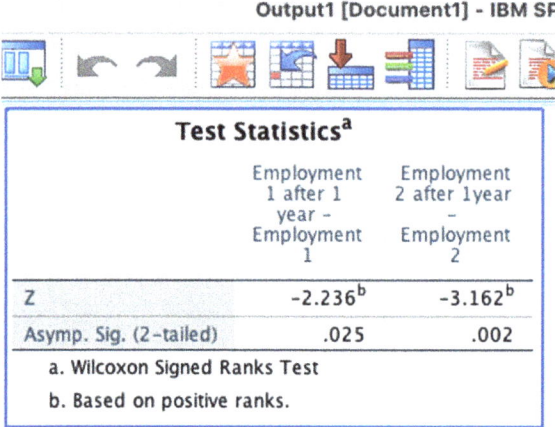

Interpretation: This indicates that the employment situation in both classes changed significantly after one year ($p < 0.05$). Exceeding the critical value of 1.96 relative to z also implies 'Reject the null'. The relatively low p and high z values of the night class also indicate that the night class is more different than the day class.

A Wilcoxon signed-rank test was conducted to determine if there were changes in the employment status of the 30 students of the Day class. Employment statuses were taken last year and this year.

Data are medians unless otherwise stated. Of the 30 participants recruited to the study, there was a statistically significant median increase in this year from to this year, $z = -2.236, p = 0.025, < 0.05$.

12.4 Analysis of Variance-ANOVA

The ANOVA table decides whether the model is significant. The model is compared to a 'null' model, where every observation is predicted to be the same.

(a) One-way ANOVA, when the dependent variable is continuous: see the example discussed under F-test in the previous chapter.
(b) Kruskal–Wallis test: The non-parametric equivalent to the one-way ANOVA test for non-parametric variables.
(c) Friedman test: the non-parametric equivalent to repeated measures ANOVA for non-parametric variables with repetitive measurements.
(d) Two-way ANOVA.

12.4.1 Kruskal–Wallis Test

The Kruskal–Wallis test (non-parametric equivalent to one-way ANOVA) is a non-parametric method for testing samples with the same distribution. The independent classification variable (two or more categorical variables or factors) is used to compare two or more independent samples of similar or different sample sizes in the rankings. It is an extension of the Man-Whitney U test used to compare only the two groups discussed above.

Friedman test (non-parametric equivalent to one-way ANOVA and repeated measures ANOVA) is used for one-way recurrence measurements of variance according to ratings. In its ratings, it is similar to the Kruskal–Wallis single-track analysis of variance in terms of ratings.

Test: Comparison of categorical variables such as age groups, vaccinated and employed in day class and night class. The hypothesis here is that there is a similarity.

[**SPSS**: Analyse > Nonparametric tests > Legacy dialogs > K Independent samples > select variables > Click OK].

Test: Kruskal-Wallis test   **Results:**

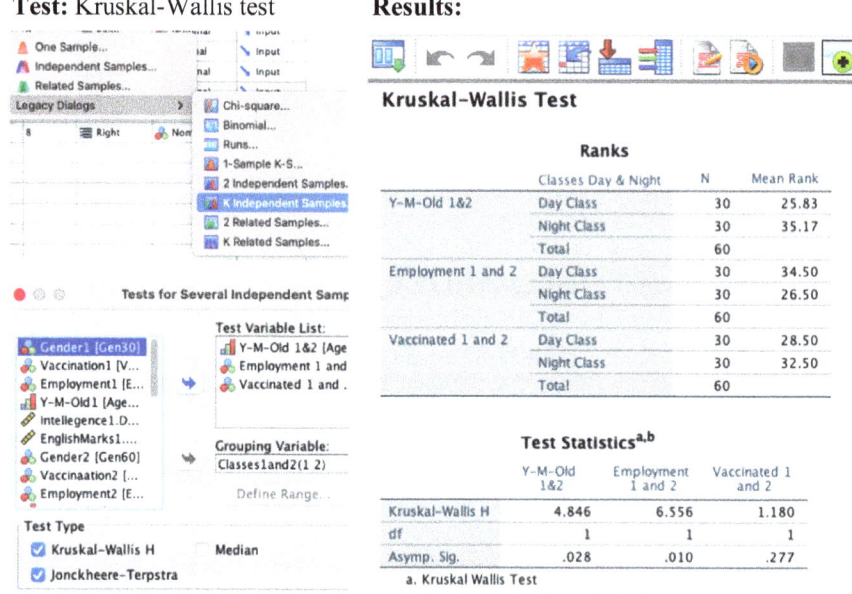

Kruskal–Wallis Test

Ranks

	Classes Day & Night	N	Mean Rank
Y–M–Old 1&2	Day Class	30	25.83
	Night Class	30	35.17
	Total	60	
Employment 1 and 2	Day Class	30	34.50
	Night Class	30	26.50
	Total	60	
Vaccinated 1 and 2	Day Class	30	28.50
	Night Class	30	32.50
	Total	60	

Test Statistics[a,b]

	Y–M–Old 1&2	Employment 1 and 2	Vaccinated 1 and 2
Kruskal–Wallis H	4.846	6.556	1.180
df	1	1	1
Asymp. Sig.	.028	.010	.277

a. Kruskal Wallis Test

b. Grouping Variable: Classes Day & Night

Interpretation: The above results show that although there is no significant similarity between age groups and employment relationships (Reject the null, < 0.05), there is a similar relationship with respect to the vaccine (Retain the null, > 0.05).

A Kruskal–Wallis test was conducted to determine if there were differences in distributions of Age groups, Employment and Vaccination status that differed in between Day (n = 30) and Night (n = 60) classes. Distributions of Age groups and Employment were not similar for the two classes. Age groups and Employment were statistically significantly different between the classes, $\chi^2(1) = 4.846$, $p = 0.028$ and $\chi^2(1) = 6.556$, $p = 0.010$. A Bonferroni correction for multiple comparisons was made with statistical significance accepted at the p < 0.05 level. This post hoc analysis revealed statistically significant differences in Age groups between the Day class (mean rank = 25.00) and Night class (mean rank = 35) ($p = 0.028$) and Employment Day class (mean rank = 34) and Night class (mean rank = 26) ($p = 0.010$), but not in the Vaccination status (mean ranks = 28 and 32 for the Day and Night classes respectively) ($\chi^2(1) = 1.180$, $p = 0.277$).

12.4.2 ANOVA (Two-Way ANOVA)

The two-way ANOVA compares the mean differences between groups divided on two independent variables (called factors). The primary purpose of the two-way ANOVA is to assess whether there is an interaction between two independent variables on the dependent variable. The two-way ANOVA test is a group comparison test called F-test for statistical significance. Suppose the variance within the groups is smaller than the variance between groups. In that case, the F-test will find a higher F-value, thus suggesting that the observed difference is realistic and not accidental.

Intelligence level (Intellegence) using gender (male/female: Gen60) and occupation (employed/unemployed) or age groups (youth, middle-aged and older: Age60) as independent variables among night class students ($n = 60$). To examine whether there is an interaction between gender and employment/unemployment on the dependent variable. (Two way ANOVA step by step guide: https://statistics.laerd.com/spss-tutorials/two-way-anova-using-spss-statistics.php).

[**SPSS**: Analyze > General Linear Model > Univariate > Univariate: Profile Plots "Employment*Gender" added to the Plots: box, Univariate: Options dialogue box > Check descriptives > OK].

Test:

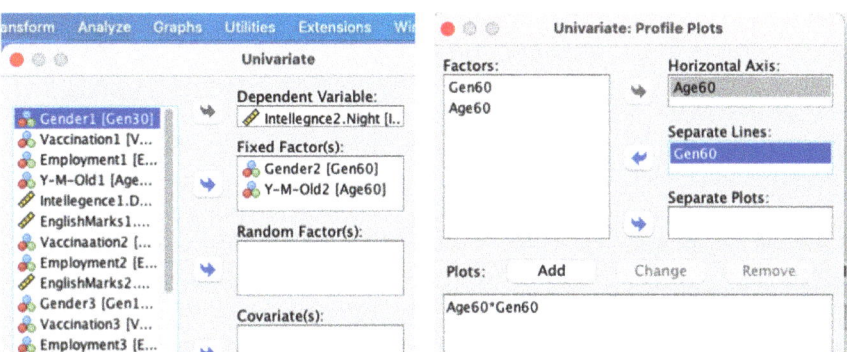

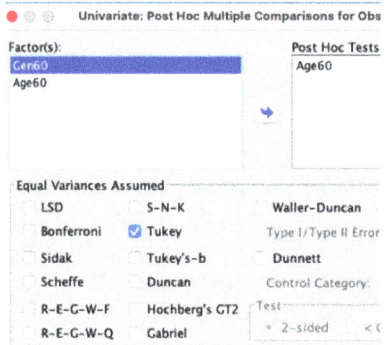

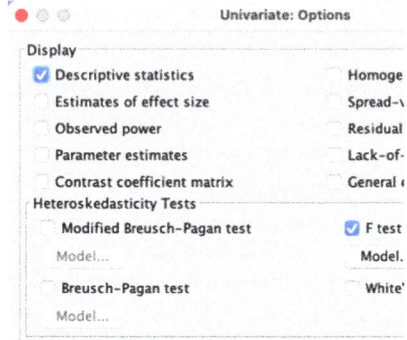

Results:

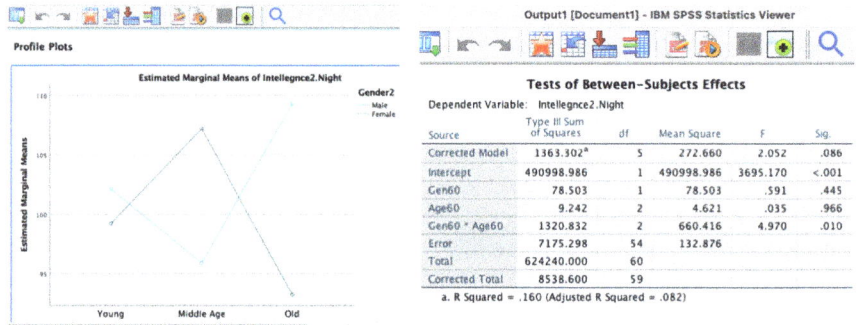

Interpretation-Graph: An interactive effect can be seen as a group of lines that are not normally parallel. The lines actually intersect. Therefore, there must be a statistically significant interaction, which is also confirmed by test statistics.

Interpretation-Statistics: It is important to look at the 'Gen60 * Age60' interaction first as this determines the interpretation of the results. The Sig. column shows statistically significant interaction with $p = 0.010$. The results of 'Gen60' and 'Age60' must be defined in the interactive context. The interaction between males and females (p = 0.445) and age groups (p < 0.966) showed that there was no statistically significant difference with the mean level mean.

A two-way ANOVA was conducted that examined the effect of gender and Age groups on Intelligent levels. There was a statistically significant interaction between the effects of gender and Age groups on Intelligence level, F (2, 54) = 4.970, p = 0.010.

You must examine each hypothetical sequence in interpreting the results. If the interaction is significant, the main effects cannot be defined by the ANOVA table. Here we have to do separate group ANOVA or use mean using graphs to explain the effects.

12.5   Correlation

Used to measure the strength of a relationship between two variables that show a correlation (r). The correlation coefficient ranges from -1 (perfect negative correlation) to $+1$ (perfect positive correlation). The absolute value of the Cohen (1992) correlation is defined in Table 12.3.

[**SPSS:** Analyse > Correlate > Bivariate Correlation > OK].

12.5.1   Simple Associations

Is there a correlation between intelligence values and English marks in the above example sample? What is the probability of having a correlation? What is the effectiveness of the correlation?

H_0: No correlation between intelligence values and English scores (hypothesis expected to be rejected)
H_1: There is a correlation between intelligence values and English scores (expected hypothesis)

Test: Correlation test: What is the probability that the alternative hypothesis is true? (At a significance level of less than 5%; alpha $= 0.05$). Let's test this separately for Day class and Online class.

[**SPSS:** Analysis > Correlate > Bivariate > Select variables into Variable pane > check Pearson Box] Let's choose as shown in the two screenshots below.

Table 12.3 The absolute value of correlation (r)

Correlation coefficient value	-0.3 to $+0.3$	-0.5 to -0.3 or 0.3 to 0.5	-0.9 to -0.5 or 0.5 to 0.9	-1.0 to -0.9 or 0.9 to 1.0
Association	Weak	Moderate	Strong	Very strong

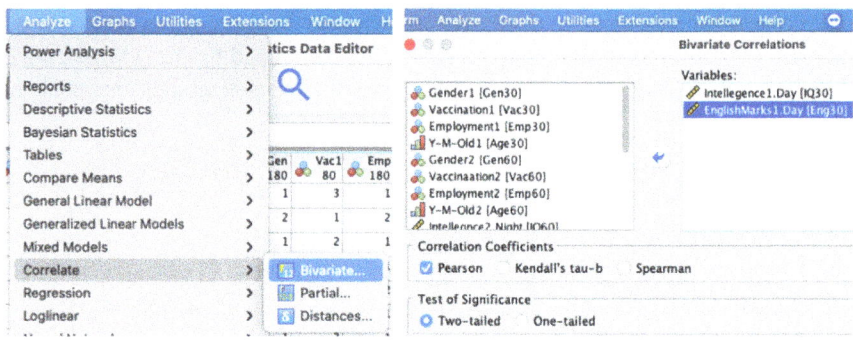

Results: the results of the Day class and the Online class are shown in the following screen shots

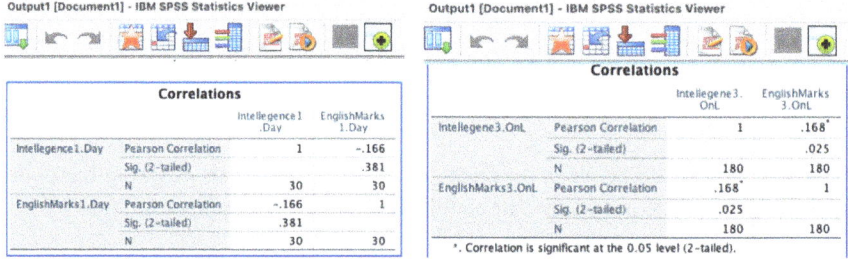

Interpretation: The result of the test is given as a correlation coefficient and Sig. (p). If the p value is less than 0.05, the H_0: no correlation between intelligence values and English scores can be rejected with 95% of confidence. The value of r indicates the intensity of the correlation as a quantitative value. This r value can range from -1 to $+1$. -1 is the perfect inverse relationship and $+1$ means the perfectly proportional relationship between intelligence and English marks.

The results show that the correlation between Day class intelligence and English scores is not significant ($p = 381 > 0.05$); however, the Online class shows a significant correlation of $r = 0.168$ with $p = 0.025 < 0.05$) between intelligence scores and English marks. A Pearson's product-moment correlation was run to assess the relationship between Intelligence values and English marks in Day class ($N = 30$) and Night class ($N = 180$) together.

The APA format;

Preliminary analyses showed the relationship to be linear with both variables normally distributed, as assessed by Shapiro–Wilk's test ($p > 0.05$), and there were no outliers.

There was a statistically significant, moderate positive correlation between Intellegence values and English marks in Night class, r(180) = 0.168, $p <$ 0.05, but the relationship between Intellegence values and English marks was not statistically significant in Day class, r(30) = −0.166, $p = > 0.05$).

A test for practice: What is the correlation between stock market activity and economic growth in your country? Try to do this research by following Example 12.1.

12.5.2 Correlation Between Non-parametric Variables

Spearman's rank correlation coefficient is a non-parametric statistical measure of the strength of a monotonic relationship between paired data. The notation used for the sample correlation is rs.

Dependent variable: Continuous/Ordinal.

Independent variable: Continuous/Ordinal.

The age group data in our example sample is a calibrated variable. Let's look at the effect of age groups on English scores.

[**SPSS:** Analysis > Correlate > Bivariate > Select variables into Variable pane > check Spearman] As shown in the two screenshots above, this time we will choose Spearman instead of Pearson.

Results:

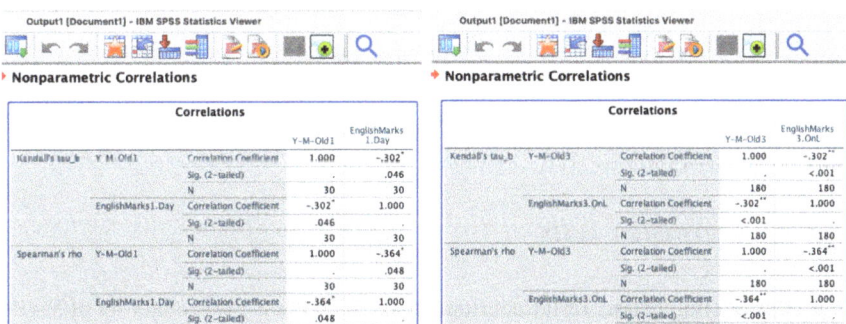

Interpretation: The Spearman results above show a significant correlation between the age group and English scores in the Day class and Online class (rs1 = −0.364, p1 = 0.048 < 0.05) and (rs3 = −0.364, p3 = 0.001 < 0.05) in Day Class and Online Class respectively.

A Spearman's rank-order correlation was run to assess the relationship between Age groups and English marks in the Day class (N = 30) and Online class (N = 180). There was a statistically significant, moderate negative correlation between Age groups and English marks in both Day and Online classes as (r_{s1} (30) = −0.364, p_1 = 0.048 < 0.05) and (r_{s3} (180) = −0.364, p_3 = 0.001 < 0.05) respectively.

12.5.3 Kendall's Tau Rank Correlation Coefficient

A *tau* test is a non-parametric hypothesis test for statistical dependence based on the tau coefficient. Specifically, it is a measure of rank correlation, i.e. the similarity of the orderings of the data when ranked by each of the quantities. Values of Tau-b range from −1 (100% negative association or perfect inversion) to +1 (100% positive association or perfect agreement). A value of zero indicates the absence of association.

Dependent variable: Continuous/Ordinal.

Independent variable: Continuous/Ordinal.

[**SPSS**: Analysis > Correlate > Bivariate > Select variables into Variable pane > check Kendall's Tau] This test can also be performed by selecting (checking) Kendall's Tau instead of Pearson/Spearman, as shown in the two Pearson and Spearman screenshots above.

Results: Kendall's Tau Rank Correlation Coefficient is also shown in the previous screenshot.

Interpretation: The Kendall's Tau Rank Correlation Coefficient results show a significant difference between the age groups and English scores in the Day class and Online class (rk1 = −302, p1 = 0.046 < 0.05) and (rk3 = −0.302, p3 = 0.001 < 0.05) in Day Class and Online Class respectively.

12.5.4 Regression Tests

How does intelligence scores affect English mark? What is the probability of having an impact? How magnitude is that impact?

H_0: has no effect on English marks (hypothesis expected to be rejected).

H_1: Intelligence has an effect on English scores (expected hypothesis).

Dependent variable: English marks (En_3).

Independent variable: Intelligence score (In_3).

Test: Regression model: $En_3 = B + (\beta \times In_3)$.

This test is not valid for Day class as the above tests have shown that Day class intelligence and English scores do not show a normal distribution.

Results:

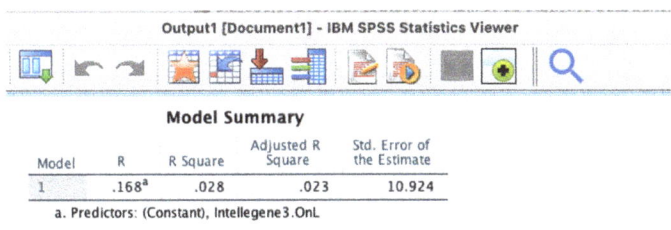

Output1 [Document1] - IBM SPSS Statistics Viewer

Model Summary

Model	R	R Square	Adjusted R Square	Std. Error of the Estimate
1	.168[a]	.028	.023	10.924

a. Predictors: (Constant), Intellegene3.OnL

ANOVA[a]

Model		Sum of Squares	df	Mean Square	F	Sig.
1	Regression	613.980	1	613.980	5.145	.025[b]
	Residual	21239.820	178	119.325		
	Total	21853.800	179			

a. Dependent Variable: EnglishMarks3.OnL

b. Predictors: (Constant), Intellegene3.OnL

Coefficients[a]

Model		Unstandardized Coefficients B	Std. Error	Standardized Coefficients Beta	t	Sig.
1	(Constant)	40.135	10.684		3.757	<.001
	Intellegene3.OnL	.227	.100	.168	2.268	.025

a. Dependent Variable: EnglishMarks3.OnL

Interpretation: Although these results are significant ($p < 0.05$), the r^2 value is only 2.3%, suggesting that the explanation covered by the sample is very small. Refer to Examples 12.1 and 12.2 to study the presentation of results in accordance with the APA rules. Examples 12.1 and 12.2 are for multiple variables, but the definition of one variable is as follows.

Example 12.2 and Two relevant sentences.

The results showed that Intelligence has a significant positive influence on English Marks ($\beta = 0.168$, $p < 0.05$), suggesting that, for every one-unit increase in Intelligence, there were 0.168 units increase in English Marks (prediction)…
this variable explained 2.3% (R-squared) of variance in English Marks.

12.5.5 Multivariate Regression

How do business turnover and assets affect profits? What is the probability of having an impact? How magnitude is that impact?
H_0: Business turnover has no effect on profit (hypothesis expected to decline).
H_0: Business assets have no effect on profit (hypothesis expected to be rejected).
H_1: Business turnover has an impact on profit (expected hypothesis).
H_1: Business assets have an impact on profit (expected hypothesis).
Test: Regression: Here the effect of two (or several) independent variables on the dependent variable is assessed simultaneously.
Model Equation: Profit $= a + \beta_1$ (Turnover) $+ \beta_2$ (Assets).

Results and interpretation: The result of the test is obtained as regression coefficients $\beta1$ and $\beta2$. The β values and the p values are given separately for each of the independent variables. The value β indicates the magnitude of the change in the dependent variable when the independent variable changes by one unit. If it is close to zero (0), then there is almost no correlation. Here again, if the p value is greater than 0.05, these results are considered not significant.

The value of R^2 indicates the ratio of the variance of the dependent variable covered by the model. In certain fields such as psychology, even a small R^2 of 6%–9% is considered sufficient. The higher the R^2 value, the higher the predictive power of the test. Table 12.4 (Reproduced below): Regression Results ($p = 0.109$) show R^2 as 0.089 (8.9%). Although that value is small, this test has been determined to be significant due to psychological factors such as attitudes.

Table 12.4 (reproduced): Regression Results: The Four Dimensions on IB

	Standardized coefficients	t	p	95.0% confidence Interval for IB	
	Beta			Lower bound	Upper bound
(Constant)		7.283	0.000	1.697	2.96
Attitude	0.135	1.580	0.116	−0.012	0.109
Power	−0.246**	−2.595	0.010	−0.062	−0.01
Social Norms	0.159*	2.004	0.047	0.001	0.138
Knowledge	0.041	0.489	0.625	−0.044	0.072

Dependent Variable: IB: Investment Behavior
Note * $p < 0.05$ and ** $p < 0.01$, Significant at the 5 and 1 percent levels respectively
$R^2 = 0.089$

Adjusted R2 (adjusted R2) is the ratio calculated to the degree of freedom of the model (Degree of freedom-d.f.).

The multiple regression model statistically significantly predicted IB, F (4, 158) = 2.393, $p < 0.05$, adj. $R^2 = 0.089$. The variables Power and Social Norms added statistically significantly to the prediction, $p < 0.05$. Regression coefficients and Confidence Intervals can be found in Table 12.4).

Box 12.2 Regression interpretation

For further analysis, based on CFA results, we computed composite measures for each factor by adding constituent items. Table 12.4 shows the results of the regression of the IB and the predictors: Attitude, Power, Social Norms and Knowledge. The …results showed that Perceived Power has a significant negative influence on IB ($\beta = -0.246$, p < 0.01) suggesting that, for every one unit increase in Perceived Power, there was a 0.245 units decrease in IB. Social norms also showed an influence on IB ($\beta = 0.159$, p < 0.05) suggesting that every one unit increase in the Social Norms percentage has resulted in a rise in IB by 0.159 units, after taking the influences of factors reflecting dimensions of Attitude and Knowledge into account. Attitudes and Knowledge showed no significant influence on IB directly. Overall, it appears that these observed associations are specific to the Fiji socioeconomic context. All the variables explained 8.9% (R-squared) of variance in IB.

12.5.6 Other Factors

Multicollinearity Diagnostic test: Multicollinearity test is a state of the high inter-relationship with independent variables, and from the results of this test, it helps to identify the regression model's predictive ability (Daoud, 2017). According to Pallant (2011), the Pearson correlation coefficient between any two independent variables is higher than 0.9, indicating a risk of multicollinearity.

Tolerance values and Variance Inflation Factor-VIF: In order to avoid the risk of low multicollinearity, it is necessary to show low tolerance values (T < 0.2) and high VIF values (VIF > 5) in the results.

12.6   Model Building and Path Diagrams

The formulation of hypothesis tests is currently being severely criticized, and researchers suggest that the process be abandoned altogether (Schmidt, 1996; Rogers, 2010; Newton and Rudestam, 2013). They are of the opinion that the relationship patterns of the variables can be presented more clearly by modeling with diagrams that represent reality.

The above examples can be presented as follows by modeling and path diagrams.

12.6.1   Single Variable Regression

H_1: Business turnover has an impact on profit.
Model Equation: Profit $= a + \beta$ (Turnover) + Error.

Path diagram:

12.6.2   Multivariate Regression

H_1: Business turnover and assets have an impact on profits.
Model building and path diagrams.
Model Equation: Profit $= a + \beta_1$ (Turnover) $+ \beta_2$ (Assets) + error.

Path diagram: Observable variables are represented by squares.

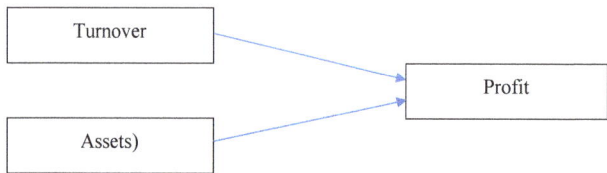

The above model shows the relationship between turnover, assets and profit. This single-headed arrow indicates a cause-and-effect relationship and the direction of influence.

A double-headed arrow indicates an association without a causal direction of influence. This two-headed arrow is used in the form of a curve to represent a correlation.

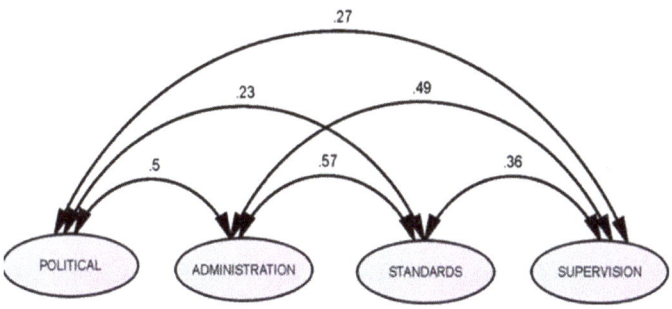

12.6.3 *An Interrelationship is Presented as Follows*

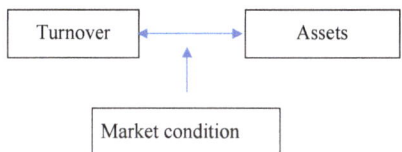

The above diagram shows that the correlation between turnover and assets is examined only under market conditions as moderating variable (assuming other factors are stable).

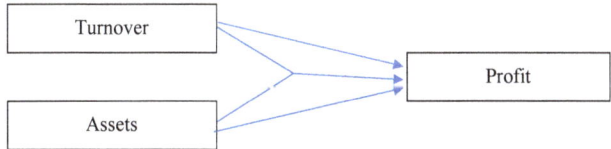

The diagram above represents a hypothesis that turnover and assets directly affect profits and that the combination of turnover and assets affects profits indirectly.

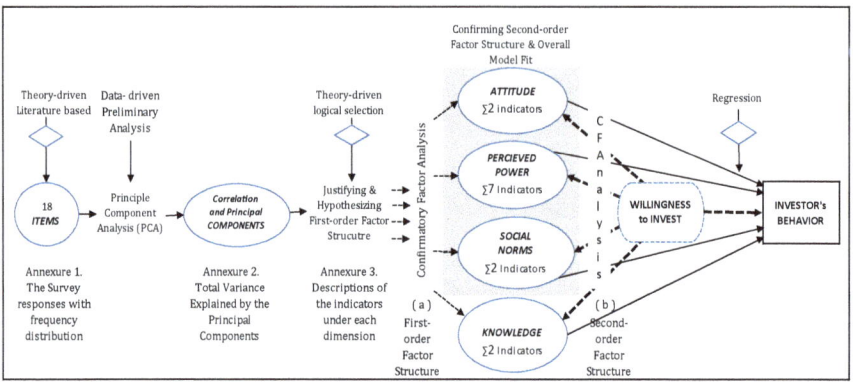

Fig. 12.6 The theoretical framework and hypothesized pathway analysis. *Source* Saliya (2021b)

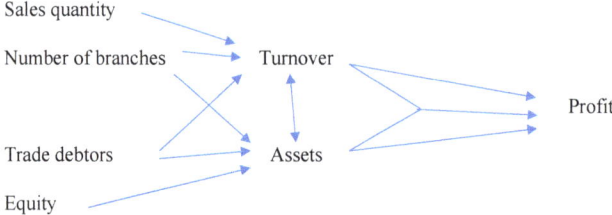

As the diagram above illustrates, many suggestions can be hypothesized to easily model a research strategy using path diagrams, avoiding scary, complex and unnecessary complications. It is standard to present variables that can be directly observed and measured directly in squares and latent variables in oval shapes.

Example 12.2, the path analysis, is illustrated by Fig. 12.6, it shows a research strategy as a path diagram that combines several of the tests discussed above.

A test for practice: How do the change in the stock exchange price index and the foreign exchange rate affect the economic growth of your country? (Two predictors and one outcome).

Example 12.3 Driving Forces of Individual Investors in Stock Market Participation, Review of Economics and Finance (Saliya, 2021b).

Figure 12.7 illustrates the influence of a latent variable *enterprising personality (EP)* that cannot be directly observed using observable variables/factors on investment inclination (II) to invest in company shares (Stock market participation-SMP). SEM (Structural Equation Modeling) is used to do this study. Here, the EP variable is a second-level latent variable. Three first-level latent variables (Intuition, Knowledge & education and Socio-cultural norms) are suggested to assess the influence of the first level latent variable (EP). In addition, the dotted lines suggest that these first-level latent factors directly influence investment intent. All of these suggestions are

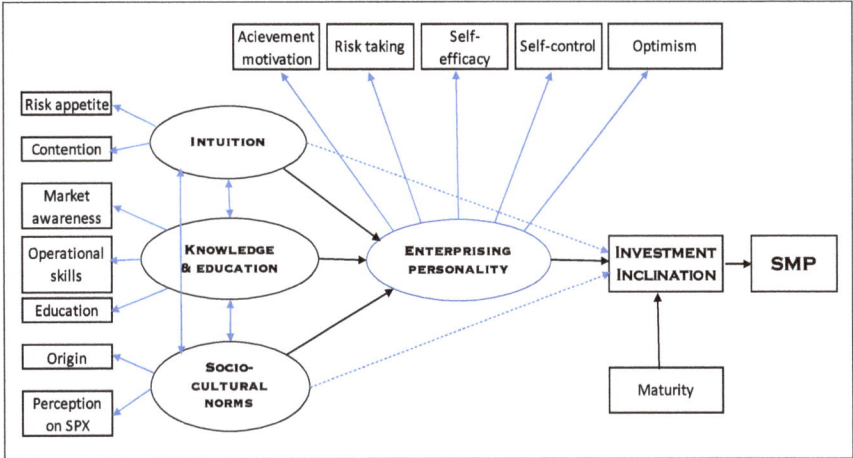

Fig. 12.7 The theoretical framework and hypothesized pathway analysis. *Source* Saliya (2022c)

hypotheses that are expected to be tested in this research process. This is discussed in Sect. 12.4 as Example 12.3.

12.7 Summary

- The statistical tests presented in this chapter should be mastered using tutorials and websites with expert guidance.
- There are two main types of analysis that can be done using statistics. They are descriptive analysis and statistical technical analysis.
- If a court case is hypothesis testing, the jury will consider the likelihood of accepting the charges (retaining the null) based on the evidence with 95% confidence if the evidence shows less than 5% of significance (Alfa: α, $p < 0.05$), they reject the null hypothesis, acquit the accused from charges.
- There are two main approaches to selecting an appropriate statistical method: according to the research problem/objectives as shown in Fig. 12.5 and according to the nature of the data/variables as shown in Table 12.2.
- There is an interactive website (https://maths.shu.ac.uk/mathshelp/WhichTest.php) to choose the basic statistical technique for the data.
- When data distribution is not normal, Wilcoxon signed ranks Nonparametric alternatives test or the Mann-Whitney U test can be used as an alternative.
- The correlation coefficient (r) is used to measure the strength of the relationship between two variables. The correlation coefficient ranges from -1 (perfect negative correlation) to $+1$ (perfect positive correlation).
- Path diagrams allow you to easily model a research strategy on a variety of hypotheses.

- It is standard to present directly measurable/observable variables in squares and latent variables in ovals.
- Path analysis is used to format various research problems avoiding complex mathematical formulas and freeing the researcher as well as the reader from unnecessary complications.
- These statistical analyses are very easy to find and study on the internet. What is more crucial is to integrate those analyses successfully with the existing knowledge and present them logically.

References/Further Reading

Advanced Regression: https://www.youtube.com/watch?v=cJpWQkoe4BA.
Analysis of Variance (ANOVA): https://www.youtube.com/watch?v=ITf4vHhyGpc.
Choosing a method: Interactive website: https://maths.shu.ac.uk/mathshelp/WhichTest.php.
Confidence Intervals: Statistics Help: https://www.youtube.com/watch?v=tFWsuO9f74o.
Daoud, J. I. (2017). Multicollinearity and regression analysis. *Journal of Physics: Conference*.
Data files and links to books: https://stats.idre.ucla.edu/other/examples/.
Granger causality: https://www.youtube.com/watch?v=ZUv7T8iPGrc.
Interpretation of ANOVA: https://www.youtube.com/watch?v=VvlqA-iO2HA.
Marshal, E. (2021). The Statistics Tutor's Quick Guide to Commonly Used Statistical Tests. University of Sheffield.
Newton, R. R. & Rudestam, K. E. (2013). *Your Statistical Consultant*; Answers to Your Data Analysis Questions. London. Sage.
One-Way ANOVA Test in SPSS: https://www.youtube.com/watch?v=OEOeXpxSjf8.
Pallant, J. (2011). *SPSS survival manual: A step by step guide to data analysis using the SPSS. Program* (4th ed.). Berkshire. Allen & Unwin.
P Valiue: https://www.youtube.com/watch?v=KLnGOL_AUgA.
Reporting Statistics APA Style: https://www.statisticshowto.com/probability-and-statistics/reporting-statistics-apa-style/Series, 949, 012009. https://doi.org/10.1088/1742-6596/949/1/012009.
Rogers, J. L. (2010). The epistemology of mathematical and statistical modeling: A quite methodological revolution. *American Psychologist, 65*(1), 1–12.
Saliya, C. A. (2020a). Dynamics of credit decision-making: a taxonomy and a typological matrix. *Review of Behavioral Finance*. https://doi.org/10.1108/RBF-07-2019-0092.
Saliya, C. A. (2021b). Driving Forces of Individual Investors in Stock Market Participation. *Review of Economics and Finance, 19*, 73–79. https://refpress.org/ref-vol19-a8/.
Saliya, C. A. and Pandey, S. K. (2021). Financial battle against climate change—assessing effectiveness using a scorecard. *Qualitative Research in Financial Markets*. https://doi.org/10.1108/QRFM-05-2020-0087.
Schmidt, F. L. (1996). Statistical significance testing and cumulative knowledge in psychology: Implications for training of researchers. *Psychological Methods, 1*(2), 115–129.
Step by step guide for SPSS: https://statistics.laerd.com/spss-tutorials.
Single Sample Hypothesis Z-test: https://www.youtube.com/watch?v=HoqzIR8xj4s.
Single Sample Hypothesis Z-test: https://www.youtube.com/watch?v=tsPv-ffN-0M.
t test: https://www.youtube.com/watch?v=pTmLQvMM-1M.
The use of SPSS software: https://www.youtube.com/watch?v=99fGYHGyO5U.
Time Series Analysis: https://www.youtube.com/watch?v=Aw77aMLj9uM.
Two sample t- and z-tests for unmatched data in SPSS: https://www.youtube.com/watch?v=RGMaN4EWIkc.
Z and t test examples given here using the link: https://www.youtube.com/watch?v=YsalXF5POtY.
Z critical value: https://www.youtube.com/watch?v=8Aw45HN5lnA.

Chapter 13
Structural Equation Modeling (SEM)

Structural Equation Modeling (SEM) is a complex statistical technique used to estimate the effect of observable factors on a variable that cannot be directly observed. Structural Equation Models (SEM) have two components; the measurement component and the structural component. The measurement component defines latent constructs that reflect study concepts with multiple indicators.

Latent growth modelling is a statistical technique used in the SEM framework to estimate growth trajectories. It is a longitudinal analysis technique to estimate growth over a period of time. It is widely used in the field of psychology, behavioural science, education and social sciences. It is also called latent growth curve analysis. The latent growth model was derived from the theories of SEM.

A test for practice: The correlation between democratic freedom in a country and the happiness of the people.

The following is the model built into Example 10.2 for SEM. Oval shapes should represent invisible (latent) variables (see Fig. 13.1).

Here the dependent variable is shown on the right and the four independent variables that are shown on the left and how it is affected on the left through the mediator variable investment willingness (IW). All five of these variables are hidden variables that cannot be directly observed. The loading factors (shown as 1–13, represented by little boxes) for latent independent variables should point toward independent latent variables as shown in Fig. 13.1. Those observable factors 1–13 (data collection survey questions) should be represented by small boxes as exemplified in F1, F10 and F13. The interpretation of this result is given in Box 13.1: Interpretation below.

> **Box 13.1 SEM Results Interpretation**
>
> Overall, the CFA model provided evidence for a hypothesized first-order factor structure. Accordingly, Attitude, Power, Social Norms and Knowledge are four distinct dimensions of investment willingness (IW). The highest correlation between Power and Social Norms suggests a close relationship or the least

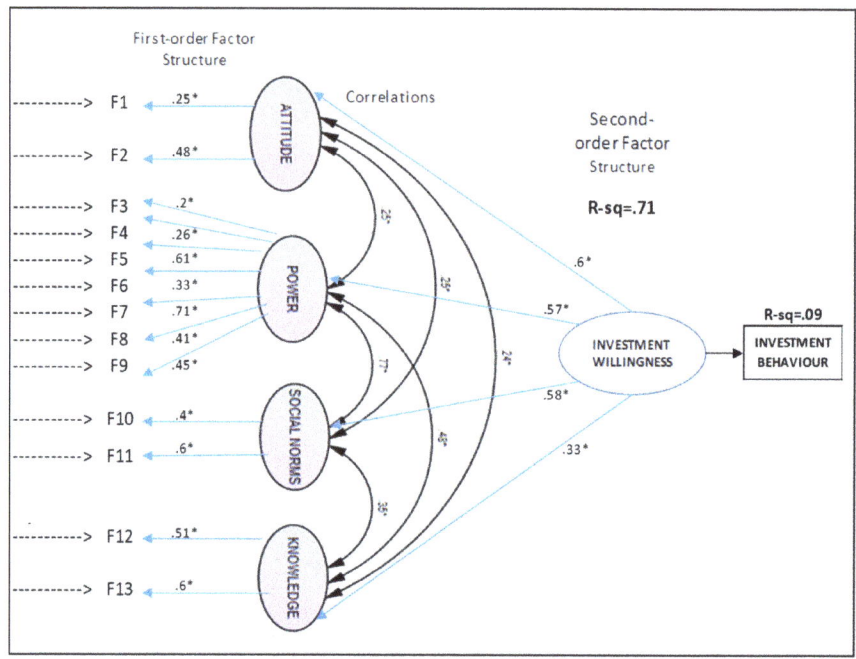

Fig. 13.1   Model Pathway analysis with results. *Source* Saliya (2021b)

distinctiveness between these two dimensions. The lowest correlations between Attitude and Power and Attitude and Knowledge suggest a weak relationship or the greatest distinctiveness between the two dimensions. However, significant correlations among the four factors suggest that there is a common variance across these factors, which may represent a higher-order factor. Thus, we tested the second CFA after incorporating the higher-order factor of investment willingness.

All 13 items showed significant substantial factor loadings to respective dimensional factors ranging from 0.2 to 0.71 ($p < 0.05$) showing the reliability and validity of these items in relation to the respective factor (Bollen, 1989). There were no significant cross-factor loadings. Four-dimensional factors showed significant and substantial loadings (0.6, 0.57, 0.58 and 0.33 to Attitude, Power, Social Norms and knowledge, respectively) to the second-order overall latent factor of the IW (significant error correlations were freed to be correlated, not shown in the figure).

Overall, the CFA model provided evidence for a hypothesized factor structure. Accordingly, Attitude, Power, Social Norms and Knowledge are four distinct dimensions of investment intention. Also, there exists an overall, higher-order factor of investment intention along with four first-order factors reflecting four dimensions.

Overall, it appears that these observed associations are specific to the Fiji socioeconomic context. All the variables explained 71% (R-squared) of variance in IW, and 9% variance in IB.

The following is an example built into Example 13.1 for modelling structural equations. Invisible (latent) variables should be represented by oval shapes Fig. 13.2: Latent Second-order Factor Structure reflecting different Dimensions of Vigilance (measurement error correlations are not shown).

Example 13.1 Determinants of Financial-risk Preparedness for Climate Change.

Figure 13.3 shows how the correlations between these four latent variables are shown. This illustration is an excerpt from Example 13.1.

Box 13.2 Correlations between dimensions

The highest correlation between administration and standard suggests a close relationship or the least distinctiveness between these two dimensions. The lowest correlation between political and standard suggests a weak relationship or the greatest distinctiveness between the two dimensions. However, significant correlations among four factors suggest that there is a common variance across these factors, which may represent a higher-order factor. Thus, we will test the second CFA after incorporating the higher-order factor of preparedness.

SPSS AMOS is software that can be used with SPSS Statistics. This software must be installed separately and can only be used with Windows OS (there is currently no

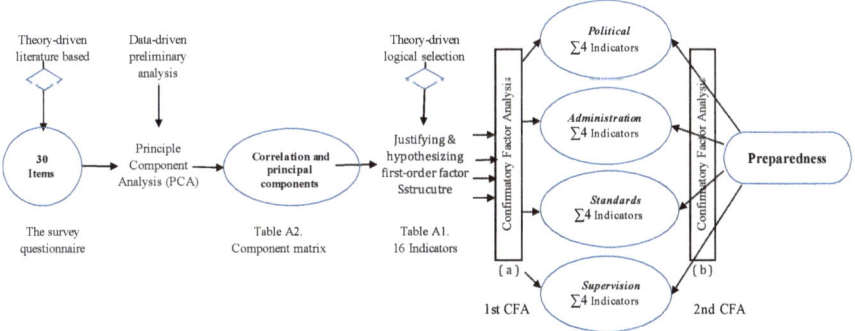

Fig. 13.2 A research strategic model that combines several experiments. *Source* Saliya and Wickrama (2021)

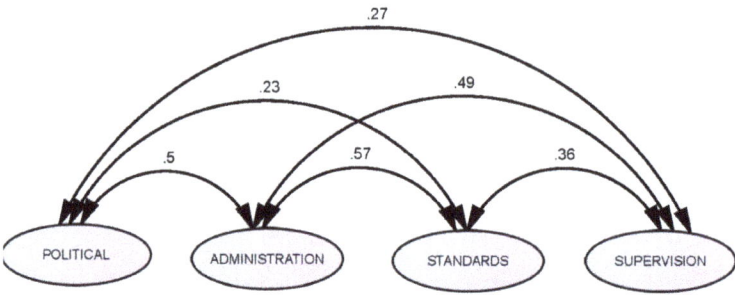

Fig. 13.3 Correlations between independent latent variables. First-order Factor Structure reflecting four Dimensions of Preparedness (measurement error correlations are not shown). *Source* Saliya and Wickrama (2021)

version for Apple). Path analysis can be used to construct SEMs attractively using the AMOS or Mplus software. The screenshot below illustrates an example of such a path analysis.

Building an SEM with AMOS: Youtube video.

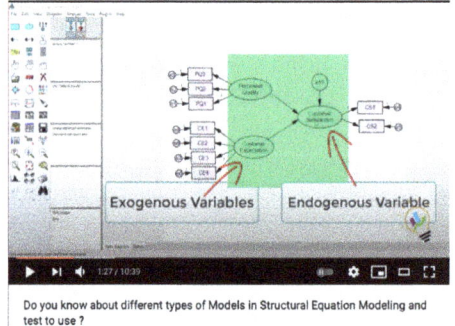

https://www.youtube.com/watch?v=qTVPT6jFKbs

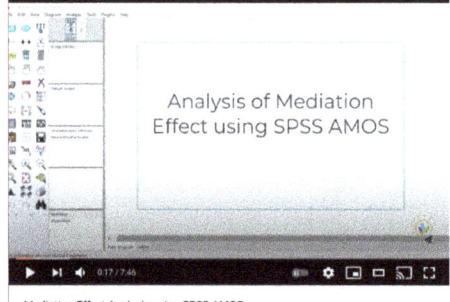

https://www.youtube.com/watch?v=XUqBlONbf7w

You can learn more about SEM from plenty of Youtube video including above links and https://www.youtube.com/watch?v=MQR0kLXDqhk.

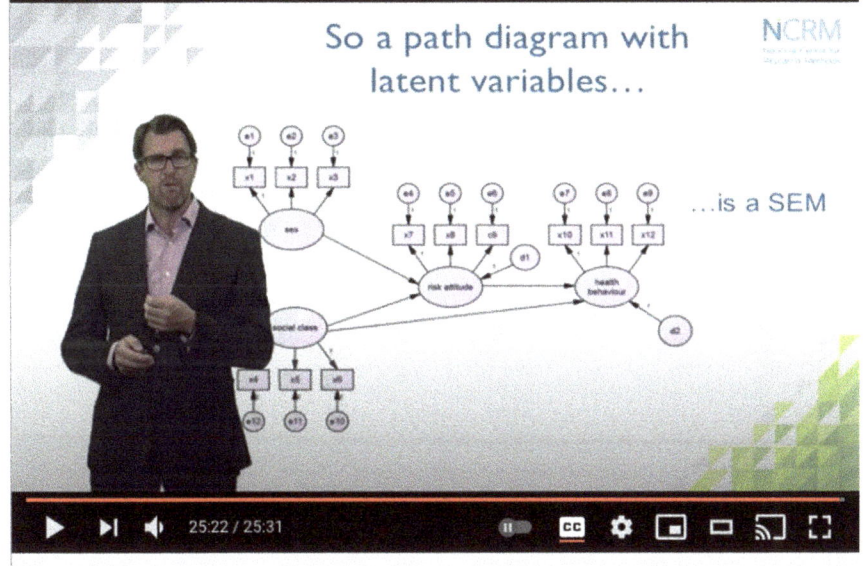

Structural Equation Modeling: what is it and what can we use it for? (part 1 of 6)

Source https://www.youtube.com/watch?v=eKkESdyMG9w

A test for practice: What is the extent to which the financial system of your country is prepared for the financial risk posed by climate change and what are the cryptocurrencies that contribute to it and their impact?

13.1 Multilevel Modeling-MLM

Statistical techniques of linear regression are used, and multi-level modeling focuses on specific static and random effects. Growth Curve Modeling (GCM) There are several statistical approaches to designing trajectories of change in a variable or variable over time. The simplest approach is to change the variance of a variable from one time point to the second time point. However, in longitudinal studies involving more than two estimates over time, more sophisticated techniques are needed not only to describe the trajectories of change but also to determine the interrelationships of critical predictive variables.

13.2 Latent Growth Curve Analysis-LGCA

Statistical techniques of linear regression are used, and multi-level modeling focuses on specific static and random effects. Latent Growth Curve Analysis (LGCA) is a powerful technology-based on Structural Equation Formatting (SEM).

Growth Curve Modeling (GCM) is an extensive set of statistical models for repetitive measurement data, and over the past decade or so, the term has primarily come to define a set of analytical approaches. The growth curve model is a statistical method that allows estimating patterns of change over time, commonly known as time trajectories, time tracks, growth curves, or cryptic trajectories. The results of this test are static (showing no change with flat-time); Gradually increasing or decreasing over time, Linear or curved form (Curran et al., 2010).

A test for practice: Analysing and forecasting the long term behaviour of the Colombo Stock Exchange and the stock market in Mauritius Island.

Modeling: General Latent Variable Modeling Framework.

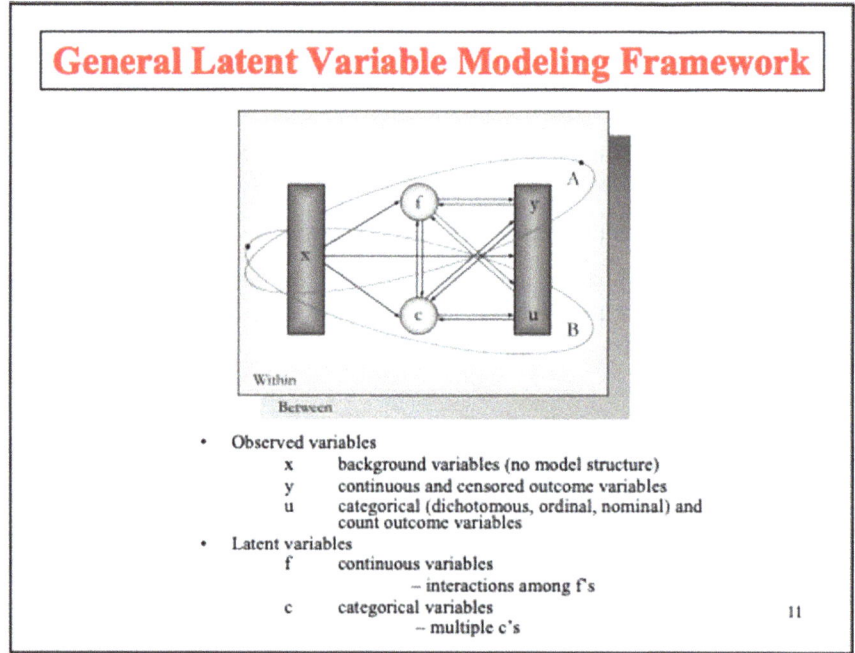

13.3 Statistical Models and Path Diagrams

You can learn more about this in the Youtube video below and other related Youtube videos.

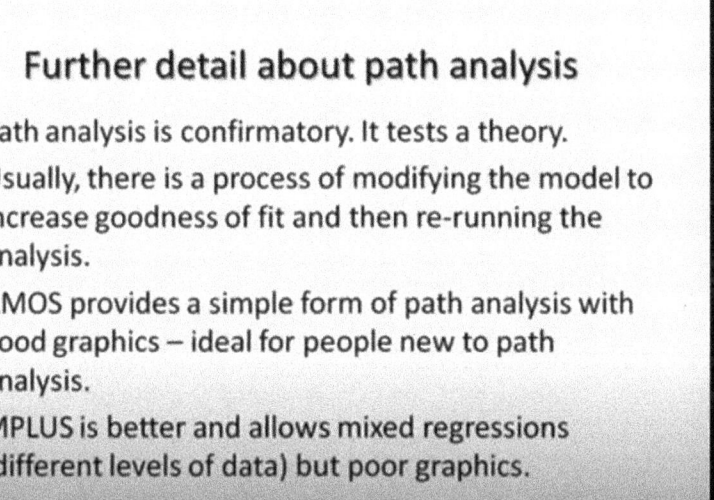

What is path analysis (for people who hate statistics) by Dr Pauline McGovern

https://www.youtube.com/watch?v=FICHixupOX4

With the help of the internet today, these statistical analyses can be easily found and studied by Google search. What is more crucial is to interpret those analyses properly and present them with cogent arguments. For this, it is necessary to successfully integrate with the existing knowledge related to the subject. Part V of this book discusses this in detail using illustrations.

13.4 Summary

- Structural Equation Formatting (SEM) is a complex statistical technique used to estimate the effect of observable factors on a variable that cannot be directly observed.

- Latent Growth Curve Analysis (LGCA) is a powerful technology-based on Structural Equation Formatting (SEM).

References/Further Reading

Curran, P. J., Obeidat, K., & Losardo, D. (2010). Twelve Frequently Asked Questions About Growth Curve Modeling. *Journal of cognition and development: official journal of the Cognitive Development Society, 11*(2), 121–136. https://doi.org/10.1080/15248371003699969.

Hooper, D., Coughlan, J., & Mullen, M. R. (2008). Structural equation modelling: Guidelines for determining model fit. *The Electronic Journal of Business Research Methods, 6*, 53–60.

Hsu, H. Y., Lin, J. J. H., Skidmore, S. T., et al. (2019). Evaluating fit indices in a multilevel latent growth curve model: A Monte Carlo study. *Behavioural Research, 51*, 172–194. https://doi.org/10.3758/s13428-018-1169-6

Kline, R. B. (2005). *Principles and practice of structural equation modeling.* Sage.

Saliya, C. A. (2021b). Driving Forces of Individual Investors in Stock Market Participation. *Review of Economics and Finance, 19*, 73–79. https://refpress.org/ref-vol19-a8/.

Saliya, C. A. & Wickrama, K. A. S. (2021). Determinants of Financial-risk Preparedness for Climate Change. *Advances in Climate Change Research.* https://doi.org/10.1016/j.accre.2021.03.012.

Wickrama, K. A. S., Lee, T. K., O'Neal, C. W., & Lorenz, F. O. (2016). *Higher-order growth curves and mixture modeling with Mplus: A practical guide.* Routledge.

Youtube Links

https://www.youtube.com/watch?v=qTVPT6jFKbs.
https://www.youtube.com/watch?v=MQR0kLXDqhk.
https://www.youtube.com/watch?v=XUqBlONbf7w.
https://www.youtube.com/watch?v=eKkESdyMG9w.
https://www.youtube.com/watch?v=qTVPT6jFKbs.
https://www.youtube.com/watch?v=FICHixupOX4.

Chapter 14
Econometrics

Econometrics was pioneered by Lawrence Klein, Ragnar Frisch, and Simon Kuznets. All three won the Nobel Prize in economics in 1971 for their contributions. Today, it is used regularly among academics as well as practitioners such as Wall Street traders and analysts.

Econometrics is the use of statistical methods using quantitative data to test existing assumptions or theories in economics or finance. Econometrics rely on techniques such as regression analysis and hypothesis testing. Economics can also be used to predict future economic or financial trends. Techniques such as SEM and Path analysis discussed in the previous chapter can also be used successfully here.

An example of an econometric test is the study of the impact of income using observable data. An economist can assume that when a person increases his income, his expenses will also increase. If the data show that such a relationship exists, a policy analysis can be conducted to understand the strength of the relationship between income and consumption.

Δ Consumption $= a + \beta$ (Δ Income) $+$ Error.
Δ (Delta) $=$ increase/change; Delta represents change.

Like the feedback function discussed above, this feedback function is the same equation $[Y = mX + c]$ that we learned in school. We have studied 'c' as intercept and 'm' as slope. We derived this equation from a free hand marked line with the help of a scatter graph or by calculating c and m using the least square method. The sequence here is the value used in the reflection as the regression coefficient (Fig. 14.1: Regression line).

Econometrics adapts and uses statistical methods into problems of economic life. These adaptive statistical methods are commonly referred to as econometric methods. Such systems are adjusted to measure inferential (stochastic, randomly sequential) relationships that operate on real-world data.

Econometrics is often 'theory driven' while statistics tend to be 'data driven'. Statistical research is generally driven by data analytic issues, not by some preconceived theory.

© The Author(s), under exclusive license to Springer Nature Singapore Pte Ltd. 2022   241
C. A. Saliya, *Doing Social Research and Publishing Results*,
https://doi.org/10.1007/978-981-19-3780-4_14

Fig. 14.1 Regression line

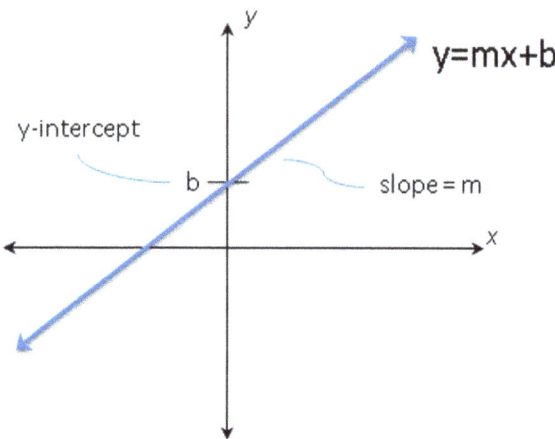

Typically, econometricians test theory using data but often do little if any exploratory data analysis. On the other hand, statisticians tend to build models after looking at data sets (Hyndman, 2014).

Eviews is one of the most widely used computer software for Econometrics. The following links provide useful tutorials. https://www.eviews.com/Discovering/dem onstrations.html; https://www.eviews.com/Learning/index.html.

Statisticians use samples to make statistical inferences about large aggregates. Economists examine contradictory case relationships to make causal assumptions. There are slight differences between the two groups in terms of terminology. The 'longitudinal data' of statistics is the 'panel data' of econometrics. Statistically, 'robust' means insensitivity to outliers, while in econometrics, it means insensitivity to heteroskedasticity and autocorrelation.

'Longitudinal data' in statistics = 'Panel data' in econometrics. And 'Survival analysis' in statistics = 'Duration modeling' in econometrics. *Robust* means insensitive to outliers in statistics but in econometrics, robust means insensitive to heteroskedasticity and autocorrelation. Almost all of the statistical techniques discussed in the chapter above are occasionally used in econometric tests. But it is important to measure and predict the relationship between long-term trends in the study of economic activity.

The intervention of Gringer and Newbold (1974), who pioneered such studies, led to a discourse that the lack of lag dynamics in retrospective tests was a spurious one. This discourse included the notion that traditional econometrics has no solution to the 'spurious regression' of non-stationary (non-stable, systematic change) time series. Although many researchers believe that the only barrier to pseudo-regression is the non-static nature of the pseudo-variables, it has been argued that invisible variables may also be another major cause of pseudo-regression.

How can this non-static nature be a hindrance? If the relevant time series does not show a systematic trend, the forecast may not be practical and reliable.

What are the invisible variables that apply? The invisible variable here is the difference between two sequential data in a variable. Although the data (mean and variance) do not change over time, the differences between the data may change. One of the objectives of the technique is to identify that change (lag) as a variable and find a solution to this problem.

There have been a number of new approaches to address this issue for a past few decades, with the ARDL (Autoregressive Distributed Lag) approach leading the way. Therefore, the ARDL strategy is described in this chapter in detail.

14.1 Autoregressive Distributed Lag (ARDL) Approach

A new approach to the problem of testing the existence of a level relationship between a dependent variable and a set of regressors, when it is not known with certainty whether the underlying regressors are trend- or first-difference stationary (Perasan, Shin and Smith, 2001).

When regressors; independent variables that influence the **behaviour** of a dependent variable are not known with certainty whether there is a tendency or stability at the first interval, it is necessary to examine what level of correlation exists between that dependent variable and the group of background regression.

14.1.1 Autoregressive Distributed Lag (ARDL) Cointegration Technique or Bound Testing

The ARDL bound test approach to cointegration is carried out for time series data in a three-stage procedure to test the direction of causality. In the first stage, the order of integration was tested using tests such as the Augmented Dickey-Fuller (ADF) and Phillips Perron (PP) unit root tests. The second stage involved testing for the existence of a long-run equilibrium relationship between variables. The third stage involved constructing standard causality tests augmented with a lagged error-correction term where the series were cointegrated.

The process of the ARDL bound testing integration technique is illustrated in Fig. 14.2.

To study this approach, it is necessary to understand some of the basic theoretical concepts such as the stationarity process and cointegration. One by one, those concepts are discussed below (https://www.youtube.com/watch?v=qovhIG_WzoE).

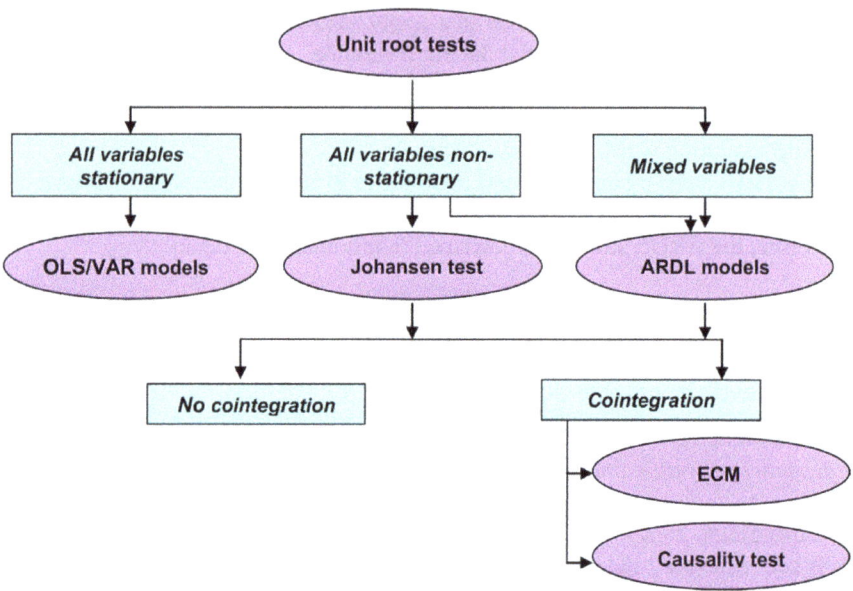

Fig. 14.2 Process of the integration technique. *Source* Shrestha and Bhatta (2018)

14.1.2 Stationary or Stable Process

A stationary process is a stochastic process whose unconditional joint probability distribution does not change when shifted in time. Consequently, parameters such as mean and variance also do not change over time.

A static process is a time series with a stochastic process whose parameters such as mean and variance do not change over time. Its mean and variance may vary for short periods, but for long periods it shows a static, unchanging nature. An example is the time series showing the stock market price-earnings ratio (P/E Ratio) shown in the first chart in Fig. 14.3. That is, the stock market PE ratio in Fig. 14.3 shows that there is a unit root. The second graph in Fig. 14.3 provides an example of stabilizing a non-stationary time series using data spacing. The black line on that chart is a time series such as GDP (GDP, Yt), and the red line indicates GDP growth, the change from one year to the next [Yt − Y (t − 1)].

Using non-stationary time series data in financial models produces unreliable and spurious results and leads to poor understanding and forecasting. The solution to this problem is to transform the time series data so that it becomes stationary. If the non-stationary process is a random walk with or without a drift, it is transformed to a stationary process by differencing.

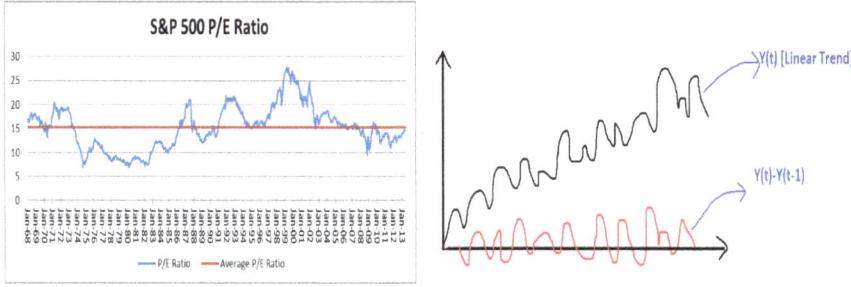

Fig. 14.3 Stock market price-earnings ratio

14.1.3 Unit Root

All you really need to know if you're analyzing time series is that the existence of unit roots can cause your analysis to have serious issues like:

Spurious regressions: you could get high r-squared values even if the data is uncorrelated.

Errant behavior due to assumptions for analysis not being valid. For example, t-ratios will not follow a t-distribution.

A unit root is a nonstationary-nonstable tendency in a time series. If there is a unit root in a time series, it shows an unexpected systematic pattern. A unit root is an inherent (stochastic, random order) tendency in a time series. Figure 14.4 shows an example of a unit root that can be averaged. The red line represents an observed drop in output. Green shows the path of recovery if the series has a unit root. If there is no unit root and the series is trend-stationary, it is indicated by a blue prototype. The blue line returns more consistently to the dashed line than the green line. The fixed autocorrelation of a unit root series is a series that recovers as slowly as the green series. The feedback assessments made with such grades are far from reality.

Unit root $\rightarrow \rightarrow \rightarrow$ Non-stationarity (Green series).

Fig. 14.4 Behavioral patterns in two time-series

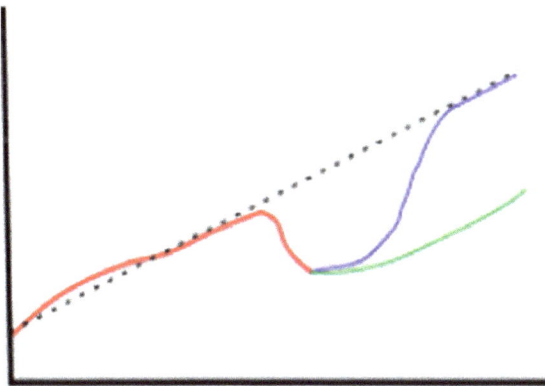

Since the unit root time series is not constant, it is tested whether stability can be observed using data intervals. For unit root time series, it is called difference-stationary and when there is no unit root, it is called trend-stationary.

Therefore, the first thing to do for the ARDL test is to make sure that the variable time series group has no unit root (at 5% significant level, $p < 0.05$).

The Dickey-Fuller test was the first statistical test developed to test the null hypothesis that a unit root is present in an autoregressive model of a given time series, and that the process is thus not stationary.

The DF test is the most popular test for the unit root test. H_0: β and H_1: $\beta < 0$.

KPSS test: Another prominent test for the presence of a unit root is the KPSS test. Kwiatkowski et al. (1992), conversely to the Dickey-Fuller family of tests, the null hypothesis assumes stationarity around a mean or a linear trend, while the alternative is the presence of a unit root.

Non-stationarity is primarily caused by the integration process. Bound tests are used to assess the variability of variables in the long run. The sequence of order (order of integration; I(0): integration at level, I(1): integration at the first difference, I(2): integration at second difference etc.) Augmented Dickey-Fuller (ADF) and PHP-Phillips Perron units are tested by root tests. Table 14.1: Stationarity tests show an example of this static test. The definition here is shown in Box 14.1.

Table 14.1 Stationarity tests

Variable	Dickey-fuller test for stationarity of all variables			
	In levels		At first differences	
	Without trend	With trend	Without trend	With trend
SMD	-3.687^{***}	-3.946^{**}	-8.979^{***}	-8.817^{***}
LQDT	-3.134^{***}	-5.635^{***}	-5.448^{***}	-5.341^{***}
INF	-2.496^{**}	-5.057^{***}	-4.567^{***}	-4.294^{**}
FDI	-0.529	-2.584	-7.483^{***}	-7.242^{***}
BNKD	1.526	-1.423	-3.212^{***}	-3.439^{*}
GDP	6.733	-0.101	-1.692^{*}	-3.900^{**}
INV	-0.895	-4.868^{***}	-7.404^{***}	-7.030^{***}
SAV	-1.351	-2.552	-5.144^{***}	-4.941^{***}
COC	-1.880	-1.551	-6.228^{***}	-6.535^{***}
GVTEF	-1.005	-2.46	-4.896^{***}	-4.861^{***}
PS_AV	-0.632	-0.278	-4.292^{***}	-4.813^{***}
RGLQ	-0.246	-1.173	-5.810^{***}	-6.241^{***}
ROL	-0.921	-2.861	-3.664^{***}	-3.811^{**}
DMCY	-1.513	-1.682	-3.179^{***}	-3.196

Notes
a: [*,**,***]Significant at the 10, 5 and 1% levels respectively
b: Lag Length based on SIC
c: Probability based on MacKinnon (1996) one-sided p-values

Example 14.1 Stock Market Development and Nexus of Market Liquidity (Saliya, 2020)

Box 14.1 Stationary test interpretation

To examine the stationarity properties of the variables, we use the Dickey-Fuller generalized least squares test. Table 14.1 reports the results of unit root tests of the variables in levels and at the first differences. It shows that variables such as SMD, LQDT, INF and INV are stationary in level while all variables are stationary at the first differences, except DMCY. The ARDL model is appropriate to estimate the long-run relationship between all the variables that are $I(0)$ and/or $I(1)$ but not $I(2)$, therefore DMCY does not qualify for the estimation.

At 5% level when the ADF (or DF) test is less than −2.86 if the regression has no trend, and when the test is less than −3.41 if the regression has a trend.

14.1.4 Cointegration Process

In statistics, the order of integration denoted I(d), of a time series, is a summary statistic, which reports the minimum number of differences required to obtain a covariance-stationary series.

There are often trends in time series. They are either deterministic or stochastic. There is statistical evidence that many macroeconomic time series (such as GDP, wages, employment, etc.) have inherent trends. If two or more series are combined individually (in the sense of time series) and, however, some of their linear combinations are combined in a different order, that series is called cointegrated.

Statistically, the time-series sequence is a summary representation of I(d), which records the minimum number of variations required to obtain a cointegrated static series. Cointegration is a statistical expression of a collection of time series variables (X1, X2, …, Xn). First, it is necessary to see if all the series are combined in order at some level [I(d): 'I'stands for 'integration', and 'd' stands for any level, the value of the difference, the first difference and the second difference etc. abbreviated as I(1), I(2)]. Then, suppose a linear combination is combined in a sequence lower than that level [I(0), I(1), I(2), 'ie at levels' (equal level), one level less, two levels less, etc. respectively]. In that case, the time series is considered cointegrated at those levels.

Engle and Granger (1987) introduced the cointegration technique as a solution of spurious regression due to non-stationary time series. According to Granger, the non-stationary time series are cointegrated if their linear integration is a stationary process. As opposed to other multivariate cointegration techniques such as Johansen and Juselius (1990) allow the cointegration relationship to be estimated by OLS once the lag order of the model is identified. Secondly, the bounds testing procedure does not require the pre-testing of the variables included in the model for unit roots, unlike other techniques such as the Johansen approach. It is applicable irrespective of whether the regressors in the model are purely I(0), purely I(1) or mutually cointegrated. Thirdly, as is the case in this study, the test is relatively more efficient in small or finite sample data sizes. The procedure will, however, crash in the presence of the I(2) series.

Therefore, an important condition here is that the integration is one of the same level [I(0)] or the first difference [I(1)] and not the second difference [I(2)]. Watch the Youtube video below.

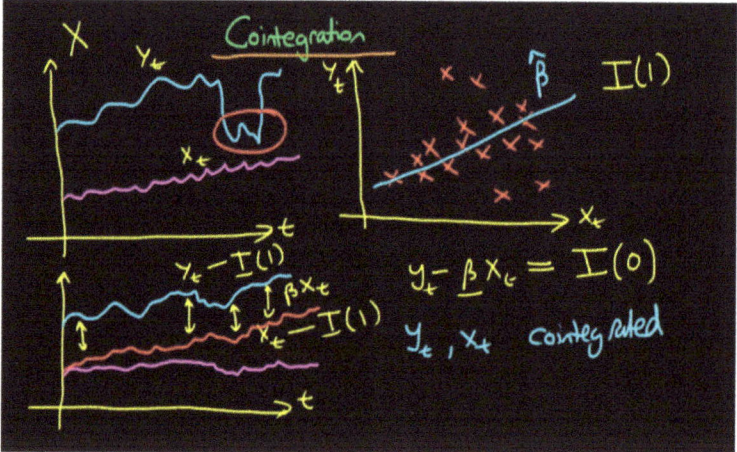

Source Cointegration: https://www.youtube.com/watch?v=vvTKjm94Ars

Johansson's test is a method of determining whether three or more time series are integrated. More specifically, it estimates the validity of a coefficient relationship using an approach to MLE maximum likelihood estimates. It is used to find the number of contacts and to evaluate those contacts.

14.1.5   Bounds Testing Approach to Cointegration

To test the combination of variables and their long-term relationships, two ARDL binding tests are often used: the AIC (Akaike information criterion) and the Schwarz information criterion (SIC: Schwarz's Information Criterion). λ–Trace (Lambda) and Max-Max-Eigenvalue are often used to determine the number of long-term associative connections. Table 14.2 shows the results of the ARDL binding test for two variables (two functions: f_1 and f_2) with the corresponding critical values for each level of maturity level. The definition is shown in Box 14.2.

Table 14.2 ARDL bounds test F-statistics for cointegration and the critical values

Null hypothesis: no long-run relationships exist

Dependent variable	Function	F-statistics	
		Akaike IC	Schwarz IC
SMD	f_1 (LQDT, INFL, FDI, BNKD, GDP INV, SAV)	5.175	5.175
SMD	f_2 (COC, GVTE, PS/AV, RGLQ, ROL, DMCY)	5.029	5.029

Critical values

	Akaike IC		Schwarz IC	
	f_1		f_2	
Level of significance (%)	Lower bound	Upper bound	Lower bound	Upper bound
10	2.03	3.13	2.12	3.23
5	2.32	3.5	2.45	3.61
2.5	2.6	3.84	2.75	3.99
1	2.96	4.26	3.15	4.43

Box 14.2 Cointegration interpretation

To examine variables' cointegration and long-run relationships, we use two ARDL bounds tests: Akaike Info Criterion (AIC) and Schwarz Info Criterion (SIC). We also used λ-Trace and λ-Max-Eigenvalue to determine the number of long-run cointegration relationships. Table 14.2 shows the results of ARDL bounds testing for two sets of variables (two functions: f1 and f2) with the respective critical values for each level of confidence.

The calculated F-statistics are 5.175 for macroeconomic variables (f1) and 5.029 for institutional variables (f2). These are higher than the critical value reported in Table 14.2 for both the AIC and SIC. Therefore, the results show that the variables in the model are co-integrated, and long-run relationships exist. Table 14.2 shows the results of the Johansen Unrestricted Cointegration Rank Test confirming rejection of the Null Hypothesis: No long-run relationships exist for all the variables.

14.1.6   *Autoregressive Model*

A statistical model is autoregressive if it predicts future values based on past values. For example, an autoregressive model might seek to predict a stock's future prices based on its past performance. A statistical model that predicts future values based on past values is a self replicating model. For example, an automated retrospective model may attempt to predict future stock prices based on its past performance. Automated progressive models implicitly assume that the future will be the same as the past. Therefore, these models can be proven wrong under certain circumstances, such as the financial crisis (financial crisis, for example, GFC-global financial crisis in 2008) or rapid technological change (online real-time trading).

14.1.7 Lagged Regression

In statistics and econometrics, a distributed delay model is a model for time series data, using a predictive equation to predict the values of a delay (previous period) of an explanatory (independent) variable and the current values of a dependent variable. For Example 10.1, which is a study of Fiji, the data were extracted from the World Bank databank but available only for 24 years from 1996, as Fiji's stock market is new. That number of data points is not enough to estimate the return between time series (experts say there should be at least 30 data points). Therefore, these annual data were converted into quarterly data using an accepted standard technique. For this, the method proposed by Boubakari and Jin (2010) was used. SMD (stock market development) refers to the year-on-year change in stock market capitalization as a percentage of gross domestic product (MCAP). Box 14.3: General regression method explanation shows the general regression procedure. Box 14.4: Lagged regression method explanation shows the delay regression process.

Box 14.3 General regression method explanation

The quantitative econometric relationship between SMD (DV: Dependent Variable) and its specific determinants can be specified based on the above hypotheses as a traditional multiple regression model in Eq. 14.1:

$$\begin{aligned} SMD = {} & B0 + \beta1\,LQDT + \beta2\,INFL + \beta3\,FDI \\ & + \beta4\,BNKD + \beta5\,GDP + \beta6\,INV \\ & + \beta7\,SAV\,\beta8\,COC + \beta9\,GVTEf + \beta10\,PS/AV \\ & + \beta11\,RGLQ + \beta12\,ROL + \beta13\,DMCY + \alpha \end{aligned} \tag{14.1}$$

To improve this model, contemporaneous levels of predictor variables (determinants: IVs: Independent Variables) will be used to predict the SMD in the following year across 24 years, 96 quarterly occasions (Boubakari & Jin, 2010), over the time period 1996 to 2018. This model predicts the level of SMD using the one-year lagged levels of determinants. By so doing, this analysis mitigates the possibility of reversed hypotheses that SMD influences macroeconomic and institutional factors.

Also, our primary thesis is that stock market determinants influence change in SMD; more convincing findings can be generated by predicting absolute change in SMD (not the level) using the lagged level of study determinants. By so doing, we examine the proportional change in SMD, not a constant change, in relation to the lagged level of determinants (Grimm et al., 2016). We believe that the investigation of proportional change in SMD, rather than the level or constant change in SMD, is more appropriate in a more vulnerable financial system such as Fiji.

Lagged means lagging behind a time interval. That is, to assume the effect of an independent variable on a dependent variable is a dynamic, such as pushing from behind. Here the effect of the 'yesterday' condition of the independent variables on the 'today' change of the dependent variable is estimated. It is now popular in economics to construct very complex equations assuming that autoregression is possible. This delay reflection process is illustrated in Box 14.4.

Box 14.4 Lagged regression method explanation

Therefore, we strengthened the hypothesized causality by (a) predicting absolute change in SMD from t_1 to t_2 (Grimm et al., 2016a, b) using lagged determinants at $t - 2$ as independent variables ensuring temporal priority (Hume, 2006). In addition, by predicting absolute change SMD from t_1 to t_2, we would be able to avoid issues related to the autocorrelation of the outcome variable.

Therefore, the time-series regression model of change (SMD $= \Delta$MCAP % of GDP) that will be used in the study is as follows (Eq. 14.2):

$$\begin{aligned} \text{SMD}(t, t-1) = {} & \text{B0} + \beta 1 \,\text{LQDT}(t-2) + \beta 2 \,\text{INFL}(t-2) + \beta 3 \,\text{FDI}(t-2) \\ & + \beta 4 \,\text{BNKD}(t-2) + \beta 5 \,\text{GDP}(t-2) + \beta 6 \,\text{INV}\,(t-2) + \beta 7 \,\text{SAV}(t-2) \\ & + \beta 8 \,\text{COC}(t-2) + \beta 9 \,\text{GVTEf}\,(t-2) + \beta 10 \,\text{PS/AV}\,(t-2) \\ & + \beta 11 \,\text{RGLQ}(t-2) + \beta 12 \,\text{ROL}(t-2) + \beta 13 \,\text{DMCY}(t-2) + \alpha \end{aligned} \quad (14.2)$$

We have used all possible combination of data for the regression across all the time points. For example, when the DV (SMD) is for 2004 to 2005, IVs (macroeconomics and institutional) are from 2003, and when DV is for 2005 to 2006, IVs are from 2004, so on, up to DV is for 1997 to 1998, IVs from 1996. The estimated regression.

Coefficients provide the average influence of lagged IVs on change in SMD across all the possible occasions. In this model, β1 to β13 regression coefficients can be interpreted as the amount of absolute change in SMD for one-unit change in the lagged determinant.

The regression analysis was performed separately for macroeconomic determinants and institutional determinants to reduce the degree of freedom due to the large number of variables. The ARDL bounds testing procedure could also add robustness for regressions with small size samples, and where the variables are not integrated in the same order (Ho, 2019).

Software called Eviews supports this technique. The statistical analysis and statistics obtained using Eviews are given in Table 14.3.

The retrospective test results of Table 14.3 are interpreted in Box 14.5.

Box 14.5 Regression analysis

Table 14.3 shows the results of the Regression of change in Stock Market Development (SMD) on Lagged Macroeconomic Determinants. The results showed that lagged GDP has a significant positive influence on change in SMD ($\beta = 2.145$, $p < 0.001$) suggesting that for every one unit increase in GDP, there was a 2.145% increase in SMD growth in a subsequent year. The influence of GDP showed the highest statistical significant ($p < 0.01$). Inflation showed a negative influence on the change in SMD ($\beta = -0.574$, $p < 0.007$), suggesting that every one percent increase in inflation percentage has resulted in a fall in SMD growth in the subsequent year by 0.574% after taking the influences of GDP and domestic credit into account. This is consistent with the findings of previous studies that showed the inflation has a negative impact on SMD. Domestic credit to the private sector (BNKD) also had a negative impact on subsequent SMD change in the ($\beta = -0.208$, $p < 0.008$) after taking the influences of GDP and inflation into account. Each one-unit increase in BNKD has resulted in a fall in 0.208% SMD growth in the subsequent year. This is contrary to the findings of previous studies that showed a positive correlation.

Table 14.3 Regression of Change in SMD on lagged macroeconomic determinants using ARDL

Variable	Coefficient	Std. Error	p-value
LQBT	–0.361	1.699	0.834
INF	–0.574***	0.182	0.007
FDI	0.094	0.171	0.591
BNKD	–0.207***	0.065	0.007
GDP	2.145***	5.100	0.001
INV	–0.246	0.188	0.213
SAV	6.496	9.948	0.525
C	8.385	5.034	0.119

R-squared 0.690455
Adjusted R-squared 0.499966
Note $* p < 0.05$, $** p < 0.01$, $*** p < 0.01$, Unstandardized coefficients

The value stocks trade, foreign direct investment, gross capital formation, and gross savings did not significantly impact subsequent SMD change. Overall, it appears that these observed associations are specific to the Fiji socioeconomic context. All the variables explained 69% (R-squared) of variance in SMD change.

The Boxes of Example 10.1 illustrate the ARDL bound testing procedure followed by the Eviews software.

14.2 Summary

- Statisticians use samples to make statistical inferences. Economists test relationships to make assumptions.
- Econometrics is often based on 'theory', and statistics are driven by 'data'.
- Statistical research is driven by data analytical problems. Econometricians test theory using data, and statisticians tend to build models after looking at datasets.
- A unit root is a non-stationary-nonstable tendency of a time series. Unit root $\rightarrow \rightarrow \rightarrow$ Non-stationarity.
- Since unit root time series are not stable, difference-stationary is obtained using data intervals, and trend-stationary is used when there is no unit root.
- Co-ordination technology is used as a solution to the pseudo-regression process caused by non-stationary time series.
- An important condition in the ARDL bound technique is that the integration is either at the same level or at the first difference and not at the second difference [I(2)].

References/Further Reading

A manual for ARDL approach to cointegration: https://nomanarshed.wordpress.com/2014/11/16/a-manual-for-ardl-approach-to-cointegration/.
Eviews training: https://www.eviews.com/Learning/forecasting.html.
Ho, S.-Y. (2019). The macroeconomic determinants of stock market development in Malaysia: An empirical analysis. *Global Business and Economics Review, 21*(2), 174–193.
Ho, S.-Y., & Iyke, B. N. (2017). Determinants of stock market development: A review of the literature. *Studies in Economics and Finance, 34*(1), 143–164.
Nkoro, E., & Uko, A. K. (2016). Autoregressive distributed lag (ARDL) cointegration technique: Application and interpretation. *Journal of Statistical and Econometric Methods, 5*(4), 1–3.
Yartey, C. A. (2008). *Determinants of stock market development in emerging economies: Is South Africa different?* Washington, DC. IMF.
Yartey, C. A. (2010). The institutional and macroeconomic determinants of stock market development in emerging economics. *Applied Financial Economics, 20*(21), 1615–1625.

Youtube Link

Cointegration: https://www.youtube.com/watch?v=vvTKjm94Ars.

Part IV
Research Traditions: Qualitative Strategies

Chapter 15
Qualitative Startegies

Even today, traditional, research papers prepared using the 'quantitative scientific method' have more potential for getting published than qualitative papers. One reason is limited availability of qualitative journals. Since the 1960s, scientific research has been divided between two competing disciplines as sociologists have taken a powerful step towards a better, more natural and subjective approach. Both strategies are traditionally based on sensory empirical data, with a tendency to use both numerical data and non-numerical data. But even if it is not perceptible, some theories can be scientifically convincing. Anthony Giddens argues that to contribute to social progress, researchers must seek knowledge beyond the scientific method.

Paulo Reglus Neves Frier's Critical Educational Movement developed concepts from critical theory and related traditions and played a vital role in the field of education and social progress. This is a noticeable contribution for qualitative approach for pragmatic social emancipation.

Human reasoning, which can be subject to pseudoscience and irrationality, is an 'imitation of real science', and social constructivism provides a quantitative answer to this confusion. Social constructivism is a sociological theory of knowledge in which human development and social knowledge are built on the interaction with others. As a philosophical approach, strong social reformism suggests that the natural world plays an important role in building scientific knowledge. In this context, even quantitative researchers tend to see the world as subjectivist, gradually breaking free from their strict objectivist positions. In the 1990s, this dichotomy led to the introduction of mixed strategies, using the same methods as both quantitative and qualitative strategies, based on certain common philosophical positions. Methods are the procedures used for data collection, description, analysis, and interpretation. This process is named by Wolcott as the D-A-I formula (D-A-I formula: Description, Analyzing and Interpreting) (Wolcott, 1994).

C. A. Saliya, *Doing Social Research and Publishing Results*,
https://doi.org/10.1007/978-981-19-3780-4_15

Table 15.1 Dimensions of comparison of five qualitative research traditions

Research tradition / Dimension	Narrative research	Case study	Phenomenology	Ethnography	Grounded theory
Focus	Exploring life experiences	Developing on in-dept analysis of cases	Understanding the essence of experience about a phenomenon	Describing, interpreting cultural and social groups	Developing a theory grounded in data from the field
Discipline, origin	Anthropology, literature, history, psychology, sociology	Political science, Business management, all social sciences	Philosophy, sociology, psychology	Cultural anthropology, sociology	Sociology, business management
Data collection	Primarily interviews and document analysis	Documents, interviews, observations, physical artefacts, experiences	Long interviews up to 10 individuals	Observations and interviews, additional artifacts during extended time in the field	Interviews with 20–30 individuals to "saturate" categories and construct a theory
Data analysis	Stories, epiphanies historical content	Descriptive, narrative	Statements, meaning theme, general description of the experience	Description analysis, interpretations	Open coding, axial coding, selective coding conditional matrix
Narrative form	Detailed picture of an individual's life	Storytelling, thick descriptions	Description of the essence of the experience	Description of the cultural behaviour of a group or an individual	Theory of theoretical model

Sources Creswell (2003, 2007), Denzin and Lincoln (2005), Saliya (2017).

This chapter discusses a brief introduction to research strategies without qualitative or statistical analysis. A detailed study of these methods requires the study of books and research papers specializing in each approach. Over the past two or three decades, the methodology of academic research has undergone revolutionary changes and has progressed to new dimensions, despite the blurring of boundaries between disciplines and obstacles such as criticism and resistance.

Cresswell (2007) classifies all types of inquiries into five 'traditions'. These include case studies, narrative research, phenomenology, ethnography, and grounded theory, and compare the fundamental differences as summarized in Table 15.1.

Needless to say, this resistance grows out of neo-conservative discourses and the recent report published by the National Research Council, which have appropriated neopositivist, so-called evidence-based epistemologies. Leaders of this movement assert that qualitative research is non-scientific, should not receive federal funds, and is of little value in the social policy arena (Lincoln & Cannella, 2004).

Qualitative inquiry is the name for a reformist movement that began in the early 1970s in the academy. The interpretive and critical paradigms are central to this movement (Denzin & Lincoln, 2005).

15.1 Other Qualitative Strategies

In addition to the traditional qualitative strategies mentioned above, some scholars identify the following strategies as different strategies. Poetic inquiry and action research strategies are discussed with examples in two separate chapters.

Heuristic inquiry is a process that begins with a question or a problem which the researcher tries to illuminate or find an answer to. The question itself is usually focused on an issue that has posed a personal problem and to which answers are required. It seeks, through open-ended inquiry, self-directed search and immersion in active experience, to 'get inside' the question by becoming one with it (Gray, 2018).

Poetic Inquiry

The use of poetry for qualitative research is nothing new. Poetry enables researchers to retain their expressions, to present them intensely, and to communicate in a more enthusiastic and accessible way. There are two versions of poetry in qualitative research; Namely 'discovered poems' (words extracted from interviews and poetically adapted to illustrate stories) and 'generated poems' (more autobiographical poems, to allow the researcher to share his/her own understanding of his/her/or others 'experiences. This methodology will be discussed later with an example in Chap. 21.

Lynn Butler-Kisber
.il 10.2 · EdD Harvard University

Use of poetry in qualitative research is not new. Poetry enables to retain the voices of research participants, and to communicate more evocatively and accessibly. There are two versions of poetry in qualitative research; 'Found poetry' (words extracted from interviews and crafted into poetic form, to depict the poignant stories) and 'Generated poetry' (...more autobiographic poetry, when the researcher uses his/her own words to share understandings of his/her own/or others' experiences) (Butler-Kisber, 2010).

Action Research

This research methodology can be termed as an experiment at your workplace or in any program or concept you intend to implement. At the end of the research, by reflection, the results of the experiment are reviewed to see how successful the experiment is, then standardised the procedures by further fine-tuning and continuous improvement. This methodology will be discussed later with an example in Chap. 21.

Action research involves close collaboration between researcher and practitioners and places an emphasis on promoting change within organizations such as offices, hospitals, schools and prisons. While the is on seeking information on the attitudes and perspectives of practitioners in the field, the way in which data are collected may involve both quantitative and qualitative methods. The main action research medium, however, is the case study, or multiple case studies. In some designs, both an experimental and a control case study may be used, so emulating the experimental approach (Gray, 2018).

Heuristic Inquiry

This category includes inquiries made from a selective or experiential decision-making approach. Through this the researcher can develop skills and ability to understand the problem and help to develop the understanding of others. Therefore, this methodology does not begin with the aim of finding an external truth. It begins and is activated by direct human experience and can only be traced back to self-inquiry, so it is a highly subjective/personalized methodology such as autobiography and weakens in terms of generalization.

Collage Inquiry

The term 'collage', which refers to a genre of art, is derived from the French verb coller, which means 'to stick', and refers to the process of cutting and sticking found materials onto a flat surface. Collage became acknowledged as art during the early part of the 10th century when artists such as Picasso and Braque used this medium to challenge the traditional conventions of art, the elitist nature of art, and the notion of a single reality. By working with this nonlinear and intuitive way, implicit assumptions can surface and/or be countered. It also discusses how collage clusters can help conceptualize dimensions of understanding that were previously unconscious, and how collage creation can be a way of making thoughts concrete, facilitating the thinking, writing and talking about inquiry. (Butler-Kisber, 2010, p. 102).

MORNING SIGHT

Cubist Photography Pablo

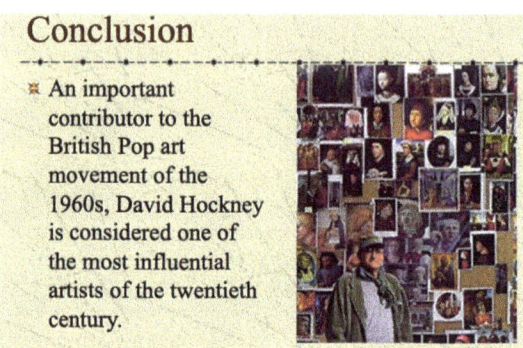

Conclusion

- An important contributor to the British Pop art movement of the 1960s, David Hockney is considered one of the most influential artists of the twentieth century.

Picasso to David: https://slideplayer.com/slide/1432469/

Despite much interest in college inquiry, it is still somewhat isolated and seems to be somewhat alienated from discussions surrounding other forms of visual inquiry, such as photo inquiries.

Photographic Inquiry

Society is sinking into an increasingly visual world. Technology has become increasingly user-friendly, reducing the need for camera equipment and/or allowing telephones to be used for better quality photography. 21st century life is fast, visible and an open to exploration in every domain. It is misleading to talk about the world or a single world. *It makes more sense to think of various versions of the world that individuals may entertain, various characterizations of reality that might be presented in words, pictures, diagrams, logical propositions, or even musical compositions. Each of these symbol systems captures different kinds of information and hence presents different versions of reality. All we have, really, are such versions; only through them do we gain access to what we causally tern 'our world' (Gardner,* 1980).

The scientific revolution of the 17th century further underestimated images because they promoted the idea that the senses perceived by the senses were deceptive. With the development of aesthetics in the middle of the 18th century, the concept of mediated understanding emerged. (Seigesmund, 2008*: 940).*

Performative Inquiry

Performative Inquiry is the exploration of a topic or issue through a play or drama show. It opens up interpersonal spaces where social responsibility and interpersonal and ethnic responses can be investigated. Provides temporary access to 'other' worlds through concert and retrospective/reflection. It is a research methodology that recognizes and respects invisible realities, social evolution trajectories, and opportunities for understanding through concerts.

MacLeod (1987) has suggested that meaning is constructed via word, number, image, gesture and sound and that drama integrates all five. Experts who follow this research methodology are of the opinion that that these five dimensions are co-ordinated by contextual presentations. They say, therefore, that it is not surprising that in our multimedia world there is a growing interest in contemporary inquiries as we explore the most comprehensive research methodologies.

The current acknowledgment of the message of the performance inquiry is that it refutes the criticism that the performing arts in general ignore suffering, injustice, and silenced grievances. This is because the contextual inquir is able to absorb the integral elements of meaning, opening up opportunities for the approaches, interventions, gains and participation dimensions that the play or concert highlights. The performative inquiry highlights the liberating potential of anthropology and suggests how the politics of resistance and ability can be created by giving street performers the opportunity to speak out on silenced grievances and accusations.

Performances such as readers theatre and other similar forms which transform data into a script that is then read aloud to an audience often using the participatory as 'actors' and/or involving the audience, have become frequent at large research conferences and shown the evocative and pedagogical potential of performative inquiry. (Butler-Kisber, p. 135).

Current thinking is that performative inquiry counteracts the criticism that representation neglects oppression, injustice, and silenced stories, for it can exploit the integrative aspects of meaning, and permit the engagement, accessibility, and participatory dimensions that drama or performance elicits (Denzin, 2003).

The current acknowledgment of the message of the performance inquiry is that it refutes the criticism that the performing arts in general ignore suffering, injustice, and silenced grievances. This is because the contextual inquiry is able to absorb the integral elements of meaning, opening up opportunities for the approaches, interventions, gains and participation dimensions that the play or concert highlights.

The performance inquiry translates into a transcript using theatres and other platforms, which are then read aloud to the audience. This happens as often as in large research conferences and demonstrates the enthusiastic and educational potential of the contextual inquiry.

15.2 Summary

- There are five main types of qualitative strategies: Case studies, narrative research, phenomenology, ethnography, and grounded theory.
- In addition, there are other quality research strategies such as poetic inquiries, contextual inquiries, collage inquiries, photographic inquiries, heuristic inquiries, and cryptographic research.

References/Further Reading

Bazeley, P., & Jackson, K. (2013). *Qualitative data analysis with NVivo*. London: Sage.
Butler-Kisber, L. (2010). *Qualitative inquiry: Thematic, narrative and arts-informed perspectives*. London: Sage.
Creswell, J. W. (2003). *Research design: Quantitative, qualitative and mixed methods approaches* (2nd ed.). Thousand Oaks, CA: Sage.
Creswell, J. W. (2007). *Qualitative inquiry and research design: Choosing among five traditions*. Thousand Oaks, CA: Sage.
Denzin, N. K. (2003). *Performance ethnography: Critical pedagogy and the politics of culture*. Thousand Oaks, CA: Sage Publications.
Denzin, N. K. & Lincoln, Y. S. (2005). Introduction: The discipline and practice of qualitative research. In N. K. Denzin & Y. S. Lincoln (Eds.), The landscape of qualitative Research: theories and issues. Sage: London. pp (1–6).
Gardner, H. (September 1980). Gifted world makers. *Psychology Today*, 92–94.
Gray, E. D. (2018). *Doing research in the real world (4th ed.)*. Thousand Oaks, CA: Sage.
Lincoln, Y. S., & Cannella, G. S. (2004). Qualitative Research, Power, and the Radical Right. *Qualitative Inquiry, 10*(2), 175–201. https://doi.org/10.1177/1077800403262373.
Macleod, D. S. (1987). ART collecting and Victorian middle-class taste. *Art History*. First published: September 1987 https://doi.org/10.1111/j.1467-8365.1987.tb00260.x.
Mienczakowski, J. (1995). The theatre of ethnography: The reconstruction of ethnography into theatre with emancipatory potential. *Qualitative Inquiry, 1*(3), 360–75.
O'Leary, Z. (2004). *The essential guide to doing research*. London: Sage.
O'Leary, Z. (2005). *Researching real-world problems*. London: Sage.
Saldana, J. (2005). An introduction to ethnodrama. In J. Saldana (Ed.), *Ethnodrama: An anthology of reality theatre* (pp. 1–38). Walnut Creek, CA: AltaMira Press.
Saliya, C. A. (2017). Doing Qualitative Case Study Research in Business Management. *International Journal of Case studies*. 6(12), 96–111. http://www.casestudiesjournal.com/Volume%206%20Issue%2012%20Paper%2010.pdf.
Seigesmund, R. (2008). Visual research. In L. M. Given (Ed.), *The Sage encyclopaedia of qualitative research methods*, Vol. 2 (pp. 940–943). Thousand Oaks, CA: Sage.
Silver, C., & Lewins, A. (2014). *Using software in qualitative research*. London: Sage.
Wolcott, H. F. (1994). *Transforming qualitative data: Description, analysis, and Interpretation*. CA. Routledge.

Chapter 16
Narrative Research

Narrative inquiry or narrative emerged as a subject stream in the field of qualitative research in the early twentieth century. Contemporary narrative inquiry can be characterized as an amalgam of interdisciplinary analytic lenses, diverse disciplinary approaches, and both traditional and innovative methods—all revolving around and interest in biographical particulars as narrated by the one who lives them (Chase, 2005).

It is a text that combines all the important information around the characters, as described by the living.

Narrative research aims to approve or construct a theory from a narrative that analyses a set of events chronologically. The key elements of this strategy are as follows.

- Script
- Images or imaginary frames used for interpretation
- Dialogues and observations that chromatic the script
- Comparative evaluations that reveal the narrator's point of view
- Extraction of conclusions with theories that are logically constructed in combination with the knowledge that has already been published.

The main data collection method of this strategy is to engage in lengthy 'pleasing conversations' with relevant parties about personal experiences, these are called interviews. Such research largely focuses on issues related to moral, cultural, and ethical confusion.

Thus, the narrative query focuses not only on data collection and processing but also on the organization of human knowledge. It also implies that knowledge is valuable and significant, even if only one person knows it. Knowledge management and narrative inquiry share the idea of knowledge communication. Knowledge communication is a theory that seeks to communicate innumerable, qualitative aspects of knowledge, including experience. If knowledge is not communicated, it becomes useless knowledge, a waste.

C. A. Saliya, *Doing Social Research and Publishing Results*,
https://doi.org/10.1007/978-981-19-3780-4_16

Philosophers speculate that it is indistinguishable from the way narratives (secondary information- told by others) and reconstruction of own experiences (first cognition- memories). Therefore, it is the belief of the scholars who follow this research strategy that storytelling is an effective and powerful way of communicating knowledge.

16.1   Participatory Narrative Inquiry (PNI)

Participatory narrative inquiry (PNI) is a participatory approach for groups to gather and work on the experience of individual groups aimed at making better decisions and gaining an understanding of complex situations. PNI focuses on a deeper consideration of values, beliefs, feelings, and perspectives through the reconstruction and interpretation of living experiences. Events, 'truth', evidence, ideas/opinions, arguments, and proof can be used as the raw material for PNI's empathy, but they are always used from personal perspectives and influenced by ontological beliefs.

The three basic elements of PNI are participation, narration, and inquiry. These three elements should be balanced with special emphasis as shown in the Fig. 16.1.

Narrative inquiry is a research strategy suitable for mature scholars with a wealth of experience. Stories can also be constructed through the reconstruction of one's own experiences. Researchers interested in politics can create a research paper using narrative strategy based on data from already published books (secondary data).

16.2   Biography

The following are some biographies based on data collected from significant surveys. The first two biographies are two internationally acclaimed books, while the second two biographies are two of the most popular in Sri Lanka.

$E = mc^2$: *A Biography of the World's Most Famous Equation by David Bodanis. Berkley Publishing Group, New York.*

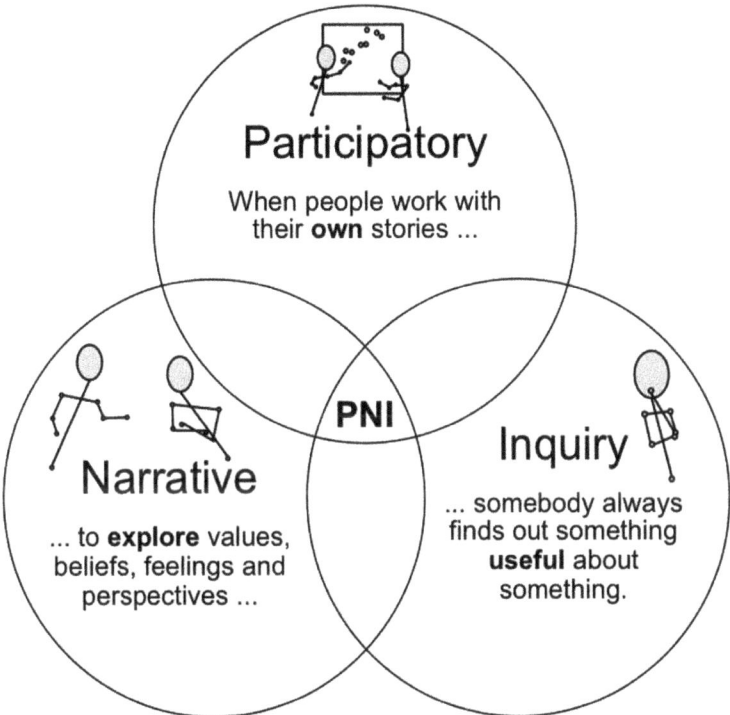

Fig. 16.1 The three basic elements of PNI. *Source* http://www.storycoloredglasses.com/p/partic ipatory-narrative-inquiry.html

This "biography of the world's most famous equation" is a one-of-a-kind take on the genre: rather than being the story of Einstein, it really does follow the history of the equation itself. From the origins and development of its individual elements (energy, mass, and light) to their ramifications in the twentieth century, Bodanis turns what could be an extremely dry subject into engaging fare for readers of all stripes. Bodanis (2000).

It is a history of where the equation ($E = mc^2$) came from and how it has changed the world. Bodanis presents its five symbolic ancestors in sequence, each with its own chapter and each with rich human stories of achievement and failure, encouragement and duplicity, love and rivalry, politics and revenge. Readers meet not only famous scientists at their best and worst but also such famous and infamous characters. Bodanis includes detailed, lively and fascinating back matter as well (Bodanis, 2000).

Mao: The Unknown Story

The most authoritative life of the Chinese leader ever written, Mao: Unknown Story is based on a decade of research, and on interviews with many of Mao's close circle in China who have never talked before, and with virtually everyone outside China who had significant dealings with him.

It is full of startling revelations, exploding the myth of the Long March, and intricate relationship with Stalin went back to the 1920s, ultimately bringing him to power. He welcomed Japanese occupation of much of China and he schemed, poisoned, and blackmailed to get his way.

After Mao conquered China in 1949, his secret goal was to dominate the world. In chasing this dream he caused the deaths of 38 million people in the greatest famine in history. In all, well over 70 million Chinese perished under Mao's rule in peacetime (Chang and Holiday, 2006).

J.R. Jayewardene of Sri Lanka and the Prabhakaran Saga

"Writing a biography of a living political figure and one still very much in office, has its pitfalls. We are well aware of these. Naturally, we may tend to look at problems through our subject's eyes, and events and issues in which he has taken part or was interested may assume a significance and proportions which they may not have had in a general history of the period we cover, or in an analysis of political systems and structures. We are at once advocates and critics, and well aware that the closer one is personally to the subject of one's biography, there is some danger of the critical role being overshadowed if not actually undermined by that of the advocate" (Silva and Wriggins, 1988, p. 8).

Although all the works are biographies, the book "Biography of the World Famous Equation" and "The Story of Prabhakaran: The Rise and Fall of an Eelam Warrior" are either case studies (about the Eelam War or $E = mc2$ equation) or can be applied to phenomenology (perceptions of war victims) for a scientific academic research.

16.3 Testimonios

Testimonios, as emergency narratives, can mobilize a nation against social injustice, repression, and violence. Collective stories can form the basis of a social movement. Telling the stories of marginalized people can help to create a public space requiring others to hear what they do not want to hear (Denzin & Lincoln, 2005).

A testimonio is a novel or novella-length narrative, produced in the form of a printed text, told in the first person by a narrator who is also the real protagonist or witness of the events she or he recounts. Its unit of narration is usually a "life" or a significant life experience….the production of a testimonio generally involves the tape recording and then the transcription and editing of an oral account by an interlocutor who is a journalist, ethnographer, or literary author (Beverley, 2005).

16.4 Examples D: The Role of Bank Lending in Sustaining Income and Wealth Inequality

Narrative D presented in Part 2 under qualitative data collection for a case study selected from a private bank in Sri Lanka (although short because it was done for a case study, these narratives are constructed in the form of long novels for narrative research purposes).

A summary of the case studies based on this narrative are presented as examples in the next chapter. These case studies were presented as early as 2006 at an international academic conference in Hawaii. It can also be referred to as a whistle bowing or an alarm. As revealed by these conference papers, this bank faced 'public confidence' issues, faced a run and near bankrupt position, the Central Bank of Sri Lanka had to intervene and save the bank.

Researchers interested in narrative research should read texts such as Labov and Woltzky's (1997) Structural Analysis Methodology, Clandinin (2007), Clandinin and Connolly (2007), Bruner (2002).

16.5 Summary

- Narrative query is a weaving structure that combines analytical lenses of different disciplines, different approaches, and traditional and creative methods, all around important information biographies as described by the living person.
- The main components of Narrative research are the script, Interpretations, dialogs and observations, balanced evaluations, and development of cogent arguments.
- Participatory narrative inquiry (PNI) focuses on recreating living experiences and in-depth consideration of values, beliefs, feelings, and perspectives through interpretation, and case facts, 'truth', evidence, ideas / opinions, arguments, and proof are always used from personal perspectives and based on ontological beliefs.
- An author who is personally close to the speaker runs the risk of undermining the role of criticism under the influence of representing the speaker.
- It is essential to look at the speaker from different angles and discuss with different people to ensure the credibility of a biography.

- Biography is a medium that can be used to speak out to the world against social injustice, oppression and violence, to lay the foundation for a social movement, to tell the stories of victims of injustice and to hear what others do not want to hear.
- The case studies related to the private bank were presented at an international academic conference held in Hawaii as early as 2006. It can also be referred to as a whistle blowing to create awareness. The government had to intervene to save the bank in 2008.

References/Further Reading

Adams, T. E. (2008). A review of narrative ethics. *Qualitative Inquiry*, *14*(175), 175–191.

Beverley, J. (2005). Testimonio, Subalternity, and narrative authority. In N. K. Denzin, & Y. Lincoln (Ed.), *The Sage Handbook of Qualitative Research*. London: Sage.

Bruner, J. (2002). *Making stories: Law, literature, life*. Cambridge. MA: Harvard University Press.

Chase, E. S. (2005). Narrative inquiry; multiple lenses, approaches, voices. In N. K. Denzin & Y. S. Lincoln (Eds.), *The Sage Handbook of Qualitative Research* (3rd ed.), pp. 651–679. London. Sage.

Clandinin, D. J. (2007). *Handbook of narrative inquiry: Mapping a methodology*. Sage.

Clandinin, D. J., & Connolly, F. M. (2007). *Narrative inquiry: Experience and story in qualitative research*. San-Francisco: Jossey-Bass.

Chapter 17
Case Studies

In one sense all research is case study: there is always some unit, or set of units, in relation to which data are collected and/or analyzed (Gomm et al., 2000, p. 2).

  Case studies can be micro-level (interpersonal and interpersonal), meso-level (organizations, institutions) or macro-level (communities, democracies, societies) and may involve one or more individuals (Swanborn, 2010).

According to case studies expert Yin (2003), 'building theory before data collection' is one factor that differentiates case study strategies from other qualitative research strategies, which is a feature of quantitative methodology. Thus, starting from a theory and accepting or rejecting that theory according to empirical evidence is an empirical approach, such as a hypothesis testing leading to conclusions, a method of drawing conclusions. Case studies with an interpretive or critical approach first collect data and then theorize.

© The Author(s), under exclusive license to Springer Nature Singapore Pte Ltd. 2022   273
C. A. Saliya, *Doing Social Research and Publishing Results*,
https://doi.org/10.1007/978-981-19-3780-4_17

O'Leary (2004) suggests that one approach may not be better than another, and more importantly, that all researchers work for reflexive awareness and informed choice. She explains, since 'cases' in a case study can involve individuals, cultural groups, communities, phenomena, events and, in fact, any unit of social life organization, virtually all methodologies and/or data collection tools can be called upon dependent on the case at hand.

In the case study strategy, the accuracy of the researchers' arguments is ensured by a considerable in-depth investigation of the number of cases studied and/or the amount of data. Clearly one good case study can demonstrate the functioning of a social system. A case study is an appropriate research method not only to explore or describe a situation but also to establish causal relationships..

Holiday (2002) suggests "… you do not have to choose between case studies, anthropology and grounded theory. Case studies may or may not be anthropological and should never be quantitative (p. 118).

17.1   Case Study Definition

Case study is an empirical test that investigates a contemporary phenomenon in its real-life context. In a broader sense, this is an analytical way of describing, understanding, predicting, and/or controlling any unit or group, such as a process, animal division, individual/household, organization or group, industry, culture, or nationality.

17.2   Types of Case Study Strategies

Different types of case studies are identified based on the number of cases involved, the number of analytical units covered, the purpose and the nature. According to Stake (2005) there are three types of case studies: Intrinsic; unique cases, not representative, Instrumental: to provide insights or enhance an existing theory, and Collective: to generalize. According to Yin (2003), three types of case studies have been identified: Exploratory, Descriptive, and Explanatory. In addition, Schwandt & Gates introduced in 2018 another type of case study aimed at what should happen (values, norms, or ideals) rather than what happened (empirical phenomenon), which is called Contributory case studies.

All case study research can be done using a single-case type or a multi-case type. Each type of case study research can include one complete analysis unit or several analysis units and use a different design. Table 17.1 summarizes the characteristics, purpose, approach, research problems, number of cases, nature of analysis units, and related theories in different types of case studies.

Table 17.1 Types of case studies

Case type Item	Exploratory	Descriptive (Interpretive)	Explanatory	Contributory
Approach	Hypothesis generation and theory development	Study of commonalities	Hypothesis and theory testing	What is and should be valued
Aim	Field work is done prior to the definition of research problem. Aimed at defining the issues of subsequent study.	Aimed at a complete description of a phenomenon within its context.	Aimed at presenting causal relationship; explaining which causes produced which effects.	Justify ends and outcomes, specifically what is right or wrong, desirable, or undesirable.
Research Issues	Broad design determined well ahead, Emphasize on actual behavioral events rather than perceptions.	Encounter enormous problems in limiting the scope of the study.	Emphasize on How and Why do Research findings get into practical use?	Normative (Value-laden) matters. Who gains and who loses?
Theories	Search for causal theories.	Theory driven data collocation.	Search for explanatory theories.	Use ethical judgements
Nature	Instrumental	Intrinsic	Instrumental	Supportive
No. of cases	Single	Single	Collective	Single/collective
Unit/s of analysis	Holistic unit of analysis	Sub-units/Holistic unit of analysis	Sub-units/Holistic unit of analysis.	Holistic unit of analysis

Sources Yin (1993, 2003), Stake (2005) and Schwandt and Gates (2018)

17.3 Credibility, Dependability and Conformability

It has been argued that a traditional 'thick description' alone is not sufficient to ensure the reliability of academic research because it can be limited to depth and detail at different levels. Whether the description is 'long' or 'short', the description is considered credible if it provides sufficient evidence to substantiate the argument presented.

Whether the description is 'thick' or 'thin' if it provides adequate evidence to the claim, the description is considered as dependable (Bailey, 2007). *They say instead of studying a thousand rats for one hour each, or a hundred rats for ten hours each, the investigator is likely to study one rat for a thousand hours Skinner (1966).*

Therefore, case studies should be supported by other validation methods and collaborative evidence, such as financial accounts and documents, and ultimately inter-rater reliability checks with peer-scholars etc. The 'usefulness' of data can be considered as an additional indicator that can be used to assess the reliability of research aimed at social progress. Therefore, the decisive factor in any decision about the quality of research is the honest quality of the methods used.

"…*method is thereby the crucial factor in any judgment made about the quality of research*" (Smith and Hodkinson, 2005, p. 917).

In planning a case study, which is a very flexible methodology, it is useful to consider the following points.

- Unit of analysis (Individuals, organisations, sector etc.)
- Criteria to be used to select cases
- How many cases, how many participants in each case etc.

Example 17.1 is a case study research. The cases were developed from Participant observation PO.8.1, PO.8.2., PO.8.3 & PO.8.4 and Narrative 8.1, 2, 3 & 4 as explained earlier an shown in Case 17.1.

Example 17.2 The role of bank lending in sustaining income and wealth inequality.

Case 17.1 Tony's garment manufacturing business.

ITEM	Credit applicant: Tony—Tony Group	Credit decision-maker: Chairperson Mr. Perera
Background	Tony is in the garment manufacturing business. He accounted for the largest single share of country's exports and 30,000 workers. Tony's credit application had been rejected by the credit officer of the Soft Bank based on an official credit-default investigation report. He had won several awards for his contribution to the country	Mr. Perera became the all-powerful chairperson of the Soft Bank with his loyalists as board members and weak Management. The Chairperson was in a business development drive as he had no faith in the bank's Management especially the credit officers. This powerful businessperson had become a fame seeker after a "near death" experience
Approach	Tony and Mr. Perera knew each other as leading businesspersons in Sri Lanka. Tony's request for an appointment with Mr. Perera was accommodated promptly even though the Chairperson was at a meeting at that time. The approach was casual and formal rules were bypassed	
Influencing factors	The business that Tony brought was lucrative and significant. The credit decision was justified by the positive cash and profit forecast by professional accountants. The Chairperson was not aiming at any personal gratification but showed some "patriotic" feelings and "sympathy" over the workforce who were without wages facing redundancy because of the foreclosure of factories called by the other banks to recover their debts. The state policies on such issues were not helpful to borrowers or to lenders. The Chairperson had an egoistic motive to boost his image and wanted to strengthen his social power base by getting another powerful businessperson tactfully hooked into his social network. Tony promised the publicity warranted by this ego-maniac Chairperson	

(continued)

(continued)

ITEM	Credit applicant: Tony—Tony Group	Credit decision-maker: Chairperson Mr. Perera
Decision-making	Tony was very impressive. The Chairperson got a quick assessment of Tony's integrity and the risk and return of the credit involved. The Chairperson had the power and he was confident that the Board would not decline what he recommended. Therefore, he instantly granted US$4 million. This casual decision was made under the enormous power entrusted to the Chairperson in a weak organizational structure with a poor management support. The decisions were defended under the façade of "patriotic" attitudes and influenced by egoistic motives especially when the decision-maker is from the top echelon of the bank and more importantly backed by projected cash flows presented by accountants	
Post-decision events	Tony's accountant complained about not bringing all the sale proceeds to Sri Lanka. This situation had caused liquidity problems therefore, Tony frequently demanded enhancement of credit. The workforce had been reduced to 15,000 and monthly sales had reduced by 25% and were making a negative contribution by 2003	Further credits were granted but later the loan was classified as bad and provisions were made to comply with the prudence principle of accounting and regulatory requirements. The interest loss was around US$0.3 million and the capital loss too was significant

Case 17.2 Silva's janitorial business

ITEM	Credit applicant: Silva—Superclean Ltd	Credit Decision-Maker: Recoveries Department
Back-ground	The threats to Silva's janitorial business include: the new entrant who was a friend of the then top Management, harassment by the credit officer due to a personal grudge. Silva lost business to the newcomer and was facing liquidity problems and financial barriers for business development and got into the bad books as his loans were not serviced	Fernando, the branch manager who nurtured Silva and his business Superclean, was promoted and transferred (Decision 1). The Superclean account was classified as non-performing and transferred to the recoveries department. First it was positively handled by the M–R (Manager Recoveries) and then negatively by the Chief M–R (Chief Manager Recoveries)

(continued)

(continued)

ITEM	Credit applicant: Silva—Superclean Ltd	Credit Decision-Maker: Recoveries Department
Approach	Decision 1. Under normal banking practices Decision 2. M–R comprehended Silva's plight by extending several bank facilities outside the normal credit rules. M–R discovered some errors of the branch staff and blamed them but did not report to the Management rather used it to intimidate the branch staff. Silva was happy with M–R and had praised him with research participants Decision 3. CM–R applied the set credit rules strictly and stopped all the facilities extended by M–R and requested Silva to settle existing outstanding immediately for him to consider further facilities Silva searched for avenues to approach the Chairperson as advised by people who are aware of the Chairperson's approach and, as Mr. Perera was boasting of helping small businesspeople via his TV shows, but without success because he was not influential and not in as powerful circle as Tony or did not have such an appealing project as Yousef had	
Influencing factors	Decision 1. Under normal banking practices Decision 2. M–R's performance was measured by the reduction shown in the bad loans handled by him. He could transfer restructured bad loans back to branches as performing loans and that process enhanced M–R's performance for which he was rewarded. But M–R did not have the necessary authority to approve such new credit facilities. Even the staff at the branch at the mercy of M–R due to their blunder committed for interest calculation had no alternative but to carry out M–R's unauthorized orders Decision 3. CM–R was an orthodox banker and he strictly applied the rules and tried to impress the Management about his "good work". Silva could not afford get the services of professional accountants to project his prospects convincingly	
Decision-making	Decision 1. Under normal banking practices Decision 2. M–R abused his authority and was at an advantage that the branch staff was under obligation. M–R was promoted and encouraged by the Management for his performance though the decisions were not duly approved Decision 3. CM–R followed normal banking practices, did not believe Silva's appeal about his business prospects and declined further credits. He was contemplating serving a letter of demand threatening legal action against Silva and his house which was mortgaged to the Bank	
Post-decision events	Research participants were of the view that if Silva had enough power or adequate influence to reach the Chairperson with the backing of accounting expertise, he could have got the necessary facilities and M–R would have got another promotion instead of losing his job. Unfortunately Silva was not from that class and social network and could not afford professional or experienced accountants	

Source Saliya (2010)

17.4 Summary

- Data collection is always for all research case studies as it is common to all research strategies to be for a particular unit or group of units.
- Case studies can be micro, meso or macro and can involve one or more individuals. Case studies also help to raise a voice for people who are sometimes unfair, privileged, marginalized, and vulnerable.

  - Descriptive case study plan aims to present information that is useful to policy makers, scholars as well as the general public.
  - Exploratory case studies involve building or proposing theories of interest.
  - Explanatory case studies use hypotheses to test hypotheses and theories.
  - Contributing to normative theory: This case study plan aims at what should happen rather than what happens (empirical phenomenon) (values, norms or ideals). These types of case studies (contributing to values/norms)

- The case studies on the based-on credit culture in in Sri Lanka provided important alarming information to regulators such as the Central Bank should have acted.
- Case study strategy provides a flexible methodology which consists of the following features; unit of analysis, criteria to be used to select cases, how many cases, how many participants in each case etc.

References

Hammersley, M., Gomm, R., & Foster, P. (2000). Case Study Method, Key Issues, Key Texts. London. Sage.

O'Leary, Z. (2004). *The essential guide to doing research.* London: SAGE.

Saliya, C. A. (2010). *The Role of bank lending in sustaining income/wealth inequality in SriLanka.* PhD Thesis, Auckland University of Technology, New Zealand. https://openrepository.aut.ac.nz/handle/10292/824.

Saliya, C. A. (2021). Conducting Case study research: A concise practical guidance for management students. *International Journal of KIU, 2*(1), 1–13. https://doi.org/10.37966/ijkiu20210127.

Schwandt, T. A. & Gates, E. F. (2018). Case Study Methodology. In N. K Denzin and Y. S. Lincoln (eds). The SAGE Handbook of Qualitative Research (5th ed.). London: SAGE.

Smith, J., & Hodkinson, P. (2005). Relativism, criteria and politics. In N. Denzin, & Y. Lincoln (Eds.), *Handbook of qualitative research, 3*, 915–932. London: Sage.

Stake, R. E. (1995). *The art of case study research.* Sage.

Stake, R. E. (2000). The case study method in social inquiry. In R. Gomm, M. Hammersley & P. Foster (Eds.), *Case Study Methods.* London, UK: Sage.

Stake, R. E. (2005). Qualitative case studies. In N. K. Denzin & Y. S. Lincoln (Eds.), *The Sage Handbook of Qualitative Research*, pp. 443–466. Sage.

Swanborn, P. (2010). *Case study research: What, why, and how? Thousand Oaks.* CA: SAGE.

Yin, R. K. (1993). *Applications of case study research* (Vol. 34). Sage.

Yin, R. K. (2003). *Case study research: Design and methods* (3rd ed. Vol. 5). London, UK: Sage.

Yin, R. K. (2014). *Case study research* (5th ed.). Sage.

Chapter 18
Phenomenology

Phenomenology is a school of philosophy related to the study of the mind. Any experience that is perceptible to the senses (five senses and mind) is recognized as a phenomenon. Simply put, epistemology is a description of a person's immediate experience. The philosophical word for what comes to mind from such an experience is 'phenomenology' and the study of that is called phenomenology. A phenomenon includes not only the cognition of the mind about a physical object such as a book or a pencil, a trip or a meal or a drink, but also illusions, dream-objects, and our memories, expectations, and aspirations.

The concept of 'phenomenon/phenomenon' identified in the context of phenomenology is broader than the terminological meaning of 'scientific' usage, which refers only to something extraordinary. A general understanding of phenomenology can be obtained from the following link.

Source https://prezi.com/_ip-lyvrxk0q/phenomenology/

© The Author(s), under exclusive license to Springer Nature Singapore Pte Ltd. 2022   281
C. A. Saliya, *Doing Social Research and Publishing Results*,
https://doi.org/10.1007/978-981-19-3780-4_18

The phenomenology was born from Edmund Hazerl's philosophical position that the starting point of knowledge is the concept of self - experience of phenomena, that is, the sensations that arise from one's conscious cognition and life experience.

Phenomenologists argue that the relation between perception and objects is not passive—human consciousness actively constructs the world as well as perceiving it... Phenomenology seeks to understand the world from the participant's point of view. This can only be achieved if the researcher 'brackets out' their own preconceptions (Gray, 2018, p. 167).

Phenomenology does not consider historical or social contexts. Phenomenologists argue that the relationship between cognition and objects is not passive inactivity, but that human consciousness actively builds the world. Phenomenology seeks to understand the world from the perspective of participants. This can only be achieved if the researcher 'isolates' his own conclusions. This means an empathy-based inquiry strategy. An essential element of this strategy is to forget the researcher's understanding, feelings, attitudes, and beliefs for a moment and to sink into the cognitions of the researcher. Buber (1958) states that human beings cannot understand others in the way they perceive objects, but that human understanding requires an openness, participation, and emotional connection that goes beyond that.

A unique and final definition of phenomenology is dangerous and perhaps even paradoxical as it lacks a thematic focus. In fact, it is not a doctrine, nor a philosophical school, but rather a style of thought, a method, an open and ever-renewed experience having different results, and this may disorient anyone wishing to define the meaning of phenomenology (Gabriella, 2014).

There are several hypotheses that can help explain the foundations of phenomenology (Orbe, 2009):

- *Phenomenologists reject the concept of objective research. They prefer subjective assumptions.*
- *They believe that one can gain a better understanding of nature by studying daily human behaviour.*
- *They insist that individuals should be explored. This is because individuals can be understood in unique ways that reflect the society in which they live.*
- *Phenomenologists prefer to collect conscious experiences rather than conventional data.*
- *They consider that phenomenology is focused on new discoveries and, therefore, they do research using much less limited methods than other sciences.*

Creswell (2007) suggests, that regardless of orientation there are four basic philosophical perspectives in phenomenology: The return of philosophy to the search for wisdom instead of a focus on science. The need to suspend all judgements about what is real. The understanding of intentionality, that reality is related to one's consciousness of it. The recognition that reality is perceived within the meaning of individual experiences (p. 58–59).

These philosophical thoughts seem to coincide largely with the interpretation of 'reality' taught in Buddhism. That is, the perceptions we acquire through the senses are not absolute reality. That is, the 'truths' in the world are relative.

Assuming that the conscious experiences mentioned in the above narrative research paragraphs (Narration D.1–4) and the case studies paragraphs (Case D.1 and Case D.2) are perceived by the people concerned, they become 'knowledge'. A typology has been created by introducing four creditors from a taxonomy that has been developed (induced) by matching the perceptions of the participants in the process (Table 18.1).

Example 18.1 Dynamics of credit decision-making: a taxonomy and a typological matrix.

This theory suggests that loan approving officers and loan applicants, under three dimensions, assume phenomena that can be perceived through their own experience. These dimensions are formed by matching the cognitions of the participants in this process.

1. Procedure of evaluating loan applications; (a) formal or (b) judging by experience
2. The relationship between loan applicants and loan officer; (a) Personal level or (b) Official level relationships and
3. Justifying loan approval; (a) logical or (b) irrational/situational.

Accordingly, the four name tags are assigned for the four categories of loan approving officers and presented as Typology.

Box 18.1 A taxonomy and a typological matrix

The **Bosses** make situational decisions employing heuristic methods. The Bosses (heuristic–situational) make heuristic decisions because they are powerful with legitimate authority to break the rules but they provide "solace" to the "needy" of course to people who are "close" to them; therefore, they are more like situational in business sense.

Table 18.1 Taxonomy of typology: The four decision-makers

Variable		3. Justification/logic		
		Rational	Irrational	
1. Evaluation procedure	Formal	BUDDY	ROBOT	2. Nature of Relationship
	Heuristic	REBEL	BOSS	

Source Saliya (2019)

The **Robots** (formal–irrational) strictly follow rules; formalities are adhered but they lack common sense; therefore, they are likely to be mechanical/irrational.

The **Rebels** (heuristic–rational) break rules for the sake of justice and fairness but lack enterprising logic. They make decisions that are beneficial to both the decision-maker and the credit applicant; therefore, they make rational decisions, but it is likely that heuristic methods are followed since formal rules forbid such decisions.

The **Buddies** (formal–rational), on the contrary, find ways within the formal system without breaking rules to support decision-seekers as well as develop business: therefore, they make rational decisions following formal procedures.

Example 18.2 Enterprising mothers in Fiji need comprehensive support.

The following is a summary of the themes of Tables 24.7 and 24.8 presented under data Analysis in Chap. 24, and the conceptualization of categories/codes created using grounded theory strategy, using the abductive inferencing, is shown in Table 18.2: Theorization of women's expectations and destiny in Fiji. Thus, the themes faced by women in relation to their professional development and the fate that befell them in society have been reduced to theoretical themes.

This knowledge construction is discussed further in the next section under Data Analysis.

One of the most important data collection methods used in phenominology is open-ended interviews. It is important to ask the questions such as 'what' as well as 'how/why' of life experience (Butler-Kisber, 2010). Researchers interested in this subject should consult the works of pioneers such as Edward Husserl (1970), Martin Buber, Van Manen and Shutz A., as well as scholars such as Moustakas C., Pieterson, C., Riemen, D. J.

Table 18.2 Theorization of women's expectations and destiny in Fiji

Demands perception	Seeking Regulations 27/32 = 84.4%	Fair share for women 25/32 = 78%
Multi-skilled 22/32 = 68.7%	Need for a legal boost, Rebuff self-effacing, Reject male dominance, Discourage toleration, Work-life balance	
Sociocultural factors 23/32 = 71.9%		

Source Saliya (2022b)

18.1 Summary

- The philosophical term for what is created in the mind by experience is 'phenomenon' and the study of it is called phenomenology.
- Phenomenologists argue that human consciousness actively builds the world.
- Phenomenologists reject the concept of objective research.
- The essential element of this strategy is to forget the researcher's understanding and sink into the cognitions of the researchers.
- Human understanding requires openness, participation and a compassionate relationship for human understanding.
- It is important to ask 'what' as well as 'how/why' cognitions about life experiences based on open-ended questions.

References

Buber, M. (1958). *I and thou*. R. G. Smith (ed. And Trans.), New York: Scribner.

Butler-Kisber, L. (2010). *Qualitative Inquiry: Thematic*. Sage.

Edward Husserl, H. (1970). *The crises of European sciences and transcendental Phenomenology*. Northwestern University.

Edmund Husserl. (1962). *Ideas: General introduction to pure phenomenology*, tr. W. R. Boyce Gibson (New York: Macmillan).

Gabriella, F. (2014). Some reflections on the phenomenological method. *Dialogues in Philosophy, Mental amp; Neuro Sciences, 7*(2), 50–62. 13 p.

Gray, D. E. (2018). *Doing research in the real world*, Thousand Oaks, CA: Sage.

Max, V. M. (2014). *Phenomenology of practice: Meaning-giving methods in Phenomenological research and writing*. Left Coast Press Inc.

Orbe, M. P. (2009). Phenomenology. In S. Littlejohn, & K. Foss (Eds.), *Encyclopaedia of communication theory*, (pp. 750–752). Thousand Oaks, CA: Sage.

Saliya, C. A. (2022b). Enterprising mothers in Fiji need comprehensive support. The Qualitative Report, Under review.

van Manen, M., & van ManenFirst, M. (2021). Doing Phenomenological Research and Writing. *PubMed*. https://doi.org/10.1177/10497323211003058.

Chapter 19
Ethnography

The Greek word 'ethno' refers to a people, race or cultural group. Ethnography is a branch of anthropology that systematically studies individual cultures. Ethnography explores cultural phenomena from the subject perspective of the study, and it involves exploring a cultural group to understand, discover, describe, and interpret a way of life from the perspective of its participants.

Ethnography is a type of social research related to examining the behaviour of participants in a particular social situation and understanding the interpretation of participants' behaviours (such as those under investigation). The goal of ethnographic research is to gain a fuller understanding of how people from different cultures and subcultures gain an understanding of their living reality (Hess-Bieber et al., 2011).

As an investigative strategy, Ethnography largely relies on the researcher participating in the background or attempting to document observations by engaging in certain roles with the people being studied. In this process, participants seek to understand social interactions, patterns, and perspectives in detail in their local context.

https://www.interaction-design.org/literature/book/the-encyclopedia-of-human-computer-interaction-2nd-ed/ethnography

C. A. Saliya, *Doing Social Research and Publishing Results*,
https://doi.org/10.1007/978-981-19-3780-4_19

19.1   Critical Ethnography

Traditional ethnography captures only the phenomena of everyday life without identifying the processes and mechanisms that cause them. ethnography has a unique
ability to approach sites of exploitation and oppression, thus giving the ethnographers
a unique opportunity to develop liberating habits.

Ethnography has a unique capability for getting close up to sites of exploitation
and oppression and hence offers the ethnographic researcher a unique opportunity
for constructing emancipatory practices (Lather, 1986). Conventional ethnography
tends to grasp only the phenomenon of everyday life without apprehending the causal
processes and generative mechanisms that drive them. Ethnography should articulate
identifiable and political issues including injustices based on race, class, gender and
sexual orientation. 'It should criticize how things are and imagine how they could
be different' (Denzin 2017).

Critical ethnography, also known as 'radical ethnography', is a method of research
that deals with a political agenda that exposes unequal, unjust influences. Critical
anthropology reveals identifiable and political issues, including injustice based on
race, class, gender, and sexual orientation. It should critique how those things are so
and explore how they change.

19.2   Auto-ethnography

Autoethnography is a research method that: Uses a researcher's personal experience
to describe and critique cultural beliefs, practices, and experiences. Acknowledges
and values a researcher's relationships with others…. Shows 'people in the process
of figuring out what to do, how to live, and the meaning of their struggles' (Adams,
2015).

Autoethnography is an approach to research and writing that seeks to describe
and systematize personal experiences in order to understand cultural experiences.
This approach challenges traditional methods of research and representation, while
research is political, socio-just, and socially conscious.

Social life is messy, uncertain, and emotional. If our desire to research social
life, then we must embrace a research method that, to the best of its/our ability,
acknowledges and accommodates mess and chaos, uncertainty and emotion (Adams,
2015).

The Examples 17.1 and 18.1 employed this self-anthropological strategy.

19.3 Performance Ethnography

Performance ethnography is vitally important as a pedagogical tool for precisely the same reason it is a potent conceptual and methodological one (Alexandar, 2006; Denzin, 2003, 2006). It exposes the dynamic interactions between "power, politics, and poetics" (Madison, 2008). Performance ethnography is such a method, incarnating critical interventions so they live in the flesh as well as on the page or the screen, though we must continually resist the temptation to conflate all performance with utopian space (Hamera, 2018). In ethnography, however, greatest risk arises at the time of publication (Murphy & Dingwall, 2001). Research participants may feel wounded or offended by published ethnographic material, often in ways that were unanticipated by the ethnographer (Ellis, 1995). Ethonography has particular problems when it comes to guaranteeing anonymity because transcripts invariably record sufficient detail to make participants identifiable (Gray, 2018).

19.4 Summary

- The Greek word 'ethno' refers to a people, race or cultural group.
- *Ethnography* involves exploring a cultural group to understand, discover, describe, and interpret a way of life from the perspective of its participants.
- *Ethnography* is a branch of anthropology that systematically studies individual cultures.
- As an investigative strategy, *ethnography* relies on participation.
- Critical ethnography, also known as 'radical anthropology', is a method of research with a political agenda to expose unequal, unjust influences.
- Self-*ethnography* is an approach to research and writing that seeks to describe and systematize personal experiences in order to understand cultural experiences.

References/Further Reading

Adams, T. E. (2015). Autoethnography and Family Research. *Journal of Family Theory & Review,* 7(4), 350–366. https://doi.org/10.1111/jftr.12116.

Bochner, A., & Ellis, C. (2016). *Evocative autoethnography: Writing lives and telling stories* (p. 87). New York: Routledge. ISBN 9781629582146.

Denzin, N. K. (2017). *A manifesto for performance autoethnography.* https://doi.org/10.1525/irqr. 2017.10.1.44.

Gray, E. D. (2018). *Doing Research in the Real World* (4th ed.). London: Sage.

Hammersley, M., & Atkinson, P. (2019). *Ethnography: Principles in practice.* www.taylorfrancis. com.

Hesse-Biber et al. (2011). *The practice of qualitative research.* London: Sage.

Lather, P. (1986) Research as Praxis, *Harvard Educational Review, 56,* 257–277.

Rathnapala, N. (2008). *The beggar in Sri Lanka.* Colombo: Sarvodaya Vishva Lekha Publivcation.

Saliya, C. A. (2019). Credit capital, employment and poverty, *International Journal of Money, Banking and Finance, 8*(1), 4–12. https://search.proquest.com/docview/2249669069?pq-ori gsite=gscholar.

Spry, T. (2001). *Performing autoethnography: An embodied methodological praxis.* https://doi.org/ 10.1177/107780040100700605.

Chapter 20
Grounded Theory

The purpose of grounded theory is to find out from participants 'what is going on' in a significant field. Here the researcher should provide space/facilities for the participants themselves to identify their main need. They have to tell the researcher themselves what the researcher needs to study. When the researcher understands the main need of the participants, the researcher constantly explores how they can solve it. This (how they solve the main issue) is the basic category, concept, or final theory of a grounded theory strategy. Everything must be allowed to emerge spontaneously.

This strategy is called grounded theory (GT) because it is a data-based theory. Glaser and Strauss (1967) are considered the pioneers of this strategy. However, the philosophical perspective of grounded theory has changed according to Glaser's positivist stance and Strauss & Cobbin's post-positivist constructivism stance. Grounded theory can be identified as the emergence of additional philosophical perspectives that have contributed to the development of research strategies over time.

The different strategic variants used for GT are threefold; Glaser's traditional or classic GT; evolved GT of Strauss, Corbin and Clarke; constructivist GT of Chamaz. Although each variant of the GT has common features, the use of literature, including the researcher's philosophical position (positivist, interpretivist and constructivist, etc.) there are distinguishing factors between approaches and procedures involved in coding theorisation (deduction, induction and abduction). The various terms used to describe the coding of these three-dimensional approaches are described in the following paragraph.

Glaser says the goal of the traditional GT is to generate a conceptual theory for the research participants that is relevant and problematic. The evolving GT is based on the sociological perspective of symbolic interactionism, which relies on individualized interpretations of human interactions for social interactions. Symbolic interaction addresses the subjective or personal meaning that people place on objects, behaviors, or events based on what they believe to be true. The third variant, developed and explained by the symbolic interlocutor Chamas, was called the Reformed GT, based on the roots of Reformation.

© The Author(s), under exclusive license to Springer Nature Singapore Pte Ltd. 2022   291
C. A. Saliya, *Doing Social Research and Publishing Results*,
https://doi.org/10.1007/978-981-19-3780-4_20

Table 20.1 Stages and related functions of grounded theory strategy

Activities	Stage 1	Stage 2	Stage 3	Stage 4
Theoretical sampling	Selection of participants for interviews	Theoretical sampling 2	Theoretical sampling 3	Theoretical sampling 4
Data collection	Interviews observation and surveys	Data collection 2	Data collection 3	Data collection 4
Inducing/abducting (modify, verify and reject)	Conceptualizing	Induction 1	Induction 2	Induction 2 abduction
Coding, memoing and tentative mini theorization	Initial coding, catagorizting, funding gerunds	Line by line coding, combining catagories to new dimensions	Focused coding, reducing dimensions exploring relationships between catagories	Classic GT: Core concept OR Constructivist GT: Most elegant, best explanations
Interrogating data and checking for deducting	Deduction 1: Check with another theoretical sample	Deduction 2: Check with another theoretical sample	Deduction 3: Check with another theoretical sample	

Essentially, grounded theory methods are a set of flexible analytic guidelines that enable researchers to focus their data collection and to build inductive middle-range theories through successive levels of data analysis and conceptual development (Charmaz, 2005, p. 507).

Grounded theory is a research tool which enables you to seek out and conceptualise the latent social patterns and structures of your area of interest through the process of constant comparison. Initially you will use an inductive approach to generate substantive codes from your data, later your developing theory will suggest to you where to go next to collect data and which, more-focussed, questions to ask. This is the deductive phase of the grounded theory process.

Since it starts with data, it follows the induction method at the beginning and categorizes it by constructing data and concepts using re-validation using deductive and abductive methods. Finally, alternative theories are proposed based on assumptions using the deductive method. Table 20.1 summarizes this process and presents a list of tasks related to each stage.

20.1   Theoretical Sampling

When data is collected and a theory begins to emerge, the researcher may feel that parts of the theory are unclear or lack data, so the researcher must return to the field and inquire about it. This is called theoretical sampling. Sample based on an emerging theory. That is, selective selection of samples that contain missing or missing data in order to build a theory.

20.2   Glaser's Traditional or Classic GT

A classical grounded theory study begins with a broadly open question. Glaser (2002) insists that GT is a form of latent structure analysis, which reveals the fundamental patterns in a substantive area or a formal area and suggests that conceptualization is the core category of GT.

The most important property of conceptualization for GT is that it is abstract of time, place, and people. This transcendence also, by consequence, makes GT abstract of any one substantive field, routine perceptions or perceptions of others, since there is always a perception of a perception, and an abstraction from any type of data whether qualitative or quantitative. Concepts in general, whether conjectured, impressionistic or carefully generated by GT, have instant "grab." (Glaser, 2002, p. 7).

Example 24.7 (Part V) presents the basic coding, categorising and conceptualisation etc., of this traditional GT strategy in Analysis 24.8.

Example 20.1 Enterprising mothers in Fiji need comprehensive support

Analysis 20.1 Chapter 24 (Data Analysis) presents a summary of 32 interviews with 11 participants. The responses obtained were basically categorized into 116 codes as shown in the first column, and their validation was done through regular discussions with the participants. Thematic analysis was then subdivided into 11 categories, categorized as shown in the second column. The conceptualization of those responses is then shown in Analysis 24.8. Following the traditional GT strategy, the core concept of the final concept stage has evolved into a theory called the Grounded Theory of Enterprising Mother. This analysis is presented in Part V.

20.3   Constructivist GT of Chamaz

According to Charmaz, Strauss and Corbin, data collection, conceptualizing, coding, memoing, and inducing tentative mini-theories were re-examined, check those theories using deductive techniques using theoretical sampling until saturation. Then conduct constant comparison conceptualised by using induction, deduction and abduction of theorising techniques.

The process of doing a GT research study is not linear, rather it is iterative and recursive. While GT studies can commence with a variety of sampling techniques, many commence with purposive sampling, followed by concurrent data generation and/or collection and data analysis, through various stages of coding, undertaken in conjunction with constant comparative analysis, theoretical sampling and memoing. Theoretical sampling is employed until theoretical saturation is reached. These methods and processes create an unfolding, iterative system of actions and interactions inherent in GT (Corbin & Strauss, 2008).

Saturation is the process of having to ask the same questions and get the same responses when testing tentative theories with theoretical samples. Then the realization that it makes no sense to ask the same question over and over again is considered saturation. Professor Chamas' discussion in the following Youtube video was also very helpful to clarify this further. This strategy is a research strategy that can be easily implemented by amateur researchers. This strategy can be understood by studying Fig. 20.1 as four steps consisting of ten basic activities.

20.4   Codification

Normally, open and axial codification are done in a simultaneous way. Firstly, codes are created through open codification and, after, they are grouped by similarity of meaning, originating the first categories.

During the codification process and data analysis the researcher developed some memos and diagrams. These analytical tools support the information record and in the research process in a general way. Memos, according to Corbin & Strauss (2008), are written records of analysis that can vary in type and format. Memos can be like codification notes, theoretical notes or operational notes.

"First cycle" coding (Miles et al., 2014), "initial coding" (Charmaz, 2006), and "open coding" (Corbin & Strauss, 2008) refer to the initial assigning of codes to data chunks. The second cycle is bundling, condensing, integrating and laying the first cycle codes into broader and more coherent categories and themes (Miles et al., 2014), identifying patterns, relationships and explanations. This is referred to as "pattern coding" (Saldana, 2013), "focused coding" (Charmaz, 2006) and "axial coding" (Corbin & Strauss, 2008). At this stage, advanced coding is essential to produce a theory that is grounded in the data and has explanatory power. During the advanced coding phase, concepts that reach the stage of categories will be abstract, representing stories of many, reduced into highly conceptual terms. Finally, themes are created to reflect "big" ideas that emerge from data (Aurini et al., 2016) conceptualising elegant themes and theorising a core category by combining categories into broader dimensions.

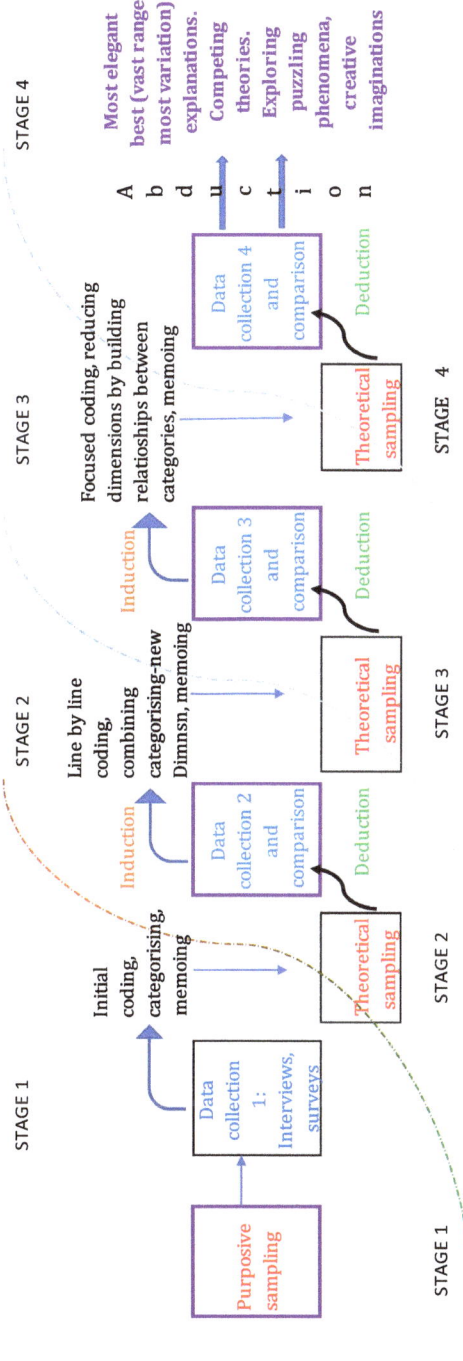

Fig. 20.1 The process path analysis of grounded theory. *Source* Saliya (2022)

Codes rely on interaction between researchers and their data. Codes consist of short labels that we construct as we interact with the data. Something kinaesthetic occurs when we are coding; we are mentally and physically active in the process (p. 5).

To develop a grounded theory for the twenty-first century that advances social justice inquiry, we must build upon its constructionist elements rather than objectivist leanings. …Contemporary grounded theorists may not realize how this tradition influences their work or may not act from its premises at all. Thus, we need to review, renew, and revitalize links to the Chicago school as grounded theory develops in the twenty-first century (Charmaz, 2005, p. 507).

In the advanced coding stage, concepts can be abstract, and many stories are concentrated into very elegant a few concepts. Finally, themes are designed to reflect the "big" ideas that emerge from the data (Aurini et al., 2016). By doing so, the theory is applied to a core concept.

In the video below you can watch an important discussion with Professor Kathy Chamasz about the evolution of grounded theory strategy. You can also easily learn data coding and thematic analysis using MS Word by watching the Youtube video at the other link.

A Discussion with Prof Kathy Charmaz on Grounded Theory

https://www.youtube.com/watch?v=D5A HmHQS6WQ

https://www.youtube.com/watch?v=4KO pSG7myOg

Glaser (2002) stated that "inviting participants to review the theory for whether or not it is their voice is wrong as a 'check' or 'test' on validity. They may or may not understand the theory, or even like the theory if they do understand it. Many do not understand the summary benefit of concepts that go beyond description to a transcending bigger picture. GT is generated from much data, of which many participants may be empirically unaware" (p. 5).

Figure 20.1 shows a comprehensive framework that seeks to incorporate design, repetitive and recursive processes, interactions between methods and motion, and every aspect of the processes that underlie GT generation. It shows the process of gradual transcendence from its conceptual level and its various cognitive levels (e.g. 3rd and 4th level cognitions).

20.5 Summary

- In the first stage, code is generated from data using terms such as "first round" coding (Miles et al., 2014), "initial coding" (Charmaz, 2006), or "open encoding" (Corbin & Strauss, 2008), by different genres.
- Building theory will suggest where to go next to collect data and ask questions that need more attention.
- In the second round, merging the first-round codes into broader and more "corporate categories and themes" (Miles et al., 2014) and Saldana calls it "pattern coding" (Saldana, 2013), Charmaz (2006) refers them as "centric coding", and Cobbin and Strauss (2008) call it as "axial coding". This is the deductive stage.
- Since it starts on data, it is initially categorized by building data and concepts using the induction and then get them confirmed by deductive and abductive reasoning.
- Finally alternative theories are proposed based on assumptions using the deduction and abduction methods.
- Grounded theorr research can be conducted in either objective methodological or reformist principles.
- In the advanced coding stage, concepts can be abstract, and many stories are concentrated to very conceptual/elegant terms.
- Finally, the themes are designed to reflect the "big" ideas that emerge from the data (Aurini et al., 2016), to conceive of Charmaz' "elegant themes".
- According to Glaser's classis GT variant, the theories are applied to a core concept by combining them.
- Inviting participants to review whether their opinion is correct is not necessary (actually wrong) as a 'query' or 'test' of validity, because they may or may not have the expertise to understand the theory, and may or may not like it if they understand it.

References/Further Reading

Aurini, J. D., Heath, M., & Howells, S. (2016). *The How To of Qualitative Research*. London: Sage

Birks, M., & Mills, J. (2015). *Grounded theory: A practical guide* (2nd ed., p. 2015). London: Sage.

Charmaz, K. (2005). Grounded theory in the 21st century: Application for advancing social justice studies. In: N. K. Denzin, Y. S. Lincoln (Eds.), *The Sage handbook of qualitative research* (pp. 207–236). Thousand Oaks, CA: Sage [Google Scholar].

Charmaz, K. (2006). *Constructing grounded theory: A practical guide through qualitative analysis*. Thousand Oaks, CA: Sage [PubMed] [Google Scholar].

Chametzky, B. (2020). Becoming an expert: A classic grounded theory study of doctoral learners. *The Grounded Theory Review, 19*(2), 20–35.

Chun Tie, Y., Birks, M., & Francis, K. (2019). Grounded theory research: A design framework for novice researchers: *SAGE Open Medicine Volume, 7*, 1–8. https://doi.org/10.1177/205031211 8822927.

Corbin, J., & Strauss, A. (2008). *Basics of Qualitative Research: Techniques and Procedures for Developing Grounded Theory* (3rd ed.). Thousand Oaks, CA: Sage.

Glaser, B. G. (2002). Conceptualization: On theory and theorizing using grounded theory. *International Journal of Qualitative Methods*, *1*(2), 23–38. https://doi.org/10.1177/160940690200100203.

Glaser, B., & Strauss, A. (1967). The Discovery of Grounded Theory: Strategies for Qualitative Research. Mill Valley, CA: Sociology Press.

Glaser, B. (2020). Getting started. *The Grounded Theory Review, 19*(1), 3–6.

Miles, M. B., Huberman, A. M., & Saldana, J. (2014). *Qualitative Data Analysis: A Methods Sourcebook*. London: Sage.

Saldana, J. (2013). *The Coding Manual for Qualitative Researchers* (2nd ed.). London: Sage.

Saliya, C. A. (2022). Enterprising mothers need comprehensive support: A case of Fiji, under review.

Ward, K., Gott, M., & Hoare, K. (2017). *Analysis in grounded theory—How is it done? Examples from a study that explored living with treatment for sleep apnea*. https://doi.org/10.4135/9781473989245.

Youtube Links

Charmaz, K. (2013). *A discussion with Prof Kathy Charmaz on grounded theory*. With G. R. Gibbs. https://www.youtube.com/watch?v=D5AHmHQS6WQ.

Coding data. https://www.youtube.com/watch?v=4KOpSG7myOg.

Chapter 21
Action Research

Active research has become an important methodology to intervene for improvement and change in organizations and communities. It is promoted and implemented by many international development institutes and university programs as well as by local community organizations around the world. In action-based research methodology, it seeks transformational change through the parallel process of action-packed programme and research intertwined by critical retrospection. Kurt Lewin coined the term "action research" in 1944. According to him, action and research are spiralling steps, each of which is a spiralling cycle of planning, action, and retrospection of facts.

There are many different types of action research. The illustration in this chapter discusses action-based research that has been implemented for productivity development within an organization.

21.1 Action Research in Organization Development

French and Bell (1973) define organizational development as 'improving organizations through action research'.

© The Author(s), under exclusive license to Springer Nature Singapore Pte Ltd. 2022   299
C. A. Saliya, *Doing Social Research and Publishing Results*,
https://doi.org/10.1007/978-981-19-3780-4_21

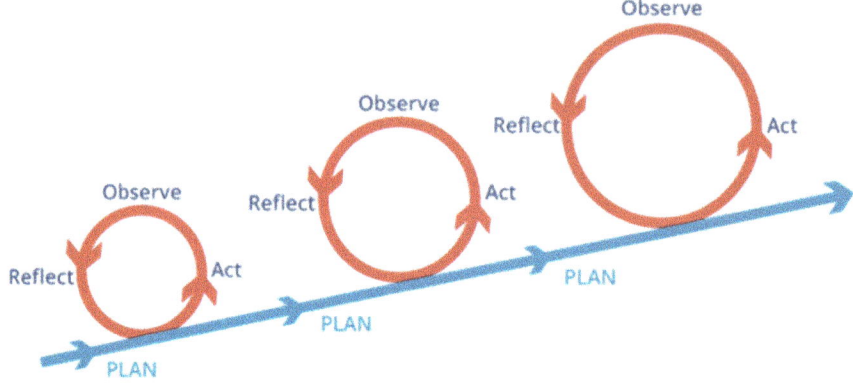

Action Research method (O'Byrne, 2016)

The first step, design, begins with identifying the problem. This requires separate research. Second, a problem-solving strategy must be developed and approved by the Management. Third, the implementation methodology should be well communicated and informed to all participating teams. Fourth, the first cycle ends with activation and retrieval. The planning of the second cycle is implemented by overcoming the weaknesses of the first cycle.

Let's look at this action research with a real example (Example 21.1: Balanced Scorecard for Commercial Banks).

21.1.1 Identifying and Defining the Problem

This was implemented to maximize shareholder's value by increasing Seylan Bank's operational productivity (optimizing productivity). To this end, Seylan Bank's productivity has been compared with that of other nearest competing banks. The screenshots below show that comparison. At the time, Seylan Bank's Cost/Income ratio, etc. were very low and the reports issued by the rating agencies were unsatisfactory. In 1999, Seylan Bank's ROAA (Return on Average Assets) was as low as 0.39% compared to other banks. The objective of the strategy was to increase it by 100% in three years, by 2002, to 0.8%. Screenshots (Fig. 21.1) show how Seylan Bank's direction, as well as analytics such as Seylan Bank's position in the life cycle, were used in the design of this strategy, using theories such as Porter's five force strategies and Ansof's market analysis.

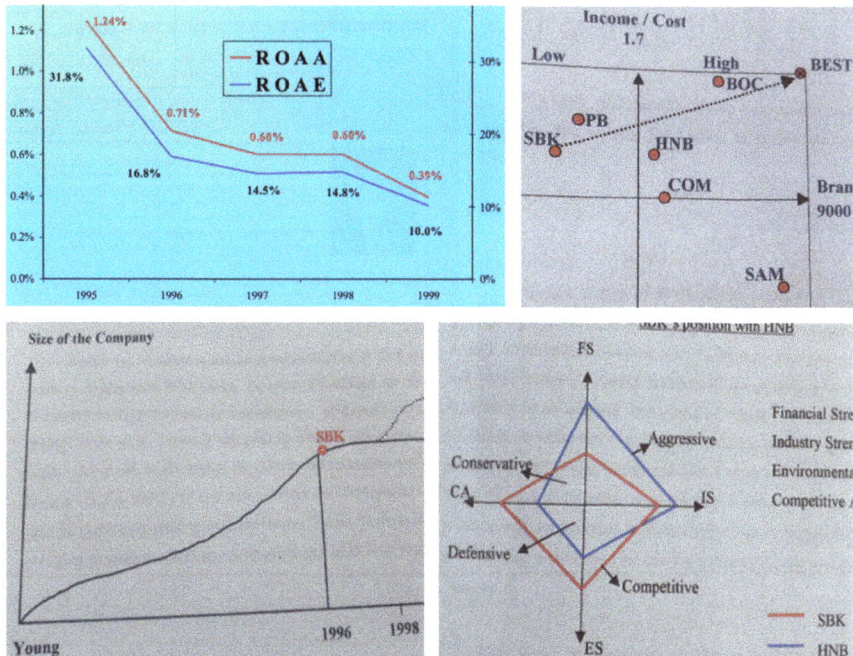

Fig. 21.1 Seylan Bank's direction

21.2 Strategy Formulation

The strategy proposed was to introduce the Balanced Scorecard (BSC) methodology to create competition among with Strategic Business Units (Branches and other profit and cost Centres such as treasury, Credit card unit, Leasing unit and International unit etc.) and motivate the staff. For this purpose, a balanced scorecard (BSC) system was developed to suit the bank. The action research was designed based on the policy paper submitted to Postgraduate Institute of Management (PIM) by the researcher (the author) as a partial fulfilment of his MBA program. The dissertation was published as a research paper in the September/December 1999 issue of The Professional Banker (Fig. 21.2).

Example 21.1 Balanced Scorecard for Commercial Banks. (Saliya, 1999)

21.3 Implementation

The following steps were followed in implementation.

1. Preparation of the proposal and draft staff circular for board approval.

Fig. 21.2   Association of professional bankers

2. Communicating the process of performance measurement and rewarding system to the staff.
3. Use BSC for budgetary control process, quarterly reviews.
4. Set up a team for branch visits for inspections (image building purposes).
5. Set up a subcommittee to get customer feedback and staff feedback, branch-wise.
6. Design MS Excel based data processing system.
7. Quarterly review management conferences (out of Colombo) for all the branch managers.
8. Annual grand award ceremony.

Aligning with Seylan Bank's Vision and Mission with the BSC's Four Perspectives (Shareholder's perspective is added), the summary strategy and detailed procedure were summarized and presented to the board of directors for approval, as shown in the screenshot below (Fig. 21.3).

21.4   Reflection

Oversight played a crucial role in the implementation of the above strategy. Since the plan was aimed at long-term goals such as 3 years, progress was reviewed through retrospective programs in each quarter of the first year (Fig. 21.4). The methodology

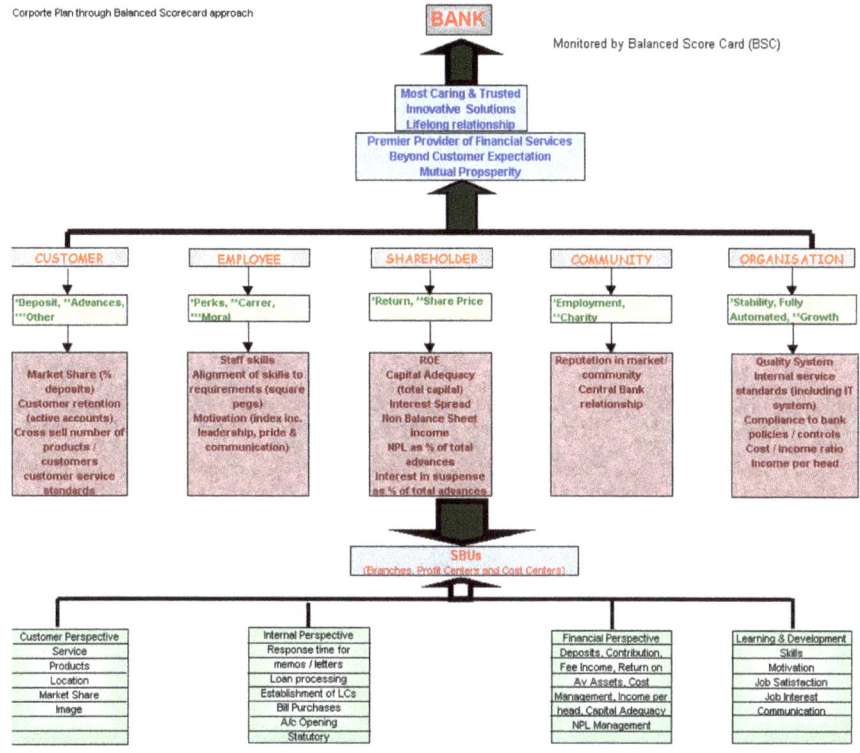

Fig. 21.3 The five perspectives. *Source* Productivity conference. 23 June 2002. BMICH. Colombo (Prepared in 1999)

was well established by 2002 with the Annual Awards/Trophies/Prizes Ceremony and Best Managers Approved. The take was successfully topped.

21.5 Balanced Scorecard—Wide Recognition

Following the presentation of the results of this research at a Productivity Conference (Productivity Conference. 23 June 2002. BMICH. Colombo) in 2002, the researcher was invited to implement this methodology in several other banks. Accordingly, the model BSC was implemented following the same methodology in another commercial bank in a short period of six months in 2004.

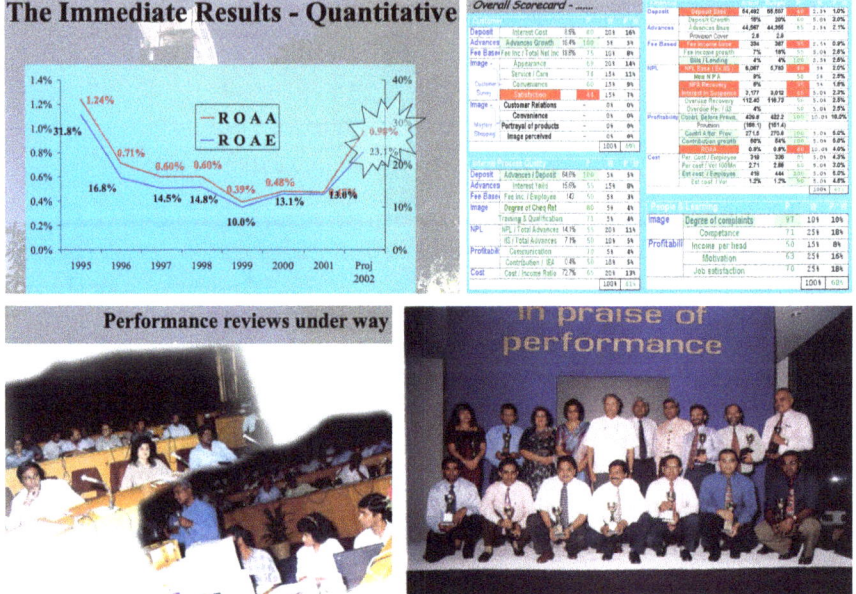

Fig. 21.4 Progress review and reflection

Productivity Conference. 23 June 2002. BMICH. Colombo.

When Professor Allan (Harvard Business School), an associate of the founder of the Balanced Scorecard at Harvard Business School, visited Sri Lanka, Seylan Bank invited him to exploit the opportunity to review the Seylan's BSC, the first successful BSC in Sri Lanka. He was pleased to witness successful implementation of his concept in Sri Lanka for the first time and appreciated the effort and provided very useful feedback.

21.6 Summary

- Action research has become an important methodology to intervene for improvement and change in organizations and communities.
- The first cycle ends with problem identification, problem solving strategy, implementation methodology and retrospective reflection.
- The planning of the second cycle is implemented by overcoming the weaknesses of the first cycle.
- Surveillance plays a crucial role in the implementation of the above strategy.
- For the first time in Sri Lanka, Seylan Bank successfully tested and implemented the Balanced Scorecard concept as a pilot project.

References

French, W. L., & Bell, C. (1973). *Organization development: Behavioral science interventions for organization improvement* (p. 18). Englewood Cliffs, New Jersey: Prentice-Hall. ISBN 978-0-13-641662-3. OCLC 314258.

Kaplan, R. S., & Norton, D. P. (1992). The balanced scorecard – measures that drive performance. *Harvard Business Review.*

Kaplan, R. S., & Norton, D. P. (1999). The balanced scorecard – translating strategy into action. *Harvard Business School Press.*

O'Byrne, I. (2016). *Four steps to conducting action research in the classroom.* W. Ian O'Byrne.

Saliya, C. A. (1999). Balanced scorecard as an approach to the digital nervous system of a commercial bank. *The Professional Banker: Journal of the Association of Professional Bankers - Sri Lanka, Colombo, 9*(2), 21–33.

Chapter 22
Poetic Inquiry

The use of poetry for qualitative research is nothing new. Today, poetry is published in academic journals (IPRAJ) in a variety of fields. However, the sociologist Richardson (1992) translated the words extracted from the interviews into verses and used them as 'found poetry' to present the grievances and stories of her research participants.

My use of the term poetic inquiry to describe the various approaches and occasions for poetry in qualitative research…it encompasses a form of work that includes much more than 'experimental' writing and production of poetry; it is an artful way of being as a researcher (Butler-Kisber, 2010, p. 98).

Butler-Kisber (2010) suggests two types of classical poetry, according to the methodology of composing poetry. They are 'found' poetry and 'generated' poetry. Poetry created from words found in research participants 'transcripts' (recorded or documented) is identified as 'found' poems, and poems created by the researcher to share his/her or 'others' experiences as autobiographies are identified as 'generative' poems.

22.1 Found Poetry

Butler-Kisber (2010) suggests two types of classical poetry, according to the methodology of composing poetry. They are 'found' or 'found' poetry and 'generated' poetry. Poetry created from words found in research participants 'transcripts' (recorded or documented) is identified as 'found' poems, and poems created by the researcher to share his/her or 'others' experiences as autobiographies are identified as 'generative' poems.

Poetry has forever had the power to attract humankind because of its ability to convey poignancy, musically, rhythm, mystery and ambiguity. It appeals to our senses and opens up our hearts and ears to different ways of seeing and knowing…it's an artful way of being as a researcher (Butler-Kisber, 2010, pp. 82–92).

© The Author(s), under exclusive license to Springer Nature Singapore Pte Ltd. 2022   307
C. A. Saliya, *Doing Social Research and Publishing Results*,
https://doi.org/10.1007/978-981-19-3780-4_22

Based on Richardson's works, Glesne (1997) says in her recorded interview that linear and progressive data do not reveal connections and subtle differences. Therefore, she suggests that the poetic creations that combine the expressions of participants and researchers are more powerful than the author-dominated individual expressions found in traditional works. As a result, the following poem published by her seems to send a very strong social message. It should be noted that this poem was created while doing research on the island of Puerto Rico, in an island context (Puerto Rican context). Puerto Rico is a small island in the United States near Cuba.

That Rare Feeling

> I am a flying bird
> moving fast
> seeing quickly
> looking with the eyes of god
> from the top of trees
>
> How hard for country people
> picking green worms
> from fields of tobacco
> sending their children to school,
> not wanting them to suffer
>
> as they suffer
> in the urban zone,
> students worked at night
> and so they slept in school
> teaching was the real university
>
> So i came to study
> to find out how i could help
> i am busy here at the university,
> there is so much to do
> but the university
> is not the island
>
> I am a flying bird
> moving fast, seeing quickly
> so i can give strength,
> so i can have that rare feeling
> of being useful

Glesne (1997, p. 215)

Much can be said poetically under the guise of consensus, in the form of satire or coercion. Poetry has always had the potential to captivate mankind because of its mysterious connotation, its musical, rhythmic, intriguing expression of the innu-merable sorrows of the world animal. It appeals to our senses and opens our hearts and senses to see and know in different ways… It is an artistic way of working as a researcher (Butler-Kisber, 2010).

22.2 Generated Poetry

The following illustration is an abstract from a creative genre poem published by an internationally rated (ranked A *) academic journal. The complete poem is given as Example 22.1. This poem is the author's first publication in such a PRIAJ magazine.

Poetry concisely registers on the whole skein of human emotions. It harrows, enthralls, awes, dazzles, confides…The soul is the depth of our being and poetry is one means of sounding that depth. A poem doesn't wile away time; it engages our fleetingness and makes it articulate. It seizes and shapes time (Wormser & Cappella, 2000, p. xiii).

Example 22.1 The Oldest Profession (Saliya, 2012).

Box 22.1 The Oldest Profession: Abstract.

Abstract:

Purpose—The purpose of this paper is to draw attention towards how religious beliefs are linked to today's accounting principles and legitimating exploitation

Design/methodology/approach—The poem is a critical thought

Findings—Religious concepts have been widely used to justify various histor-ical social injustices. Four basic accounting principles; matching, accrual, going concern and reporting cycle perfectly correspond with life cycles, the acts and rewards of such lives, and "balancing" the pains and gains of individ-uals and society. These concepts, which have now become the foundation of accounting, were in existence millennia ago

Research limitations/implications—Accounting is broader than has been widely perceived. Accounting researchers should pay more attention to how accounting helped sustain social injustices

Originality/value—This poem provokes thoughts on the origin of basic accounting principles and their historical roles in sustaining social injustices

Example 22.1: **The Oldest Profession**

While I was born rich, buoyant and preferred
You were born poor, compliant and unfavoured
Was it God? Destiny? Karma?
Or simply luck or by accident?

Was I good, were you bad in a previous life …
or more recent?
I care, but please please be fully aware
When I share, I always get back my lion's share

You must never ever say that
I have not been legally fair
When accounts are destiny-balanced,
then I will responsibly bear

All debits have credits and all credits have debits … an equal claim
You stupid know not the Matching principle,
while I know the worldly game
When all accounts are settled, in this fiscal Cycle
or in life's next Cycle
I follow well the accounting rules,
yes, based on Going-concern

For three millennia I balanced my gain
against your pain … ordained?
Remember you were born poor and
I was born preferred … die-casted?

On Earth, my Accrual concepts always prevail,
an expertise in legal exploitation
So which profession is the oldest?
You know now … just trust my reputation

Poetry briefly communicates all human emotions and sinks them into the heart. It is a force that can sometimes be painful, sometimes seductive, sometimes exhilarating, and at the same time terrifying, as well as astonishing and deceptive to people. The medium poetry does not waste time; It instantly leads us to state something clearly. It's time consuming and polishing (Wormser & Capella, 2000).

22.3 Summary

- There are two types of classical poetry: 'found' poetry and 'generated' or 'generated' poetry.
- Create poems from words found in research participants 'recordings or writings as' found 'poems and autobiographies identify poems created by the researcher to share his/her or others' experiences as 'generative' poems.
- Poetic creations that combine participant and researcher publications convey a much stronger social message than traditional author-dominated publications.
- Poetry, rhythmically, captivatingly, with the ability to express grievances metaphorically, appeals to our senses and opens our hearts and senses to see and know in different ways.
- Poetry is a concise communication of all human emotions and sinks into the heart.
- A poem instantly directs us to express something clearly, grasping time and polishing it.

References

Butler-Kisber, L. (2010). *Qualitative inquiry: Thematic, narrative and arts-informed perspectives.* London: Sage.

Glesne, C. (1997). That rare feeling: Re-presenting research through poetic transcription. *Qualitative Inquiry, 3*(2), 202–221.

Madison, D. S. (2008). Narrative poetics and performative interventions. In N. K. Denzin & M. D. Giardina (Eds.), *Qualitative inquiry and the politics of evidence* (pp. 221–249). Left Coast Press.

Saliya, C. A. (2012). The oldest profession. *Accounting Auditing and Accountability, 26*(6), 1072. http://www.emeraldinsight.com/doi/pdfplus/10.1108/09513571211250279.

Richardson, L. (1992). The consequences of poetic representation: Writing the other, writing the self. In C. Ellis & M. G. Flaherty (Eds.), *Investigating subjectivity: Research on lived experience* (pp. 125–137). Sage.

Wormser, B., & Cappella, D. (2000). *Teaching the art of poetry: The moves.* Erlbaum.

Chapter 23
Mixed Methods Strategy

Mixed methods research includes the mixing of qualitative and quantitative data, methods, methodologies, and/or paradigms in a research study. These approaches to professional and academic research emphasize that monomethod research can be improved through the use of multiple data sources, methods, research methodologies, perspectives, standpoints, and paradigms.

Qualitative and quantitative schools dominate the world of research because there is a sharp division between philosophical worldviews. The mixed-method strategy provides a bridge between these two schools. A mixed methodology is created by incorporating elements of another strategy into one strategy. Previous studies have shown that qualitatively driven or qualitatively dominant mixed methodological studies can better grasp the complexity of major social problems.

The modern origin of methodological combination is commonly dated to Campbell's 'multi-trait, multi-method matrix in psychology (Campbell and Fiske, 1959), which rendered the concept in highly formal terms (Fielding et al., 2008).

Meanwhile, Yin (2006) suggests that this combination does not necessarily have to be quantitative and qualitative, but can be a combination of several quantitative methods as well as a number of qualitative methods. Furthermore, according to Blaikie (1991), the use of different combinations of methods can give different answers to the same question, thus hampering clear conclusions. But the same reason seems to provide evidence to justify worldviews such as multiple truth ontology and subjectivist epistemology, which are flexible features of qualitative strategy. Thus it appears that the mixed methodological strategy is far from the positivist features and is much closer to the paradigms of reformist and critical thinking, and is therefore more closely related to the qualitative school.

23.1 Triangulation

Commonly used triangulation is a tactic in which one person does not communicate directly with another person, but instead builds a triangular connection through a third person to communicate to the other. It also involves the idea of controlling the relationship between two parties by dividing two people. Triangulation can be seen as a ploy to create rivalries between two people, to divide, to win, or to direct one against the other. The use of this tactic to verify the validity or reliability of research data is known as triangulation in the research context. That is, information obtained from one source can be verified independently using other sources. A parallel process has also been proposed in the mixed-method approach.

Approaches following Campbell's inspiration were based on 'triangulation', an objective aiming to test and prove relationships. The goal was causal explanation with predictive adequacy and mechanism was 'convergent validation'.

23.2 Mixed Methods Research Design

According to the hybrid methods proposed by various scholars (Ex; Creswell & Plano Clark, 2007; Gray, 2018; Yin, 2003) and different strategies suggested, there are four types of mixed methods designs.

Convergent

The quantitative and qualitative methods of any research that are applied concurrently and the results are interpreted as complementary convergent to draw the results separately and come to conclusions.

Example 23.1: Stock Market Development and Nexus of Market Liquidity, *International Journal of Finance and Economics*

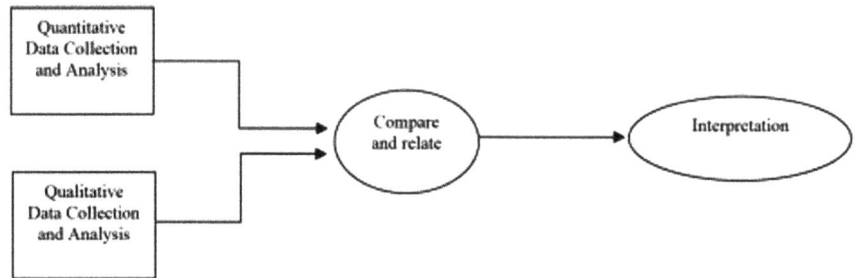

Concurrent Contradictory Interpret the various results obtained separately, perhaps paradoxically, to arrive at conclusions using methods such as deduction.

Sequential

Follow 2–3 methods sequentially. (Ivankova et al., 2006)

a. Here, qualitative methods are used to construct hypotheses, and quantitative statistical techniques are used to test theories that lead to deductions from hypotheses.
b. Final theories using qualitative methods using quantitative data and results Building theories using induction methods.

Complex

It is also possible to combine these two–three methods to develop more complex research design plans. The two examples presented here follow a mixed-method strategy. They are **Example** 10.1 (Stock Market Development) and **Example** 12.1 (Financial Battle against Climate Change).

23.3 Other Approaches

There are many other approaches suggested by various scholars (for example).

23.4 Summary

- Mixed methods research methodology can be improved by combining qualitative and quantitative methods and minimizing the disadvantages of being confined to a single methodology by mixing multiple data sources, methods, different worldviews, paradigms and theories.
- Triangulation is the independent verification of information obtained from one source using other sources to verify the validity or reliability of research data.
- Different strategies can be devised for converging, concurrent, sequential and complex mixed methodologies.
- Mixed strategy seems to be far removed from positivist features and belongs to the paradigms of reformist and critical thinking.

References

Blaikie, N. W. H. (1991). A critique of the use of triangulation in social research. *Qual Quant 25*, 115–136. https://doi.org/10.1007/BF00145701.
Campbell, D. T., & Fiske, D. W. (1959). Convergent and discriminant validation by the multitrait multimethod matrix. *Psychological Bulletin, 56*(2), 81–105. https://doi.org/10.1037/h0046016.
Creswell, J. W., & Plano Clark, V. L. (2007). *Designing and conducting mixed methods research.* Thousand Oaks, CA: Sage.

Fielding, N., Lee, R. M., & Blank, G. (Eds.) (2008). *The SAGE Handbook of Online Research Methods*. London: SAGE.

Gray, E. D. (2018). *Doing Research in the Real World* (4th ed.). London, SAGE.

Ivankova, N., Creswell, J. W., & Stick, S. L. (2006). Using mixed-methods sequential explanatory design: From theory to practice. *Field Methods, 18*(3). https://doi.org/10.1177/1525822X0528 2260.

Yin, R. K. (2003). *Case study research: Design and methods* (3rd ed. Vol. 5). London, UK: SAGE.

Yin. (2006). Mixed methods research: Are the methods genuinely integrated or merely parallel? *Research in the Schools, 13*(1), 41–47.

Part V
Methods 2: The Revelation

Chapter 24
Analysis

24.1 Introduction

Qualitative research is primarily exploratory research and used to gain an understanding of underlying reasons, opinions, and motivations. It provides insights into the problem or helps to develop ideas. Quantitative research is used to quantify the problem. It is used to quantify attitudes, opinions, behaviors, and other defined variables—and generalize results from a larger sample population. Data analysis methods for quantitative and qualitative data follow distinct strategies. They are known as statistical analysis and thematic analysis O'Leary (2004).

Quantitative data and results can be analysed using statistical techniques, whereas qualitative data analysis requires use of thematic or narrative analysis. We have discussed many quantitative statistical analysing techniques under quantitative strategies discussed in Part III. In this chapter we discuss broader approaches to data analysis with interpretations.

Basically, data analysis approaches can be classified into several types.

- content analysis
- quantitative data analysis
- thematic/interpretive analysis
- typologies and taxonomies.

24.2 Content Analysis

Fact analysis is the classical procedure for analysing text/numbers from data of any origin. Fact analysis presents an empirical methodology that aims at a systematic, interdisciplinary transparent description of significant and formal features of

C. A. Saliya, *Doing Social Research and Publishing Results*,
https://doi.org/10.1007/978-981-19-3780-4_24

information. This method is based on the use of classifications derived from theoretical models. It aims to categorize content into a systematic classification system by separating expressions, sentences, words or numbers.

24.2.1   Quantitative Content Analysis

Quantitative analysis captures important individual elements by classifying numerical and textual components. By analysing the frequency distributions statistical techniques reveal the characteristics of texts or populations under study. Information can be presented in tables or in charts. Examples: the number of times certain concepts are mentioned in a text, or how many responses there are to each preference in a survey, and so on. Example 12.1 (FBACC), Figs. 12.2 and 12.3.

The problem with quantitative analysis is that words or paragraphs can be misinterpreted out of context by isolating words or passages.

24.2.2   Qualitative Content Analysis

In this, the information related to the research question is first defined. It then analyses the data collection status, how it was generated, who was involved, who was present at the interview, and where the documents came from. In the next step, the research question is further defined on the basis of theories. This process involves three techniques:

- **Summary of the factual analysis**: That is, first, without considering the less relevant paragraphs that define the same, then group the similarly defined facts together.
- **Analysis of explicative facts**: That is, by adding contextual facts to the analysis as opposed to summarizing it, it diffuses and explains vague or contradictory passages. Glossary-based or grammatical definitions are used. Additional information, such as author details, is also available outside the text.
- **Framing the analysed facts into structures**: That is, seeks out different types of structures with a view to finding specific topics or areas relevant to the text by looking at the types or formal structures of the content; For example, 'panic' statements in interviews are always associated with issues of violence and crime.

By systematically expanding the procedure, this content analysis method is more transparent, less confusing, and easier to handle than other qualitative analysis methods. This method is primarily suitable for reducing large amounts of information and analysing their surface. However, this process is time consuming and can lead to distortion of meaning.

Table 24.1 Data entry-Excel worksheet (response feedback): There were three answer choices used; Yes, No and Neutral. Substitutes 100 for 'Yes', 0 for 'No' and 50 for 'Neutral'

	A	B	C	D	E	F	G	H	I	J	K	L	M	N	O
1	Q.1	Q.2	Q.3	Q.4	Q.5	Q.6	Q.7	Q.8	Q.9	Q.10	Q.11	Q.12	Q.13	Q.14	Q.15
2	100	100	0	0	0	0	50	50	100	100	0	0	0	0	0
3	100	100	0	0	0	0	0	50	0	0	50	50	0	0	0
4	0	100	0	0	0	0	50	50	0	100	0	0	50	0	0
5	100	100	50	50	100	50	50	50	50	50	50	100	50	50	50
6	100	100	100	50	50	50	50	50	50	100	0	100	50	0	50
7	100	100	50	0	0	50	0	0	0	100	0	0	0	0	0
8	100	100	0	0	0	50	0	0	0	50	0	0	0	0	0
9	100	100	0	0	0	50	50	50	0	50	50	0	50	100	0
10	100	100	50	0	0	50	100	100	100	100	50	50	50	50	0
11	100	100	0	50	50	0	100	100	0	100	0	100	100	0	0

SPSS DATA SET   Model   TOTAL ORIGINAL   Reliability Test   Frequencies

Average: 42.48985301   Count: 9159

24.3 Quantitative Data Analysis

In order to perform the statistical analysis discussed in Sect. 24.3, data must be prepared. The steps are as follows.

24.3.1 Constructing Data Matrix

This means compiling all the response variables for each study unit, more precisely for each question. Example 12.1: The set of questions and responses used in FBACC (Table 24.1. Data entry-Excel worksheet) and theorem building (Table 24.2. Data matrix) are as follows.

24.3.2 Cleaning the Data

Is data entered in the wrong column? It is necessary to make sure that the data is entered correctly such as whether data is missing or lost. For example, the sum of the three columns in Table 10.3 should be equal to the total number of participants 212. (Q.1, 142 + 45 + 25 = 212; Q.2, 177 + 15 + 20 = should be the same 212, etc.).

Table 24.2 Data matrix The summary of the 'Yes' count, 'No' count and the 'Neutral' count in the above survey is tabulated

Item/Questionnaire				
16 Variables logically arranged in different dymensions				
Particulars		Yes	No	Neutral
1. Are Fiji political initiatives actively working on its battle against climate change (the BACC)?	Political	142	45	25
2. Does the Fiji government acknowledges the importance of climate risk for society and business?		177	15	20
3. Are you aware of any promotions of funds other than Green bond for the BACC?		52	50	110
4. Are you aware of any special lines of credit to promote climate change mitigating projects?		62	25	125
5. Do the government agents i.e. RBF, ministries, etc provide proper active/ robust guidance to financial institutions to face the challenges of climate change?	Administrative	62	50	100
6. Do the financial institutions (FIs) have dedicated teams to implement the BACC guidelines and procedures?		39	98	75
7. Do financial institutions reward (incentives/concessions) their officers and clients who promote Climate change mitigation and adaptation action plans in their business?		54	70	88

24.3.3   Statistical Analysis

Data analysis can be performed using software such as SPSS, AMOS, Eviews and Mplus by applying the following different statistical techniques discussed in Part Two.

Univariate analyses: Frequencies, central tendency and dispersion etc.

Analyses referring to two variables: Correlations, comparison of groups etc.

Analyses with more than two variables: Multivariate analyses.

Testing associations, differences and changes.

Table 24.3 Numerical interrelationships

Variable type	Central tendency	Dispersion	Statistical test
Nominal	Mode	Frequency distribution	Chi-square test
Ordinal	Median	Range	Mann–Whitney U-test
Interval	Mean	Standard deviation	t-tests

24.3.4   Quantitative Interpretative Analysis

Interpretations also apply in quantitative analysis. The interrelationships that can be found through calculations (correlations) are defined by looking at significant explanations for these numerical relationships. Table 24.3 provides examples of this.

24.4   Quantitative Results Analysis

Quantitative data must be analysed once the results have been tested using statistical techniques. The **Analysis** 24.1 is an analysis of the results of the SEM (Structural Equation Modelling) test presented in Example 24.1.

Example 24.1 Determinants of financial-risk preparedness for climate change.

Analysis 24.1 SEM results analysis

The first-order CFA model with four latent factors representing political, administrative, standards and supervision dimensions is presented in Fig. 24.1. The responses to Q1, Q2, Q3 & Q4 were used as indicators of Political leadership. The responses to Q5, Q6, Q7 & Q8 were used as indicators of Administration while the responses to Q9, Q10, Q11 & Q.12, and Q13, Q14, Q15 & Q16 were used as indicators of Standards and Supervision dimensions respectively. The second-order factor structure with the overall preparedness factor is presented in Fig. 24.2.

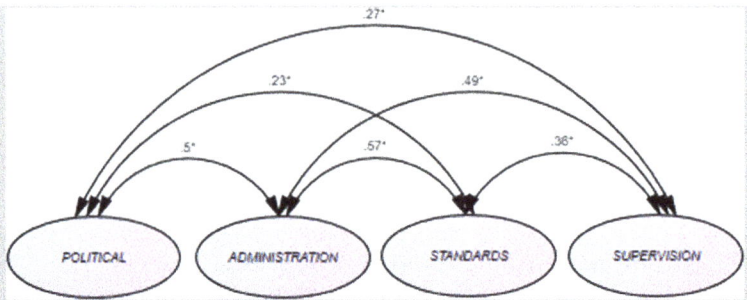

Fig. 24.1 First-order Factor Structure reflecting four Dimensions of Preparedness (measurement error correlations are not shown)

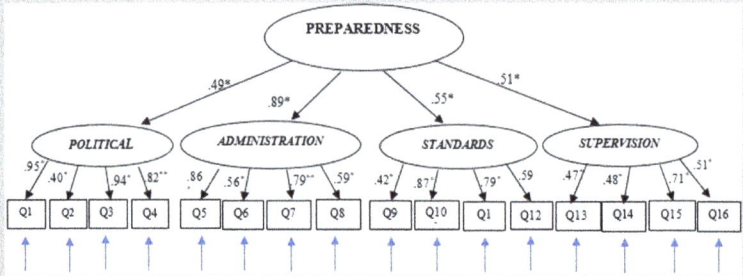

Fig. 24.2 Latent Second-order Factor Structure reflecting different Dimensions of Vigilance (measurement error correlations are not shown)

This CFA model reflected an acceptable model fit (chi-square/df = 2.12, CFI = 0.92, RMSEA-0.07, CI:0.052,0.083). Overall, the CFA model provided evidence for a hypothesized first-order factor structure. Accordingly, political, administrative, standards, and supervision are four distinct dimensions of risk preparedness.

All 16 items showed significant substantial factor loadings to respective dimensional factors ranging from 0.42 to 0.95 (p < 0.01) showing the reliability and validity of these items in relation to the respective factor (Bollen, 1989). There were no significant cross-factor loadings. Four dimensional factors showed significant and substantial loadings (0.49, 0.55, 0.51 and 0.89 to political, standard, supervision and administration, respectively) to the second-order overall latent factor of preparedness. Measurement errors of observed responses are shown in the figure (significant error correlations were freed to be correlated, not shown in the figure). This CFA model reflected an acceptable model fit (chi-square/d.f. = 2.10, CFI = 0.91, RMSEA-0.07). Overall, the CFA model provided evidence for a hypothesized factor structure. Accordingly, political, administrative, standards, and supervision are four distinct dimensions of risk preparedness. Also, there exists an overall higher-order factor of preparedness along with four first-order factors reflecting four dimensions.

For further analysis, based on CFA results, we computed composite measures for each factor by adding constituent items. For example the composite measure for Political Leadership was created by adding Q1, Q2, Q3 and Q4. These composite measures of four factors were used to produce statistics in Annexure 2. The descriptive statistics and paired t-test results for composite measures for four factors are also presented in Annexure 2. The paired t-test compared the average levels of factor scores.

24.5 Thematic/Interpretive Analysis

Thematic/interpretive analysis refers to the analysis of words, concepts, literary devices, and/or nonverbal cues.

Thematic analysis; 'narrative analysis' (Riessman, 1993) or narrative mode of knowing also referred to as the paradigmatic mode of knowing (Bruner, 1986) has become prominent because; first, as Llewellyn (1999) claims 'narrating is a mode of thinking and persuading that is as legitimate as calculating' (p. 220); second, as Czarniawska points out "the narrative mode of knowing consists in organizing experience with the help of a scheme assuming the intentionality of human action" and "'narrative' in Latin probably comes from gnarus ('knowing')" (p. 7).

Thematic analysis or 'narrative analysis' or narrative cognition mode is also known as ideal cognitive mode. They include discourse analysis, narrative analysis, conversation analysis, semiotics and techniques such as hermeneutics and grounded theory.

24.5.1 Discourse Analysis

The interpretative repertoires are analyzed. Repertoires are not spontaneous, but that people apply certain ways of talking about an issue. After selecting texts and talk accruing in natural context, which have to be described first by reading the transcript carefully. The analysis focused on the context, variability and constructions in the text, finally, on the interpretative repertoires used in the texts. The last step in this analysis is writing back to the empirical material (Flick, 2011, p. 160).

Dialogue or discourse analysis is an analysis of a wealth of meaningful knowledge. Knowledge is not spontaneous, but people use certain methods to talk about a problem. After selecting the text or paragraph and discussing it in its natural context, the script should be carefully read and explained first. The context of text or paragraph analysis focuses on constructing different genres and theories, and finally, presents theories/ideas based on the body of interpretive knowledge used in the text/paragraphs or knowledge. The final step in this analysis is to rewrite empirical information.

24.5.2   Narrative Analysis

To start with first, all non-narrative passages will be eliminated. Then followed by structural description of the content specifying the different parts of the narratives. The third part is moves away from the specifics towards intentions. Then integrate with non-narrative parts. Finally produce compared and contrasted arguments (Flick, 2011).

24.5.3   Conversation Analysis

Less interested in interpreting the content of texts than in analyzing the formal procedures with which people communicate and with which specific situations are produced. The procedure involves the following steps: Identify a certain statement or series of statements in transcripts as a potential element of order in the respective genre of conversation. Second, assemble a collection of cases in which this element of order can be found. Third, specify how this element is used as a mean for producing order in interactions and for which problem in the organization of interactions it is the answer (Bergmann, 2004).

24.5.4   Semiotics

The study of signs and symbols as a significant part of communication is one of the traditional aspects of scientific research. Unlike linguistics, the study of signs and symbols has also studied non-linguistic expressions and systems. Signs and symbols include processes, indications, designations, likeness, analogy, allegory, metonymy, metaphor, and symbolism. The study of the significance or significance and the study of communication.

24.5.5 Objective Hermeneutics

Hermeneutics is derived from the Greek word 'hermeneut' meaning 'translation' or 'interpretation'. This approach emphasizes the fundamental difference between the subjective meaning of a statement or activity for one or more participants and its objective meaning. Objective meaning is understood in terms of latent structure meaning. Semantics can only be tested using the framework of a multi-step scientific interpretation. Objectivist analysis is 'strictly sequential'.

Example 7.2 "Role of Bank Lending:" **Analysis** 24.2, 24.3, 24.5 and 24.6: **Thematic analysis**.

Analysis 24.2 Thematic analysis—Tony's case: Analytical framework.

This chapter analyses the case study stories and present additional empirical data on possible influencing factors on the behaviours of lenders, borrowers and credit mechanisms. These analyses provide the foundation for theoretical discussion of research findings in the next chapter based on Marxian critical interpretations. These analyses are developed around four basic aspects of bank-lending processes described in the case-study stories:

- The approach and method used by the borrower for credit applications.
- The approach and method used by the lender to accommodate or reject credit applications.
- The influencing factors of decision-making.
- The decision-making.

This discussion is staged case-by-case for all three cases. A combined analysis is also presented for the first two medium-sized credit applications as they share common features, and therefore, repetitions can be avoided. This analysis provides the necessary foundation for the theorization of the research outcome presented in the next chapter.

Discussion and analysis of Case Study I.

As explained in the data analysis section in the Research Method chapter, the analysis is developed through a series of questions and logical presentation of probable explanations. The common questions are: Why did the borrower approach the Chairperson directly? Why did the bank accommodate the borrower informally? How was the negotiation carried out? And what are the influencing factors and the decision?

Analysis 24.3 Thematic analysis—Tony's case: Why direct and informal?

Why did Tony approach the Chairperson of the Soft Bank directly?

For credit applications, the usual practice in Sri Lanka is that, if the customer knows an officer in a higher rank, he/she always gets that officer to introduce himself/herself to the credit officer. As discussed in Chap. 5 in detail, in Sri Lanka, the perception is that, if one doesn't know somebody in the institution concerned, the expected service is either delayed or never made available. On the other hand, Tony was a powerful businessman and it was not that awkward to contact another businessman for a business deal. Tony had the access to the Chairperson of the Soft Bank, because they were in the same social network and had been known to each other for some time. Furthermore, Tony Group had tried the formal path but had been rejected by the credit officers of the Soft Bank. Therefore, Tony had contacted the Chairperson directly for a credit facility, which can be considered as an informal personal approach.

Why did the Soft Bank accommodate Tony informally?

To uncover "Why did the Chairperson decide to lend to the Tony Group", following possible arguments are considered.

(a) Was the decision made because of the prevailing excess liquidity position causing a negative spread to the bank?
(b) Was the decision made for business development purpose because of the misleading influence of Tony and his accountants?
(c) Was it for personal gratification?
(d) Was it on sympathetic/patriotic/nationalistic grounds and/or based on social responsibility?
(e) Was it because of a personal relationship?
(f) Was the decision influenced by the ego of the Chairperson?
(g) Did the organization structure force the chairperson to make such decisions?

These possible arguments are discussed and analyzed in detail below.

Analysis 24.5 Thematic analysis—Tony's case: Why informally?

Was the decision made because of the prevailing excess liquidity position causing a negative spread to the bank?

The bank was desperate to shift their funds from low interest bearing gilt-edged Government Securities to commercial lending for higher return in June 1996. But the bank lost their additional liquidity position later, because of the strategic decision made to cut the interest rates offered for customer deposits.

The Soft Bank was comfortably canvassing deposits at lower cost in the level playing field, because the competitors also had reduced the interest rates by September 1996. The Chairperson was fully aware of the situation because, the CFO was reporting the position daily. Therefore the Chairperson's decision to extend the credit facility could not be persuasively related to the excess liquidity position or the prevailing negative return position.

Was the decision made for business development purposes because of the misleading picture shown by Tony with his accounting projections?

In previous instances the Chairperson referred such requests to the Management of the Bank for due recommendation after proper credit evaluation.

The Chairperson would have thought of getting all the banking business of Tony by bailing him out in this crucial situation. He would have been carried away by the positive cash flows projected by Tony's accountants; US$1 million per month which is over and above the required commitment of US$0.7 million to the banks. The service cost of the Soft Bank credit was around US$0.07 per month, the annual effective rate of interest (AER) charged from Tony Group was in the range of 17 to 22% as they had not been sufficiently backed by collateral. But why he did not do that previous occasions for other applications?

The Chairman was well aware of the already committed liabilities of Tony to other banks. The single-borrower limit would not allow the Soft Bank to acquire the banking business of the Tony Group with other banks by settling Tony's debts to other banks because such an exposure would be too much for a bank like the Soft Bank where the Capital Adequacy was only just at the stipulated level. Capital adequacy of a bank is calculated by weighing the assets (including loans) according to the risk of realization at a point of bankruptcy. Loans are weighed according to their collateral. The Basel Agreement, which was reached by the leading banks in the world in Basel, Switzerland, is to weigh the loans, backed by commercial properties by 50% assuming that they are at risk of price fluctuations. The Chairperson was of the view that developing countries never experience downward price fluctuations for properties. However, the volume of business that the Tony Group had offered to the Soft Bank was significant.

Was it for personal gratification?

The Chairperson refused the Mercedes Benz car offered by Tony. He had the control of over US$2 billion assets. The Chairperson rejected the offer of being the Chairperson of the Tony Group as well. Therefore, this decision could not be treated as one that was influenced by immediate personal gratification.

Was the decision made on sympathetic/patriotic/nationalistic grounds and/or based on social responsibility?

The Chairperson might have been sympathetic to the workforce of 30,000 of the Tony Group, and especially with the remarks made by the priests that "..the God has saved you to serve the people". There was not sufficient evidence to conclude that he made this decision purely on sympathetic grounds. Then, his move might have been influenced by patriotic attitudes, if we recall his criticism, quoted in the previous chapter, on banking practices and government policies towards SME development in Sri Lanka: On the other hand, the Chairperson, being the head of more than 100 companies, was fully aware of the grievances of the 9,000 small individual shareholders and 1.5 million deposit customers. But one of the colleagues argued saying that, "The fate of the 30,000 employees of the Tony Group was critical and warranted an immediate remedy, therefore, the Chairperson may have acted rationally to address first things first". However, this argument looks like eyewash to the public because the Chairman's justification to the board of directors of the Soft Bank for accommodating The Tony Group was "bringing more business to the Soft Bank". Therefore, this patriotic or social responsibility claim cannot be justified in a business sense. Also, accommodating a client who had been blacklisted in CIB is a clear indication of disregarding state intervention to protect public deposits from potential risk of default. Also it is an act of working in the opposite direction to maximizing shareholders' wealth which is the primary goal of a firm. On the other hand one could argue this act was a socially irresponsible decision with regard to public depositors, unethical in the interest of minority shareholders and totally unprofessional.

Was the decision made due to a personal relationship?

The Chairperson had great respect for Tony who is a successful premier businessperson in the country. On the other hand, most of the people who approached the Chairperson were privileged to obtain a favorable decision from him. Most cases so referred were defaulters to other banks and the Chairperson used to accommodate them with generous concessions. However, The Chairperson was very lenient in applying due diligence and acted significantly on "trust" and "gut-feeling". Therefore, this decision can be interpreted as an extending of a "favour" and getting another fellow capitalist, who is also a powerful individual (as an award-winning leading foreign exchange earner in Sri Lanka), hooked into his social network.

Was the decision influenced by motives such as ego or prestige of the Chairperson?

This could be a contributory factor since he had achieved the highest level of social status. The Chairperson was very keen on making popular decisions, followed by a good propaganda campaign. The Chairperson asked for publicity, which he got promptly from Tony.

Though the Chairperson's name Mr. Perera was a household name, he was known as only a businessperson. With the narrow escape from the bomb, now he may have thought of developing his image as a patriotic businessperson and win the hearts of the people in the latter stage of his life. He did not have any child to hand over his business and always used to say that his employees are his children. He started his own newspaper and a one-hour TV programme which showed negotiations with the clients/customers of his group. That propaganda was to convince the public that this was how generous he was in helping small entrepreneurs in Sri Lanka. All the clients who participated in that TV programme praised the Chairperson for the generous support he extended to their business, especially during the hard times.

The Chairperson was convinced that he could boost his image as a patriotic businessperson with the publicity promised. Therefore, it is more realistic to assert that he would have made this decision to fulfill his egoistic motives under the pretence of patriotism or social responsibility.

Did the organization structure force the Chairperson to make this decision?

The organizational structure was very conducive for him to make such decisions. It is evident that he had legitimate authority under a poor management support and weak organization structure. Also, he had no faith on the formal credit approval and monitoring systems. He always criticized the formal systems. He had very little confidence in the credit officers of the bank and appointed CFO, who was a non-banker, to monitor the Tony Group account. However, it was not expected of him to make such decisions at the highest level of authority and it was not the practice of the banking industry. Therefore, it is also believable that the organization structure was conducive to making such decisions especially as the decision-maker was vested with enormous power.

What were the influencing factors?

What could be inferred from the above analysis is that the arbitrary/informal decision was made under the enormous power entrusted to the Chairperson in a weak organizational structure with a poor management support. The decisions may have been influenced by patriotic attitudes and egoistic motives especially when the decision-maker is from the top slot of the bank. However, further analysis is necessary to discover the roots for such attitudes and/or to find more convincing reasons. This issue is further discussed in the next chapter: Theoretical Discussion.

How was the decision made?

Tony was very impressive and convincing on his marketing capabilities and "filling the factories with orders" for uninterrupted production if materials are supplied. The Chairperson made a quick assessment of Tony's integrity and the risk and return of the credit involved based on the cash flows presented by the team of accountants, including chartered accountants. He did not want to consult credit experts in the bank, as he had no faith or confidence in any of them. He had the power and he was confident that the Board would not decline what he recommended. Therefore, he approved this incomplete credit application and granted US$4 million over the table. **Narrative** 8.1 provides the summary of this case study.

Analysis 24.6 Thematic analysis—Silva's case.

The discussion and analysis of this case is developed in a similar manner as the previous two cases but poses different questions according to the circumstances. Initially a banker (Manager–Dehiwela Branch, Mr. Fernando) was instrumental in building up Superclean. Then a few other bankers tried to bring him down as they had personal grudges. Again another middle-level officer informally helped Superclean to rebuild, ultimately another upper-middle level manager decided to recover the money due to the bank, and declined new facilities, pushing Mr. Silva to a desperate situation.

Why do so many contradictory views and judgments exist within the formal banking systems?

The story of this case study was analysed and discussed at length in the same way that Tony's case study analysis was done. Its summary is presented as **Narrative** 8.2 in the case study chapter of the above strategy section.

24.6 Coding

A code is a term used to describe a word or phrase that captures the main essence of one small dimension of data.

Coding schemes can be developed deductively or inductively, and typically the coding schemes develops and changes as it is used. Codes may be created deductively from just thinking about a topic, from theory, or from prior studies. But codes also arise inductively from reading and interpreting the qualitative data.

The coding system generally evolves and changes as the analysis progresses. Codes can be renamed, explained, grouped together, separated, and identified in a variety of ways. In a significant sense, building and refining an encoding system sheds more light on data interpretation. Also, the task of building, modifying, and refining a coding system is greatly facilitated by quality analysis software.

'First cycle' coding (Miles et al., 2014), 'initial coding' (Charmaz, 2006), and 'open coding' (Corbin and Strauss, 2008) refer to initial assigning of codes to data chunks. Then Second cycle is bundling/condensing/integrating/grouping and laying the first cycle codes into broader and more coherent categories and themes (Miles et al., 2014), identifying patterns, relationships, and explanations (in GT, at this stage more data collection is sought, called theoretical sampling). This is referred as 'pattern coding' (Saldina, 2013), 'focused coding' (Charmaz, 2006) and 'axial

coding' (Corbin and Strauss, 2008). At this stage, initial codes may be relabelled, rearranged or eliminated. Finally, themes are created to reflect two-five 'big' ideas that emerge from data (Aurini et al., 2016).

24.6.1 Thematic Coding

To keep the reference to interviewees, as a single case when using a coding procedure, the alternative is to use thematic coding. Unlike grounded theory coding here you will go more into the depth of the material by focusing on the single case (Flick, 2011).

This means using thematic coding as an alternative to using a coding procedure to focus on the interviewers. This is not like the coding of **grounded** theory, but a deepening of the phenomenon by focusing on one phenomenon.

24.6.2 Computer Software for Qualitative Data Analysis

There are many types of computer software for quality data analysis. Unlike software for statistical analysis such as SPSS, Mplus, AMOS, Eviews, the researcher's involvement in this software is enormous. These software is very useful for grounded theory.

With advances in technology, computer software programs have been developed to help with the task of qualitative data analysis. Increasingly, qualitative data—including audio and video recordings, photos, and documents—are now readily digitized and can be stored, organized, and analysed in electronic form. The software includes tools to segment, tag, and categorize the content of these various files so that they can be sorted and analysed. While software is particularly valuable for coding and content analysis, it can also be valuable purely for data management and organized note taking. Some of the more widely used qualitative analysis software programs are Nvivo, Atlasti, and Ethno graph. There are some freeware programs as well, including answer and EZ-Text as well as ELAN and Ethno 2 (Remler & Van Ryzin, 2011).

The following is a list of such software ratings. It represents that NVivo is the most popular software today.

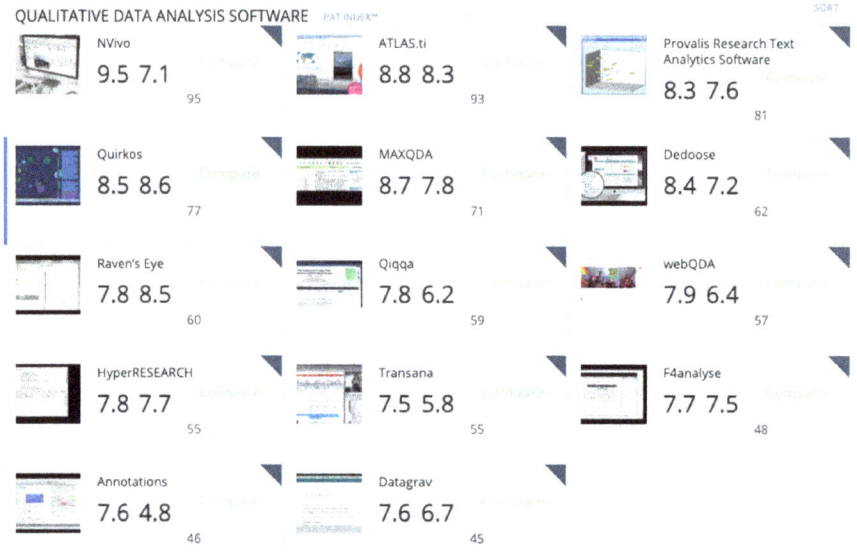

24.6.3   Nvivo

Nvivo is software that supports quality data coding. For researchers with little clue how to apply a qualitative analysis, it will not give magical results. However, it does provide a lot of facilities for managing, exploring and revealing data from questionnaires, interview scripts, focus groups and secondary data sources. It also helps to analyse audio and video recordings and digital photography. Like any software program, this type of software requires time and patience to learn. The following two resources are useful for learning NVivo.

- Bazeley and Jackson (2013), Qualitative Data Analysis with NVivo, Quality Data Analysis with NVivo: This book provides step-by-step instructions for using NVivo, as well as a set of datasets that show you how to handle and analyse tutorials.
- Silver and Lewins (2014), Using Software in Qualitative Research, Using software for quality research.

    The following is how to encode a video recorded interview using Nvivo.
    Using software for quality research.

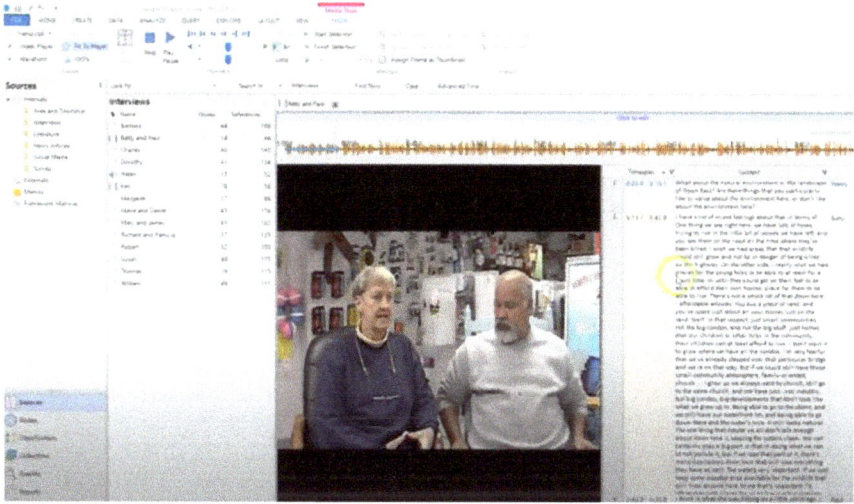

Source https://www.youtube.com/watch?v=0vxwiypXQSA

24.7 Grounded Theory

In developing the theory based on empirical data, a method of data analysis called coding is used. In the interpretation process, a number of coding procedures are used to work with text. These are called 'open encoding', 'axial encoding' and 'selective encoding'. In short, the first step in the development of grounded theory is open coding, that is, taking words/texts/paragraphs/textures and dividing them into separate categories. Axial coding is the building of connections between codes form dimensions. Selective coding involves connecting all the codes/categories/dimensions in the analysis and conceptualise a few elegant concepts (anti-positivist constructivism approach) or a core category (positivist classic grounded theory) to capture the essence of the research.

These procedures are not clearly identifiable procedures or sequential stages of a linear process. Instead, it is a process of manipulating text/theories that moves them back and forth and integrates them if necessary. Starting with data, the coding process leads to the development of theories through an abstract process.

The purpose of open coding is to express data and phenomena in conceptual form. For this, the data are first fragmented, and scattered. Interpretation-unit expressions (single words, short sentences) are classified for naming (labelling) their comments and concepts (codes). Possible sources for 'codes' in labelling are concepts derived from the sociology literature (generated code) or concepts derived from interviewer publications (in vivo codes). In open encoding, the categories (types) found in this

Analysis 24.7   Data analysis using the Grounded Theory coding strategies: Initial/Open Coding, Focus/axial and selective coding.

Open/initial coding		Axial/Focus coding creating categories based on patterns	Selective coding
Code 1: 30% is Sufficient	Code 59: Protection	**Opportunities**	**Frustration**
Code 2: 30% is a Challenge	Code 60: Traditions	Codes 1–7: %; Code 11:	**legislation**
Code 3: 30% is a Problem	Code 61: Men are leading	Law is necessary; Code	
Code 4: No %	Code 62: Changing	102: Lack of acceptance;	
Code 5: May be 30%	Code 63: Education	Code 103: Recognition;	
Code 6: 50% ideal	Code 64: Advantaged	Code 104: Lack of	
Code 7: 10% minimum	Code 65: Devotion	collaboration; Code 105:	
Code 8: Favouring males	Code 66: Career advancement	Accommodation; Code	
Code 9: Gender diversity	Code 67: Experience	106: Opportunities	
Code 10: Value addition	Code 68: Culture		
Code 11: Law is necessary	Code 69: Homemakers	**Protection**	
Code 12: Women are Juggling	Code 70: Religious beliefs	Code 12: Women are	
Code 13: Multiple roles	Code 71: Conservative attitudes	Juggling; Code 58: Women	
Code 14: Multi-task	Code 72: Think out of the box	are weaker; Code 59:	
Code 15: Multi-skilled	Code 73: Motherhood	Protection	
Code 16: Competitive	Code 74: Guilty minded	Code 76: Compromising;	
Code 17: Achievement	Code 75: Caring	Code 77: Vulnerable	
Code 18: Disadvantaged	Code 76: Compromising	Code 78: Lack of support;	
Code 19: Based on Merits	Code 77: Vulnerable	Code 111: Lack of	
Code 20: No idea	Code 78: Lack of support	Confidence; Code 41:	
Code 21: The best person	Code 79: Passion for work	Reluctant	
Code 22: Irrespective of Gender	Code 80: barriers		
Code 23: Female or male	Code 81: Glass ceiling		
Code 24: Right people	Code 82: Longer hours		
Code 25: Lack of Prominence	Code 83: Musculo-phallic		
Code 26: Gender balance	Code 84: Restrictions		
Code 27: Gender equality	Code 85: Interruptions		
Code 28: Corporate requirement	Code 86: Business needs		
Code 29: Religious principles	Code 87: Disciplined		
Code 30: Creativity	Code 88: Organized		
Code 31: Innovation	Code 89: Logical		
Code 32: Collaboration	Code 90: Diligence		
Code 33: Decision-making	Code 91: Constructive		
Code 34: Different Signals	Code 92: Dedication		
Code 35: Attractive	Code 93: Nurturing		
Code 36: Talent	Code 94: Role models		
Code 37: Intuitiveness	Code 95: Bringing up children		
Code 38: Competent	Code 96: Empowerment		
Code 39: Investors	Code 97: Family		
Code 40: Advertisement	Code 98: Expectations		
Code 41: Reluctant	Code 99: Empathizing		
Code 42: Opportunities	Code 100: Social norms		
Code 43: Discrimination	Code 101: Pre-conceived ideas		
Code 44: Practices	Code 102: Lack of acceptance		
Code 45: Male Dominance	Code 103: Recognition		
Code 46: Competence	Code 104: Sympathising		
Code 47: Perspectives	Code 105: Accommodation		
Code 48: Harassments	Code 106: Opportunities		
Code 49: Late hours	Code 107: Restrictions		
Code 50: Negative attitudes	Code 108: Outdated		
Code 51: Double standards	Code 109: God		
Code 52: Hypocrisy	Code 110: Tolerating		
Code 53: Rules	Code 111: Lack of Confidence		
Code 54: Enforcements	Code 112: Professionalism		
Code 55: patriarchal	Code 113: Performance		
Code 56: Think out of the box	Code 114: housewife		
Code 57: ego	Code 115: Musculo-centric		
Code 58: Women are weaker	Code 116: Musculo-phallic		

(continued)

Analysis 24.7 (continued)

Open/initial coding		Axial/Focus coding creating categories based on patterns	Selective coding
		Gender equality Code 9: Gender diversity; Code 26: Gender balance; Code 27: Gender equality; Code 34: Different Signals; Code 102: Lack of acceptance; Code 103: Recognition; Code 104: Lack of collaboration; Code 105: Accommodation; Code 106: Opportunities; Code 61: Men are leading; Code 62: Changing	**Gender balance** **self-effacing** **challenging**
		Restrictions Code 80: barriers; Code 81: Glass ceiling; Code 66: Career advancement; Code 18: Disadvantaged; Code 28: Corporate requirement; Code 85: Interruptions; Code 86: Business needs; Code 49: Late hours; Code 21: The best person; Code 22: Irrespective of Gender; Code 23: Female or male; Code 24: Right people; Code 67: Experience; Code 96: Empowerment	
		Multi-tasks Code 13: Multiple roles; Code 14: Multi-task; Code 15: Multi-skilled; Code 16: Competitive; Code 46: Competence; Code 47: Perspectives Code 112: Professionalism; Code 113: Performance; Code 88: Organized; Code 89: Logical	**Multi-talented** **fixing**
		Creativeness Code 35: Attractive; Code 36: Talent; Code 37: Intuitiveness; Code 38: Competent; Code 91: Constructive; Code 30: Creativity; Code 31: Innovation; Code 32: Collaboration; Code 33: Decision-making; Code 10: Value addition	

(continued)

Analysis 24.7 (continued)

Open/initial coding		Axial/Focus coding creating categories based on patterns	Selective coding
		Caring Code 93: Nurturing; Code 94: Role models; Code 95: Bringing up children; 99: Empathizing; Code 92: Dedication; Code 69: Homemakers; Code 104: Sympathising	**Caring empathising tolerating**
		Tolerating Code 97: Family; Code 98: Expectations; Code; Code 65: Devotion; Code 110: Tolerating; Code 74: Guilty minded; Code 92: Dedication	
		Male dominance cultural practices Code 50: Negative attitudes; Code 51: Double standards; Code 52: Hypocrisy; Code 53: Rules Code 54: Enforcements; Code 55: patriarchal Code 56: Think out of the box; Code 100: Social norms; Code 101: Pre-conceived ideas Code 57: ego; Code 71: Conservative attitudes; Code 115: Musculo-centric; Code 116: Musculo-phallic	**Discrimination believing**
		Beliefs Code 109: God; Code 101: Pre-conceived ideas; Code 102: Lack of acceptance; Code 103: Recognition; 74: Guilty minded; Code 70: Religious beliefs	

Source Saliya (2022)

Analysis 24.8 Categorizing and theorizing.

	New dimensions Eliminating categories		
Elegant core concepts	**Concept 1** Legislations for gender balance	**Concept 2** Multi-skilled tolerating mother	**Concept 3** Male dominating traditions and beliefs
Single Core Concept	**Enterprising mother**		

Source Saliya (2023b)

way are further grouped and the characteristics of each category are labelled and abstracted. You can specify codes to link line by line, sentence by sentence, paragraph by paragraph, or the entire structure.

For all coding strategies, it is suggested to address the text regularly with the following list of so-called basic questions: What id the issue here? Which phenomenon is mentioned? Who are involved? Which roles they play? How do they interact? Which aspects of the phenomenon are mentioned (or not mentioned)? When? How long? Where? How much? How strong? Which reasons are given or can be constructed? With what intention, to which purpose? By which means, tactics, and strategies for reaching the goal? Then, after a number of substantive categories have been identified, the next step is to refine and differentiate the categories that result from open coding (Strauss and Corbin, 1998).

Selective encoding focuses on building core concepts or variables. This leads to elaborate or processing of the story of the phenomenon. In this case, the two schools of grounded theory collide. According to Glaser's positivist school, the outcome at any given time should be only one central core concept, while the structuralize propositions of Strauss, Cobbin, and Charmers can be constructive. In any school, the process of interpreting data, such as integrating additional factors, ends when theoretical saturation is reached.

Analysis 24.7 and 24.8 of **Example** 20.1**: Grounded theory of mother enterprising.**

24.8   Summary

- Quantitative data can be analysed using statistical techniques and qualitative data may require the use of thematic or interpretive analysis.
- Basically data analysis approaches can be categorized into several types; Content analysis; Quantitative data analysis; Thematic/interpretive analysis; the use of typography and taxonomies.
- Thematic/interpretive analysis refers to the analysis of words, concepts, literary devices, and/or nonverbal cues.
- Grounded theory uses a method of data analysis called coding. This method builds theories by classifying the data several times.
- Coding basically consists of three steps. They are called first or open encoding, axial or focal encoding and selective encoding. Finally, 'themes' are created to reflect two or three 'important' ideas that emerge from the data.
- There are many computer software for qualitative data analysis. Unlike software such as SPSS and Mplus for statistical analysis, in this case, the researcher's involvement is very large.

References/Further Reading

Aurini, D. J., Heath, M., & Howells, S. (2016). *The how to of Qualitative Research*. Sage.
Bazeley, P., & Jackson, K. (2013). Qualitative data analysis with nvivo. *Qualitative Research in Psychology, 12*(4), 492–494. https://doi.org/10.1080/14780887.2014.992750.
Bollen, K. A. (1989). *Structural equations with latent variables*. John Wiley & Sons. https://doi.org/10.1002/9781118619179.
Bruner, J. (1986). *Making stories: Law, literature, life*. Cambridge, MA: Harvard University Press.
Charmaz, K. (2006). *Constructing grounded theory: A practical guide through qualitative analysis*. Sage.
Corbin, J., & Strauss, A. (2008). *Basics of qualitative research: Techniques and procedures for developing grounded theory*. Thousand Oaks, CA: Sage.
Flick, U. (2011). *Designing qualitative research*. Thousand Oaks, CA: Sage.
Llewellyn, D. (1999). The Economic Rationale for Financial Regulation.
Miles, M., Hubernman, M. A., & Salidina, J. (2014). *Qualitative data analysis: A methods sourcebook*. Sage.
O'Leary, Z. (2014). *The essential guide to doing research*. London: Sage.
Remler, D. K., & Van Ryzin, G. G. (2011). *Research methods in practice: Strategies for description and causation*. Sage.
Riessman, C. K. (1993). *Narrative analysis*, vol. 30. London: Sage.
Saldina, J. (2013). *The coding manual for qualitative researchers*. Thousand oaks, CA: Sage.
Saliya, C. A. (2023b). Enterprising mother: a grounded theory emerged from Fijian working women. Equality Diversity and Inclusion, Under review.
Silver, C., & Lewins, A. (2014). *Using software in qualitative research*. London: Sage. https://doi.org/10.4135/9781473906907.
Strauss, A., & Corbin, J. (1998). Basics of qualitative research: grounded theory procedures and techniques. Beverly Hills, CA: Sage Publications.

Youtube Links

Nvivo video recorded interview: https://www.youtube.com/watch?v=0vxwiypXQSA.

Chapter 25
Typologies and Graphics

What happens in this method is that the whole thing rises into a picture. It helps the researcher to present the idea clearly and to make it clearer to the reader. This is a very old methodology. Such offerings include the swastika of ancient Indian religions, the Dharma Chakra, which condenses the teachings of the Buddha, and the cross of Christianity.

The symbol facing the right (卐) is called the swastika and symbolizes the sun, prosperity and good fortune, while the symbol facing the left (卍) is called the swastika, symbolizing the elements of Kali or night.

https://m.facebook.com/path.nirvana/posts/1317101618357157:0?locale2=ar_AR

The following are some of the most widely used such presentations in recent times in the academic world.

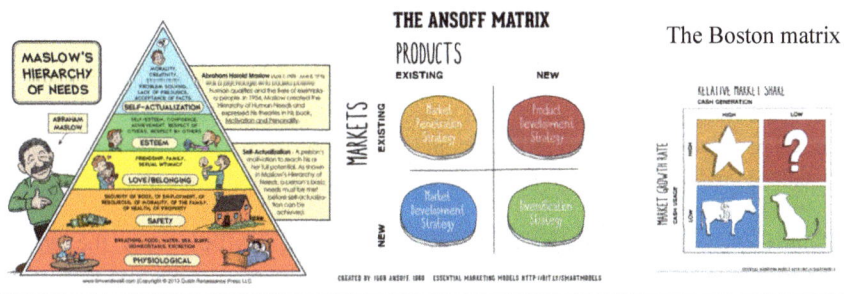

(continued)

C. A. Saliya, *Doing Social Research and Publishing Results*,
https://doi.org/10.1007/978-981-19-3780-4_25

(continued)

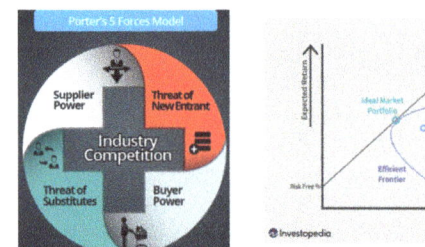 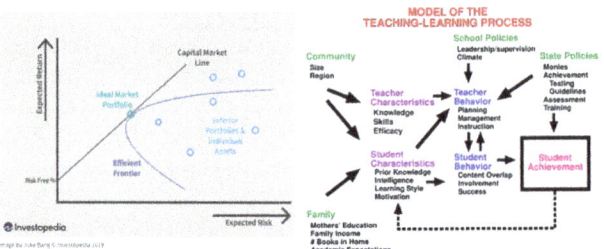

Sources Investopedia, 2019; https://www.danmartell.com/hierarchy-of-entrepreneurial-needs/; www.timevandevall.com; Dutch Renaissance Press LLC, 2013; https://www.smartinsights.com/; https://visual.ly/

Typologies are constructed in both quantitative and qualitative research. Some suggestions for how to apply this, are:

- *First, compare systematically the issues in hand referring to the data collected.*
- *Second, define relevant dimensions of comparison.*
- *Third, group the issues according to substantial dimensions and to analyse empirical regularities.*
- *Fourth, construction of a typology to analyse substantive meanings.*
- *Finally, characterize the construction types, exploring which features or combinations of features characterize the issues that have been allocated to the various types: what do they have in common, what distinguishes the issues in the different types? (Flick, 2011).*

The naming of the Boston Matrix as Dog, Cow, Star and '?' is also an attractive presentation, as characterized by these classifications. Let us try to study this procedure by example.

25.1   Four Credit Decision-Makers

Factors influencing the lending decisions discussed in Example 7.2 and how those factors change from time to time were categorized into different dimensions and a research paper was published as presented in Example 26.5. The first is a taxonomy and the second is a construction of a typology and nomenclature according to philosophy.

Figure 25.1 shows an improved variant based on Barry's Taxonomy, the factors influencing those credit decisions. Using this classification, a replication was made of the factors influencing the lending decisions discussed in Example 7.2 and how

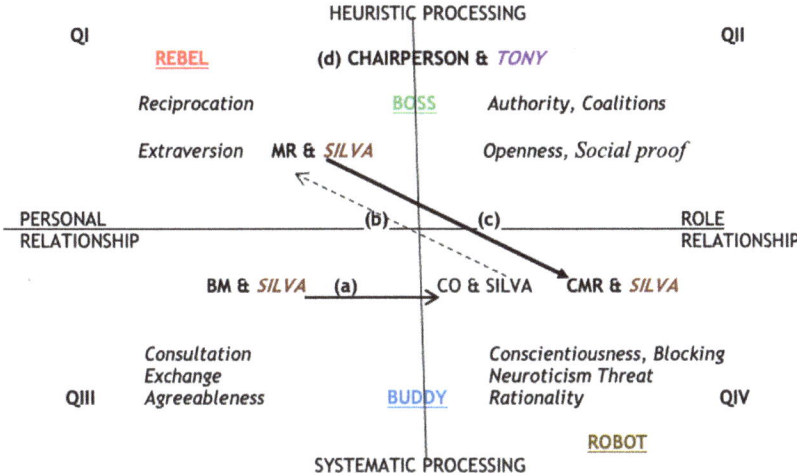

HEURISTIC PROCESSING

QI — REBEL — (d) CHAIRPERSON & *TONY* — QII

Reciprocation — BOSS — Authority, Coalitions

Extraversion — MR & *SILVA* — Openness, Social proof

PERSONAL RELATIONSHIP — (b) — (c) — ROLE RELATIONSHIP

BM & *SILVA* (a) — CO & SILVA — CMR & *SILVA*

Consultation / Exchange — Conscientiousness, Blocking / Neuroticism Threat

QIII — Agreeableness — BUDDY — Rationality — QIV

ROBOT

SYSTEMATIC PROCESSING

(a) Breaking the relationship with the branch staff after Dehiwela branch manager (BM) was promoted and transferred to head office. Credit officer (CO) started handling Silva's account.

(b) Develop a temporary relationship with Manager-Recoveries (MR) which did not last long and MR lost the job.

(c) Settle down in the due quadrant due to strict application of rules by CMR when the applicant is from the powerless-class.

(d) Powerful Chairperson was handling powerful Tony

Fig. 25.1 Barry's Taxonomy (replication). *Source* Saliya (2019b)

those factors change from time to time. (Tony and Silva in Example 7.2, as David and Sena in Example 26.5).

25.1.1 Typology

Typologies are 'organised systems of type's that enables us to better comprehend, understand and explain complex social realities (Kluge, 2000).

Box 25.1 Shows this typology development process

From the research findings, a taxonomy and a typological matrix are presented to encompass the types of processes of decision-making, the types of relationships, rationality of decisions and the personality traits using six major

Table 25.1 Six dynamics of credit decision-making

Dimensions	Dynamics
Process of decision-making	Systematic/Formal methods—policy-based approach
	Heuristic methods—the rule-of-thumb approach
Types of relationships	Personal relationship
	Role relationship
Justification of credit	Rational—sensible
	Irrational—situational

dynamics under three dimensions namely; (a) the procedures followed for credit evaluation (Systematic/Formal—policy-based approach or Heuristic—the rule-of-thumb approach); (b) the type of relationship between credit seekers and credit decision-makers (personal or role relationship), and (c) the justification of credit (Rational/sensible or Irrational/Situational). These findings provide guidance to broaden understandings on how certain credit decisions are made and why credit managers make such decisions. The Table 25.1 shows these six dynamics of credit decision-making.

Building a model and naming according to this philosophical theory, the examples are as shown in Examples 8.2 and 26.5: Box 18.1 and Box 25.2 its description is given in Table 25.1 and Fig. 25.2, Table 25.2.

Explicit dimensions / Implicit dimensions	Women's' managerial positions should be secured by law	Women are offered limited opportunities
Women are multi-talented	Hopeful-enterprising	Hapless mothers
Women behave differently due to traditional and religious conventions	Nice to have-passive	

Fig. 25.2 Taxonomy of typology: Theorization of women's expectations and destiny in Fiji. (Reproduced)

Table 25.2 Taxonomy of typology (reproduced)

Variable		3. Justification/logic		
		Rational	Irrational	
1. Evaluation procedure	Formal	BUDDY	ROBOT	2. Nature of relationship!
	Heuristic	REBEL	BOSS	

Source Saliya (2019b)

Thus, under four dimensions (two-dimensional and two-dimensional), a triad of women is defined as theoretical proposals regarding the conditions they face in relation to their professional development and their fate in society. Figure 25.2: An explanation of the taxonomy of typology is shown in Box 25.2.

Box 25.2 Theorization of women's expectations and destiny in Fiji

Hopeful-enterprising women: These women implicitly show that they have been offered less opportunities and working hard to get the positions they deserve. They do post graduate studies and devote more for work related activities.

Nice to have-passive women: These women are confined to the jobs they are doing and not much interested in advancing up in their careers, and do not have any intention of pursuing postgraduate studies.

Hapless mother: Hapless mothers show more attachment on home affairs and tend to prioritize mother's role compared to office tasks.

25.2   The Nexus of Liquidity

A graphic representation of the nexus of stock market liquidity is presented in Example 10.1 Research Paper as shown in Fig. 25.3.

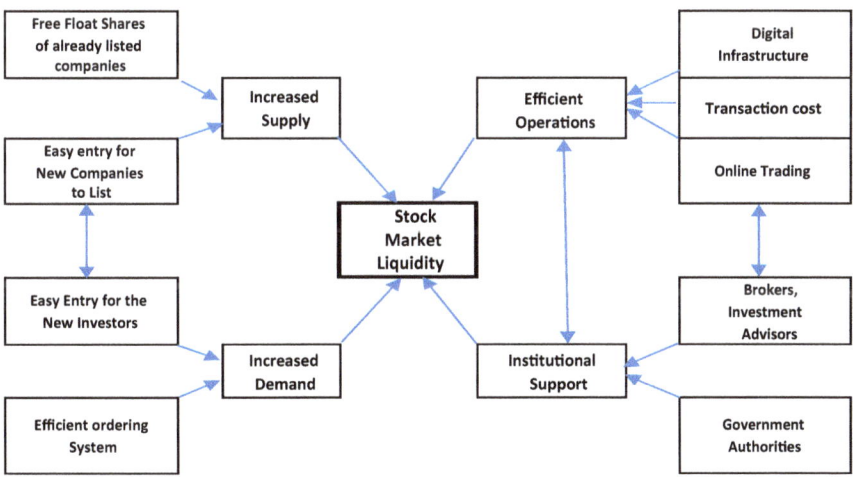

Fig. 25.3 The nexus of stock market liquidity. *Source* Saliya (2020)

The graphic 'The nexus of stock market liquidity' illustrates that stock market liquidity depends primarily on four variables. These four variables are (a) the demand for the shares, (b) the supply of shares for sale by the shareholders in the market, (c) the efficiency of the stock market activity, and (d) the support of the institutions that regulate the stock market. Nine factors affecting these four variables have also been identified and presented separately. There are two factors for each of the three variables (a), (b) and (d) and three factors for the variable (c). This opens up opportunities for future research, such as the replication of the Example 10.1 research paper (replication) for the Colombo Stock Exchange or for market comparisons.

25.3 Women's Expectations and Destiny

In Table 25.3 presented in Chap. 24 (Analysis), the example presented under the grounded theory is to analyse the data in the Example 25.1: Grounded Theory of Enterprising mother, summarize its thematic themes, and formulate a theory through visual studies. (Resubmitted below).

25.3.1 Graphical Presentation

Example 25.1 Grounded theory of enterprising mother data encoding, and classification is presented in Fig. 25.4: Data encoding and classification as shown in coding and classification of data. It presents a graphical representation of the 'coding and conceptualization stages' of data analysis to generate a theory called the grounded theory of enterprising other, the core concept of the final conceptualization stage.

Table 25.3 Data analysis using the Grounded Theory coding strategies: Emergent theories (Reproduced)

	New dimensions eliminating categories		
Elegant core concepts	**Concept 1** Legislations for gender balance	**Concept 2** Multi-skilled tolerating mother	**Concept 3** Male dominating traditions and beliefs
Single core concept	**Enterprising mother**		

Source Saliya (2023b)

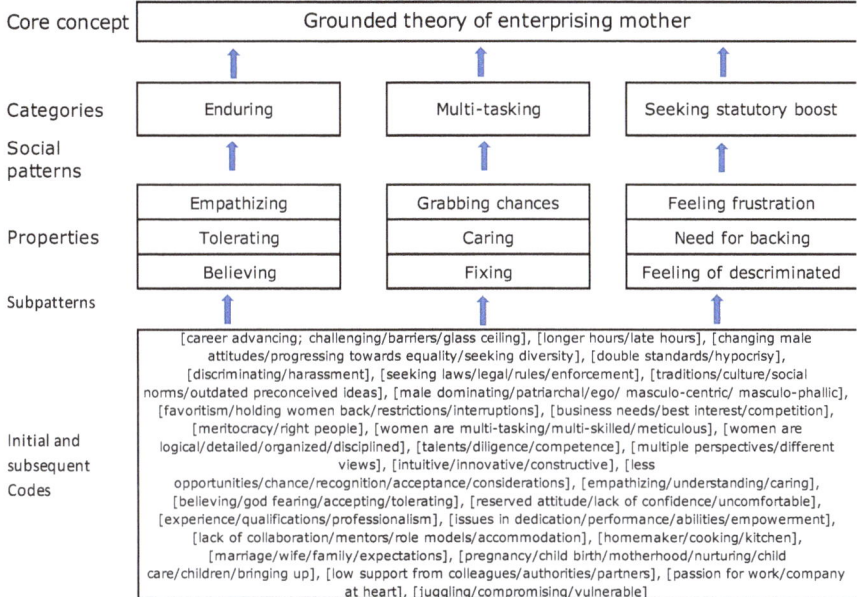

Fig. 25.4 Data analysis using Grounded Theory coding strategies and classification of data. *Source* Saliya (2023b)

25.4 Poverty and Certain Financing Decisions

Example 25.2 CREDIT CAPITAL, EMPLOYMENT AND POVERTY: A Sri Lankan Case Study.

"How can poverty and certain financing decisions be mutually reinforcing when they are interconnected?" The research problem is to find and show the interrelationships between the two variables, which are described in Box 25.3 and graphically presented in Fig. 25.5.

Box 25.3 The research problem is the interrelationship between two variables

The research question was sharpened to investigate how and why a certain powerful group in a society could have more access for finance than that of the powerless group. Similarly, how and why the financing decision-makers facilitate credit applicants from the powerful group and deny the credit applications from disadvantaged group of the poor class. Then, the researcher seeks to.

Explore what theoretical perspectives could be applied to usefully explain the rationale behind this social mechanism and, to provide evidences for existence of the mutual relationship between such financing methods and the

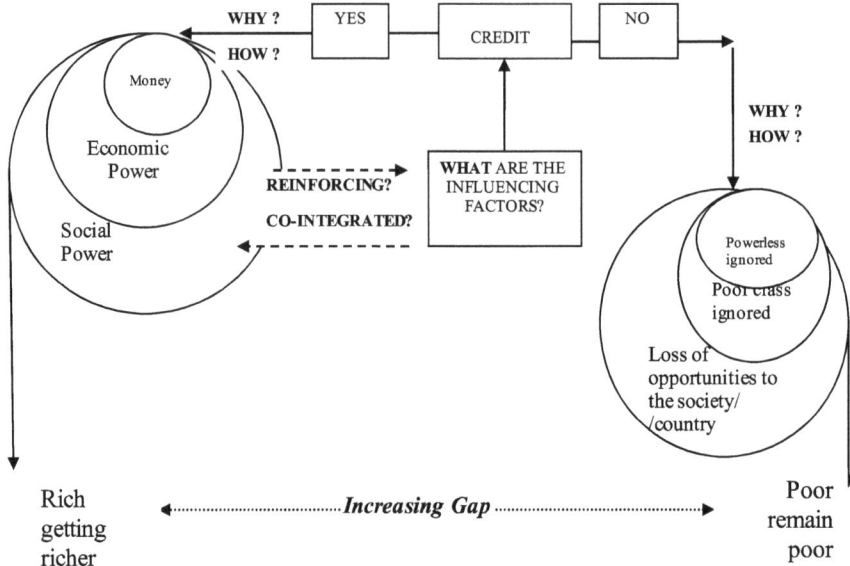

Fig. 25.5 Illustrated Integrated Research Questions and Proposition. *Source* Saliya (2019a)

poverty of the society at large. Mutual relationships between variables could generally create reinforcing-mechanism which could finally be transformed into a vicious cycle. These types of variables, which are complementary to each other's survival and enhancement, could be interpreted as a strongly co-integrated function.

Therefore, all individual, social-cultural and economic-political factors are collectively directed towards protecting and strengthening the social power of an affluent class of the capitalist or feudal society (Lapavistas, 2003). The economic power afforded by money eventually leads to social power and in turn, social power becomes the fundamental driving force of arbitrary/informal decision making in the banking/finance industry in Sri Lanka (Saliya & Yahan-path, 2016). Then, because the powerless is ignored, poor class is neglected. Therefore, opportunities are lost to the society/country as a whole, so the poor remains poor. On the other hand, because powerful class acquires more power through such informal decisions, it can be concluded that, poverty in the society and preferential lending processes are mutually reinforcing. And therefore, socially powerful rich class get richer and richer while power-less poor class remains stagnant. This is illustrated in Fig. 25.5.

25.4.1 Arbitrary Decision-Making Model

The combined impact of personal, institutional, political and socio-environmental factors (DM = decision maker) on informal decision making in the formal banking system of Sri Lanka is described in Box 25.4: Fig. 25.6. Arbitrary Decision-Making Model. The research problem is the interrelationship between two variables.

Box 25.4 Arbitrary Decision-Making explanation

Arbitrary/Informal financing methods, which is currently in operation within the formal banking system in Sri Lanka, mostly driven by social power with or without legitimate authority vested with the decision maker. In addition, personal characteristics and traits such as patriotism, ego, prestige and aspirations of the decision maker are also playing some roles in making decisions but they do not seem important. Lack of support from the government in developing the SME sector seems to be having significant impact on the formal banking systems pushing decision makers for informal decisions with regards

---

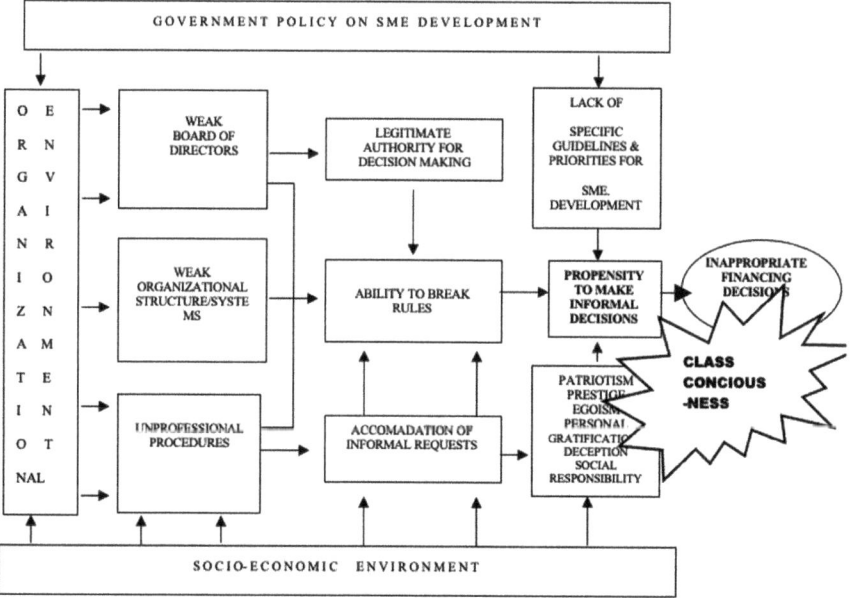

Fig. 25.6 Arbitrary Decision-Making Model: Combined Influence of Individual, Organizational, Political and Social Environmental Factors on Informal Decision Making in Formal Banking System in Sri Lanka (DM = Decision Maker). *Source* Saliya (2010)

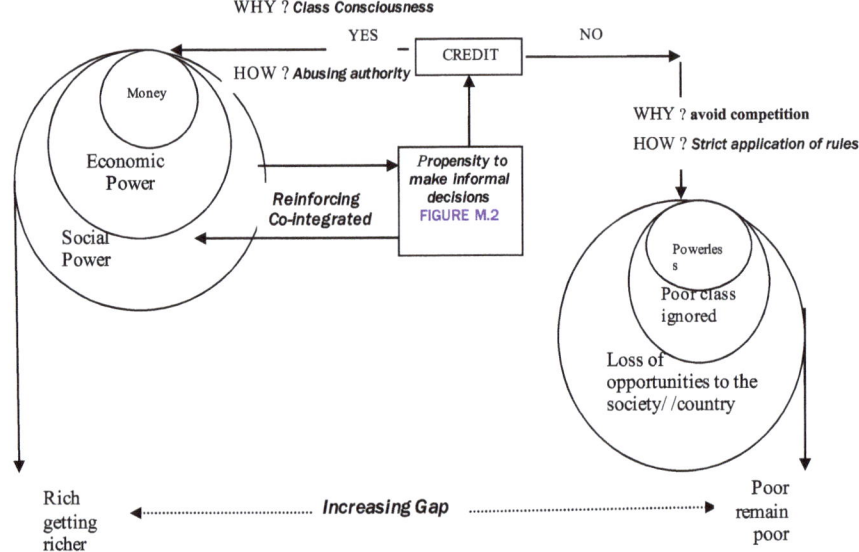

Fig. 25.7 Integrated research problem and proposition with research conclusions. *Source* Saliya (2019a)

to financing micro and small enterprises. These forces collectively influence informal decision making process especially when the organizational structure is weak. These factors are summarized and presented in a model illustrated in the Fig. 25.7: Arbitrary Credit Decision-Making Model below.

25.4.2   *Research Problem and Conclusions*

The following is a graphical presentation of the proposed explanations for this research problem in Fig. 25.7. Loan approval for its powerful applicants are based on Marxist concept of bourgeois class consciousness in this context: capitalist class consciousness abuses authority suggests that.

25.4.3 Discriminatory Credit Decision Model

Example 25.3 Conducting Case Study Research: A Concise Practical Guidance for Management Students.

> **Box 25.5 Discriminatory credit decision model explanation**
> In general, apart from the legitimate authority and the formal credit evaluating factors (such as risks, cash flows, feasibility and historical factors) the following factors have also been identified as critical in approving credit; the weight of socio-economic power of the parties involved, the strength of the credit

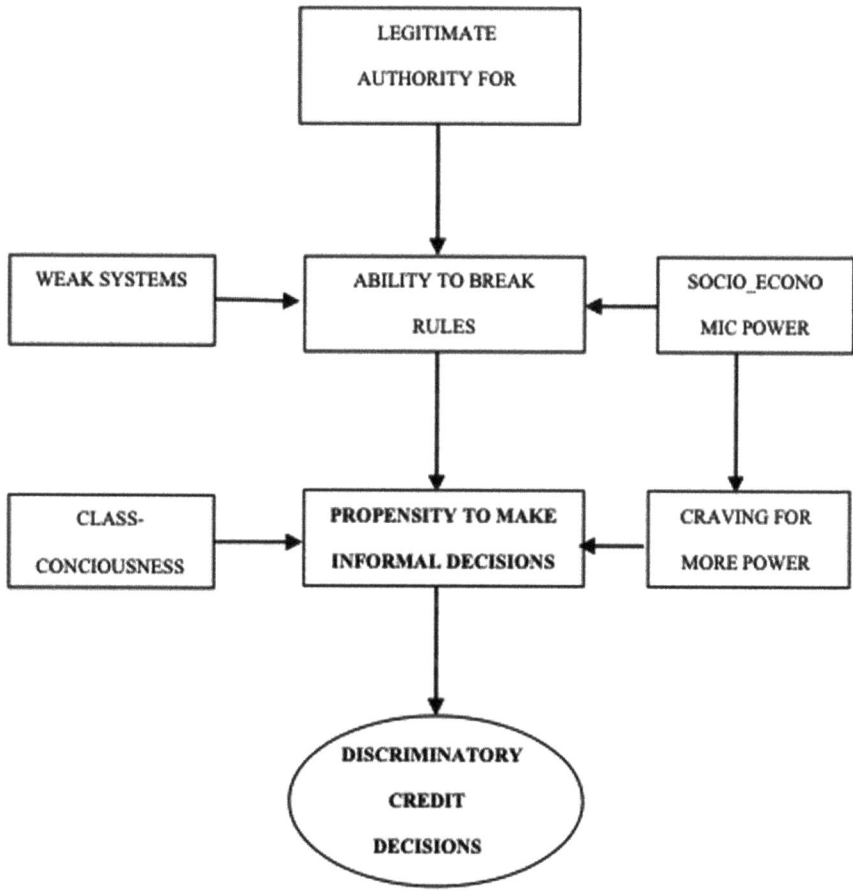

Fig. 25.8 Discriminatory credit decision-making model. *Source* Saliya (2019a)

policies & procedures and the vigour of the class-consciousness of the social classes. These factors are summarised and presented as a model illustrated in the Fig. 25.4: Discriminatory credit decision-making model below (Fig. 25.8).

25.5   Balanced Scorecard—A Graphical Presentation

Example 25.4 Financial Battle Against Climate Change—FBACC

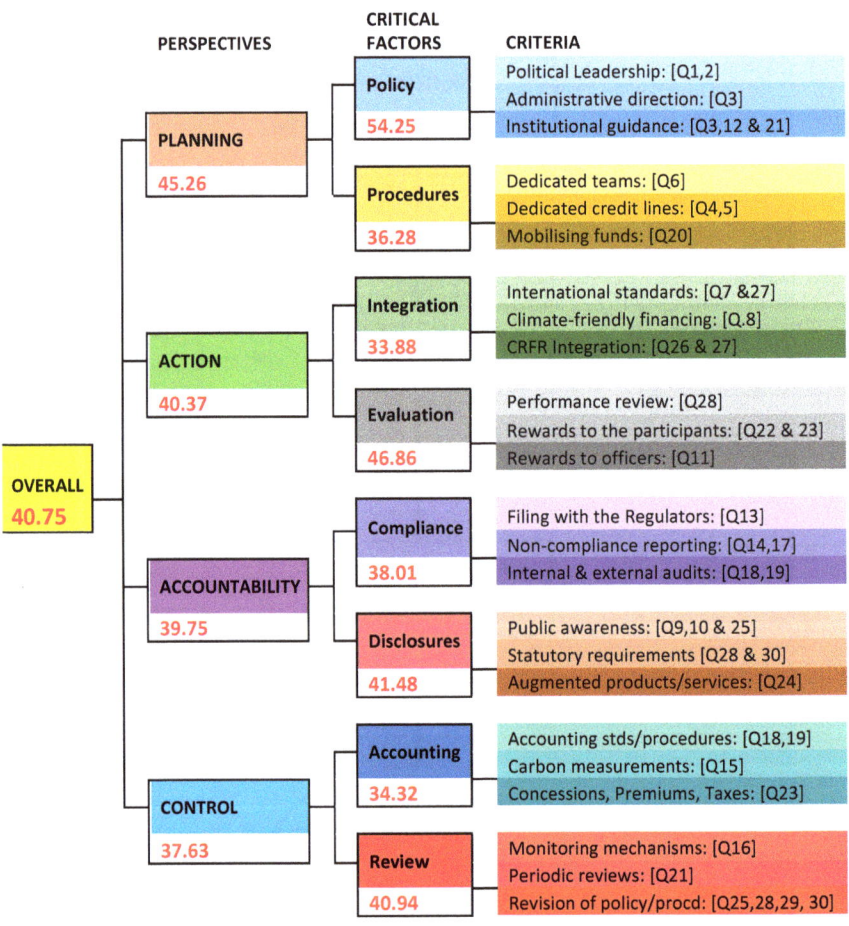

> **Box 25.6 Balanced Scorecard explained**
> The overall score of 40.75% is below the theoretical benchmark of 50%, and just above the unsatisfactory level of 40%. The Perspectives of Accountability and Control scored only 39.75% and 37.63% respectively, below the unsatisfactory level. Only the Critical Factor Policy scored an Average level with 54.25% of the score due to an excellent score of 82.3% for Political Leadership criteria. All other criteria, critical factors scored below 50% while many of them scored below 40% at unsatisfactory level, especially the Critical Factor Integration of existing standards and requirements into the Fiji financial regulatory system reporting the lowest score of 33.88%.

25.6 Methodological Graphical Presentations

25.6.1 Research Lenses on the Study Approach

See Fig. 25.9.

25.6.2 Constructivist Grounded Theory Process

See Fig. 25.10.

25.7 Summary

- Typological studies, classifications, and graphic building are found in both quantitative and qualitative research strategies.
- Typology is the 'organized category/category system' that enables us to better understand, comprehend and explain complex social realities.

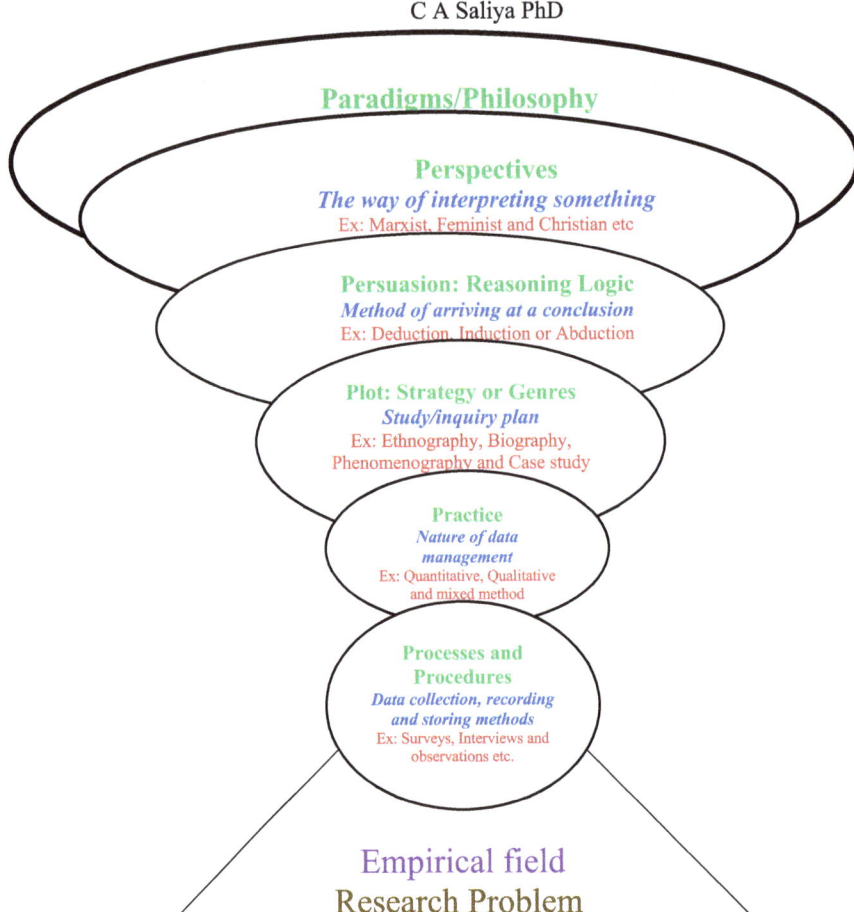

The Seven Phases (7Ps) of Academic Research Design
C A Saliya PhD

Paradigms/Philosophy

Perspectives
The way of interpreting something
Ex: Marxist, Feminist and Christian etc

Persuasion: Reasoning Logic
Method of arriving at a conclusion
Ex: Deduction, Induction or Abduction

Plot: Strategy or Genres
Study/inquiry plan
Ex: Ethnography, Biography,
Phenomenography and Case study

Practice
*Nature of data
management*
Ex: Quantitative, Qualitative
and mixed method

Processes and
Procedures
*Data collection, recording
and storing methods*
Ex: Surveys, Interviews and
observations etc.

Empirical field
Research Problem
and objectives

Fig. 25.9 Lenses of the research and application of case study approach. *Source* Saliya (2010)

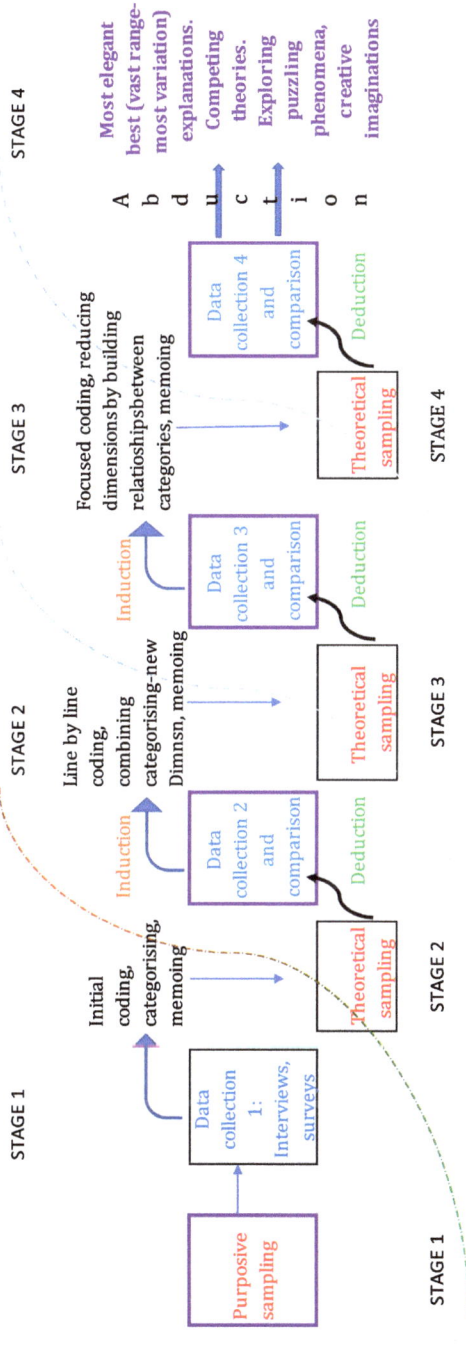

Fig. 25.10 The Process Path analysis of Grounded Theory (Reproduced)

References

Flick, U. (2011). Introducing Research Methodology: A Beginner's Guide to Doing a Research Project. Los Angeles: Sage.

Kluge, S. (2000). Empirically grounded construction of types and typologies in qualitative social research [14 paragraphs]. *Forum qualitative sozialforschung/forum: Qualitative Social Research*, (Vol. 1, No. 1), Art. 14, http://nbn-resolving.de/urn:nbn:de:0114-fqs0001145.

Lapavitsas, C. (2003). *Social Foundations of Markets, Money and Credit*. London: Routledge.

Saliya, C. A. (2010). *The Role of bank lending in sustaining income/wealth inequality in Sri Lanka*. PhD Thesis, Auckland University of Technology, New Zealand. https://openrepository.aut.ac.nz/handle/10292/824.

Saliya, C. A. (2019a). Credit capital, employment and poverty. *International Journal of Money, Banking and Finance, 8*(1), 4–12. https://search.proquest.com/docview/2249669069?pqorigsite=gscholar.

Saliya, C. A. (2019b). Dynamics of credit decision-making: a taxonomy and a typological matrix. *Review of Behavioral Finance*. https://doi.org/10.1108/RBF-07-2019-0092.

Saliya, C. A. (2021a). Conducting Case Study Research: A Concise Practical Guidance for Management Students. *International Journal of KIU, 2*(1), 1–13. https://doi.org/10.37966/ijkiu2021 0127.

Saliya, C. A. (2023b). Enterprising mothers in Fiji need comprehensive support. The Qualitative Report, Under review.

Saliya, C. A., & Yahanpath, N. (2016). Petty-Bourgeois Nationalism – A case study. *Qualitative Research in Financial Markets, 8*(4), 359–368.

Chapter 26
Discussion and Conclusions

The end of a research paper is to interpret the results and come to conclusions. Let us consider what are the essential elements that should be included in the discussions and conclusions for this. This chapter then discusses how the paragraphs should be organised to illustrate the various stages of the research process in the above chapters are constructed and the conclusions drawn.

After presenting the results or findings, the researcher then interprets the meaning of the results. This often comes in the discussion section of a report. Basically, interpretation of results involves moving away from detailed results and taking into account research issues, questions in a study, existing literature, and perhaps personal experience, and their greater meaning.

For quantitative research, this means comparing the results with the initial research questions asked to determine how the question or hypotheses were answered in the study. It also means comparing the results with prior predictions or explanations drawn past research studies or theories, which provide explanations for the researcher has found. In qualitative research, the interpretations provide similar explanations about the results but with a few differences. The qualitative researcher needs to address how the research questions were answered by the qualitative findings. Also, comparisons can also be made of the findings with past research studies in the literature. But in addition to the approaches, qualitative researchers also may bring in their personal experiences and draw personal assessments of the meanings of the findings. This last feature sets qualitative research apart from quantitative approaches, and it reflects the role of the qualitative researcher, who believes that research (and its interpretations) can never be separated from the researcher's personal views and characterizations (Creswell and Clark, 2011, pp. 209–210).

26.1   Discussion Section

The purpose of the discussion is to interpret and describe the significance of your findings in light of what was already known about the research problem being investigated, and to explain any new understanding or fresh insights about the problem after you've taken the findings into consideration. The discussion will always connect to the introduction by way of the research questions or hypotheses you posed and the literature you reviewed, but it does not simply repeat or rearrange the introduction; the discussion should always explain how your study has moved the reader's understanding of the research problem forward from where you left them at the end of the introduction (Sacred Heart University, 2020).

- This 'discussion' section is often considered the most important part of a research paper as the ability to think critically about a problem as a researcher, to come up with creative solutions to research problems based on findings, and to gain a more in-depth understanding of the research problem being studied. Because it shows.
- 'Discussion' is the exploration of the implications of your research, the possible implications for other areas of study, and the improvements that can be made to further develop your research concerns.
- This is the section where you need to present the importance of your study and how you can contribute and/or fill in the gaps in the field. If appropriate, these discussion paragraphs are also the place where you point out how the findings of your study have revealed new gaps in literature that have not previously been exposed or adequately described.
- This section of the paper is not strictly controlled by objective reporting, but instead can be applied to creative thinking on issues through the interpretation of evidence-based findings. This is where your results fill in the sense.

Unless the interpretation is to be a one-off and wholly personal exercise, you will have to engage in a more general consideration of the relevance and usefulness of your work. Such a consideration will bring you into touch with four related concepts: significance, generalizability, reliability and validity. All researchers need to review and defend their own work in this light (Blaxter et al., 2008).

26.1.1   Significance

The concept of significance has both a specific, statistical meaning and a more general, common-sense interpretation. In statistical parlance, it refers to the likelihood that a result derived from a sample could have been found by chance. The more significant a result, the more likely that it represents something genuine. In more general terms, significance has to do with how important a particular finding is judged to be.

26.1.2   Generalizability

The concept of generalizability, or representativeness, has particular relevance to small-scale research. It relates to whether your findings are likely to have broader applicability beyond the focus of your study. Thus, if you have carried out a detailed study of a specific institution, group or even individual, are your findings of any relevance beyond that institution, group or individual? Do they have anything to say about the behaviour or experience of other institutions, groups or individuals, and, if so, how do you know that this is the case?

26.1.3   Reliability

The concept of reliability has to do with how you well you have carried out your research project. Have you carried it out in such a way that, if another research were to look into the same questions in the same settings, they would come up with essentially the same results (though not necessarily an identical interpretation). If so, then your work might be judged reliable.

26.1.4   Validity

Validity has to do with whether your methods, approaches and techniques actually relate to, or measure, the issues you have been exploring.

It is essential to have an accurate understanding of some of the theories and arguments read in English in order to assimilate the research results with existing knowledge. There are many facilities for this on the internet. Google Translator is one such tool.

26.2   Google Translations and Adaptation

Many languages have the ability to translate English text as well as other languages using the Google Translate tool. To do so, visit: https://translate.google.lk/?hl=en&sl=en&tl=si&vi=c. But these transformations cannot be relied upon for four reasons. First, the translated sentence can be obscure and noisy. Second, the most appropriate word (unambiguous) is often not included in the translation. Third, translated words can be misspelled and erroneous. Fourth, for the same reasons, the idea of a more original texture can be distorted or misrepresented.

26.3 Drawing Conclusions

The conclusion is intended to help the reader understand why your research should matter to them after they have finished reading the paper. A conclusion is not merely a summary of your points or a re-statement of your research problem but a synthesis of key points. For most essays, one well-developed paragraph is sufficient for a conclusion, although in some cases, a two-or-three paragraph conclusion may be required.

The following points should be considered in making conclusions. A well-written conclusion provides some important opportunities to show the reader your overall understanding of the research problem. The following points are contained in a successful conclusion paragraph.

- Make a final judgment on the issues you have raised. Just as the 'Introduction' section gives your reader a first impression, the conclusion also provides an opportunity to make a lasting impression. This is done by highlighting the key points of the analysis or findings.
- Summarize your thoughts and present important impressions of your study. Conclusion "So what?" Provide an adequate answer to the question posed by placing the study in the context of past research on the topic you are researching.
- Demonstrate the importance of your ideas. This conclusion provides an opportunity to explain the importance of your findings. It must be done without hesitation, courage and confidence.
- Introduce new or expandable ways to think about the research problem. The issues not discussed in the paper should not be re-introduced here. But here you can come up with new insights and creative approaches to formulate/contextualize the research problem based on the results of your study.

26.4 Quantitative Discussions and Conclusions

26.4.1 Discussion of a Descriptive Statistics

Example 26.1 Stock Market Development and Nexus of Liquidity.

Discussion 26.1: Descriptive statistics and normality tests
Means and standard deviations of study variables are presented in Table 26.1 Skewness of all the variables were below 1.5 (except for LQDT which had skewness of 3.043 and Kurtosis value of 9.948) providing evidence for the normal distribution. Kurtosis statistics shown for the variables INF, FDI and INV are also higher than the critical value of 3 (Westfall et al., 2015) and all

Table 26.1 Normality tests

Variable	Descriptive and normality statistics					
	Distribution			Normality		
	Mean	Std. dev	Skewness	Kurtosis	Jarque-Bera	*p*-value
SMD	1.097	3.703	1.429	5.432	14.088	0.001
MCAP	13.603	6.784	0.432	2.649	0.871	0.647
LQDT	0.266	0.393	2.616	9.948	75.638	0.000
INF	3.960	3.304	0.290	4.714	3.273	0.195
FDI	5.927	5.620	0.899	3.334	3.346	0.188
BNKD	61.623	22.405	−0.301	1.527	2.531	0.282
GDP	5.610	2.510	0.866	2.732	3.073	0.215
INV	19.740	3.226	1.309	4.238	8.385	0.015
SAV	0.112	0.089	0.373	1.952	1.655	0.437
COC	0.149	0.314	−0.041	2.184	0.673	0.714
GVTEF	−0.362	0.288	−0.514	2.221	1.664	0.435
PS_AV	0.391	0.394	−0.146	1.730	1.699	0.428
RGLQ	−0.366	0.260	0.001	2.667	0.111	0.946
ROL	−0.277	0.394	0.080	1.801	1.464	0.481
DMCY	−0.266	0.452	−0.762	2.031	3.258	0.196

other variables fall between 1.5 and 3. According to Westfall et al. (2015), kurtosis says very little about the peak or center of a distribution, its only unambiguous interpretation is in terms of tail extremity. The Jarque–Bera test is a goodness-of-fit test of departure from normality, based on both the skewness and kurtosis (Jarque, 2011). Higher *p*-values of Jarque–Bera statistics confirm 12 variables as not normally distributed.

26.4.2 Discussions of Survey Results

The following is an example of a detailed discussion of survey data without the use of tests and analyses by statistical methods.

Example 26.2 FBACC - (Saliya and Pandey, 2021)
Graphical presentation/a of Survey results.

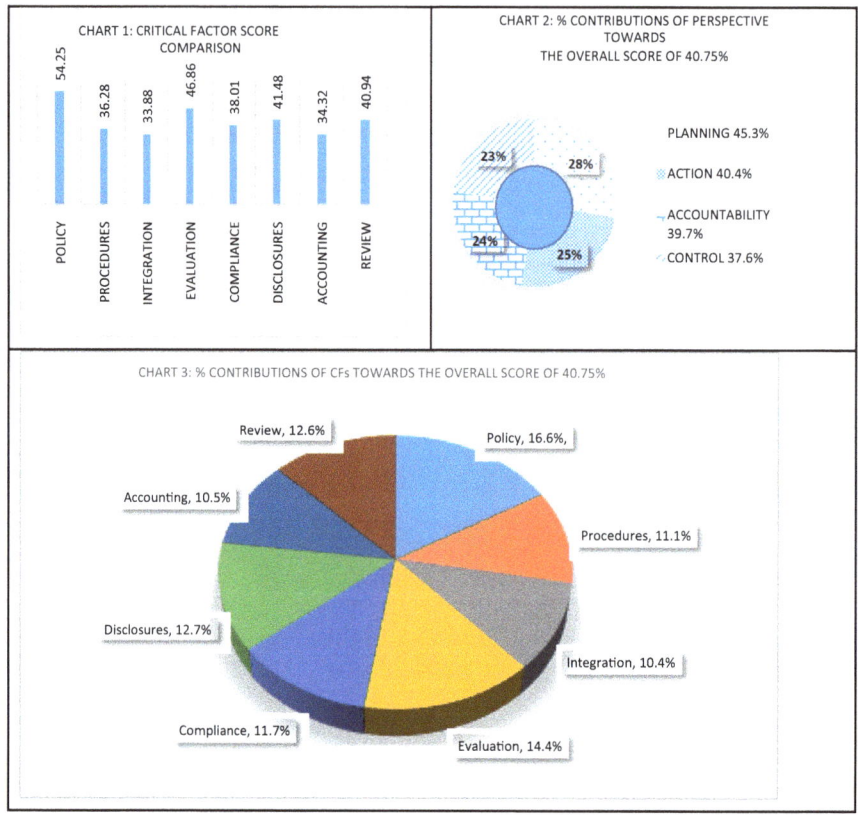

Fig. 26.1 Relative contributions and importance of CFs and Perspectives in the final score

Discussion 26.2 (FBACC)- Survey results (Saliya and Pandey, 2021)
Figure 26.1, which consists of three charts, shows the relative contributions of CFs and perspectives in constructing the overall score. Chart 1 portrays the relative importance of the CFs while Chart 3 graphically depicts the relative contributions of the CFs; Chart 2 shows the relative contributions of the Perspectives towards the overall assessment of FFBACC.

The analysis of survey data reveals several contradictions in terms of planning, implementation, and knowledge about the duties and responsibilities of administrative and regulatory authorities. With regard to the Political Leadership criteria within the Policy Critical Factor under the Planning Perspective, 177 respondents (85%) agreed that the Fiji government provides the political leadership to arrest CRFRs. This is the highest score recorded in the FBACC Scorecard with a score of 82.3%; however, 115 respondents (65%) disagree that these policy directions are not effectively implemented by the administrative agents by formulating necessary regulations, directives, and guidelines to the players in the Fijian financial system.

Another contradictory observation is that only 28.8% agree that allocation of FSGB proceeds is in compliance with its objectives, and there are 41% neutral responses while 31% disagree with the RBF's claim of expenses incurred on construction of school buildings and roads under the objective of "Resilience to Climate Change for Highly Vulnerable Areas and Sectors". Furthermore, only 14.2% agree with the external auditor's review which endorses the above allocation of proceeds as claimed by the RBF.

It is also revealed that the Public Awareness criterion scored 60% in terms of knowledge of the respondents, ranking next to the Political Leadership criterion, and none of the other criteria scored higher than 50% while the CRFR Integration criterion scored a mere 21%, and 12 criteria scoring above 40%. The mean values of the four Perspectives are 40.75% with a standard deviation of a mere 2.8 indicates that the four Perspectives are perceived by the respondents very closely in terms of their knowledge.

26.4.3 Discussion of a Simple Regression Analysis

Example 26.3 Stock Market Development and Nexus of Liquidity.

Discussion 26.3 Regression.

Table 26.2 shows the results of the Regression of change in Stock Market Development (SMD) on Lagged Macroeconomic Determinants. The results showed that lagged GDP has a significant positive influence on change in SMD ($\beta = 2.145$, $p < 0.001$) suggesting that for every one unit increase in GDP, there was a 2.145% increase in SMD growth in subsequent year. The influence of GDP showed the highest statistical significant ($p < 0.01$). Inflation showed a negative influence on change in SMD ($\beta = -0.574$, $p < 0.007$) suggesting that every one percent increase in inflation percentage has resulted in a fall in SMD growth in the subsequent year by 0.574% after taking the influences of GDP and domestic credit into account. This is consistent with the findings of previous studies that showed the inflation has a negative impact on SMD. Domestic credit to private sector (BNKD) also had a negative impact on subsequent SMD change in the ($\beta = -0.208$, $p < 0.008$) after taking the influences of GDP and inflation into account. Each one-unit increase in BNKD has resulted in

Table 26.2 Regression of change in SMD on lagged macroeconomic determinants using ARDL

Variable	Coefficient	Std. error	p-value
LQDT	−0.361	1.699	0.834
INF	−0.574***	0.182	**0.007**
FDI	0.094	0.171	0.591
BNKD	−0.207***	0.065	**0.007**
GDP	2.145***	5.100	**0.001**
INV	−0.246	0.188	0.213
SAV	6.496	9.948	0.525
C	8.385	5.034	0.119

R-squared 0.690455
Adjusted R-squared 0.499966
Note $*p < 0.05$, $**p < 0.01$, $***p < 0.01$, Unstandardized coefficients

a fall in 0.208% SMD growth in the subsequent year. This is contrary to the findings of previous studies that showed positive correlation. The value stocks trade, foreign direct investment, gross capital formation and gross savings did not show any significant impact on subsequent SMD change. Overall, it appears that these observed associations are specific to the Fuji socioeconomic context. All the variables explained 69% (R-squared) of variance in SMD change.

26.4.4 Conclusions

Example 26.4 Determinants of Financial-risk Preparedness for Climate Change.

Conclusions 26.1 Determinants of Financial-risk Preparedness for Climate Change.

This study had two main objectives. First to confirm the hypothesised theoretical framework with different dimensions of the overall financial system risk preparedness for climate change in the Fijian context, and to utilize this framework to assess the present risk preparedness in Fiji. In general, the results of the study supported the hypothesized framework for financial system risk preparedness for climate change in this context. Overall, the study provided useful findings about the varied severity of different dimensions, which may constitute important input for financial policy and programme planners.

This study revealed that four observed factors (political leadership, administrative direction, international standards and supervisory mechanisms) as distinct dimensions of risk preparedness. In addition, there exists a higher-level risk preparedness factor. These findings inform financial planners and policy makers and aid preparing for the financial risk caused by climate change. These four dimensions of risk preparedness and the overall risk preparedness can be tackled independently and also in an integrated manner in a multi-pronged integrated manner. We believe that this theoretical framework can also be used to assess financial systems in other developing countries with similar socioeconomic contexts.

The *administrative direction* and *supervision mechanism* play critical roles in designing, implementing and monitoring effective systems to ensure compliance with the standards. The regulatory returns have to be updated by incorporating appropriate benchmarks, which are suitable for underdeveloped systems in emerging economies such as Fiji's. The results also reveal that political initiatives would be futile without proper administrative direction and strong supervisory mechanisms.

Therefore, it is evident that the *political leadership* cannot be just confined to acknowledging the importance of preparing for preventing and/or mitigating the impact of climate change on the financial system. To complete the preparation process, all four dimensions have to be integrated and implemented smoothly especially because stricter standards and compliance can cause harm to small developing economies such as that of Fiji. On the other hand, if not prepared with appropriate standards and compliance procedures in a timely manner, the possible consequences could cause more painful shocks and damage to the financial system. These results confirmed that the existing political initiatives need to be effectively communicated and/or implemented in the financial system by the regulatory agencies.

The present study has used Confirmatory Factor Analysis (CFA) in Structural Equation Modeling (SEM) Framework to analyse the data. SEM allowed us to account for the measurement errors of the responses. Also, we have used several fit indices to evaluate the hypothesised model. This has enhanced the quality of estimated parameters and provided statistically more convincing results (Bollen, 1989). Further, second-order CFA confirm multi-level constructs (reflecting overall preparedness and five constituent dimensions) are provided that can be considered for policy and programme formulation.

The present study has several limitations. First, the sample size is relatively small; it would need to be larger to yield more statistical power. Second, respondents with more diverse backgrounds and from diverse geographical areas would have increased the generalisability of the study findings. Third, greater numbers of questionnaire items would have produced higher reliability of factors reflecting different dimensions. Future studies should test this theoretical framework with a larger and more diverse sample, and with a more comprehensive instrument.

Despite these limitations, the current study enhanced our knowledge about the assessment of financial system preparedness for climate change in Fijian context as well as its application for strengthening of the current financial system.

26.5 Qualitative Discussions and Conclusions

The following is just one example of a discussion provided in Example 26.5. That is, the interpretation of Barry's taxonomic model and the setting for the linguistic presentation that follows. The conclusion section is given in Discussion 26.4. (In this paper, Softbank is renamed NSL Bank, Silva is renamed Sena and Tony is renamed David).

26.5.1 Discussion

Example 26.5 Dynamics of Credit Decision-making.

Discussion 26.4 Dynamics of Credit Decision-making.

It is observed that there are four basic types of decisions have been made by the credit decision-makers in these case studies according to the situation. It is revealed that some credit decisions are made following formal rules and some are not. On the other hand, some credit decisions are seem logical but some do not have any business sense rather appear to be ridiculous. These four decision-makers are analysed below.

Case One

The Chairperson

The Chairperson is closely associated with the government and political elites. It is argued that business persons in the long run, to maintain their position of power, may accept certain responsibility to the whole society. One of the researcher's colleagues argued that, the Chairperson's decisions in the Case One are rational because they were intended to rescue 30,000 workers of David

Group. He interpreted this as a reflection of the Chairperson's perceived responsibility to the whole society. On the other hand it is argued that the social power holder might show his willingness and ability to help others to feel powerful might not bother about formalities in decision-making. However, this '*willingness to help others*' is only a mere intention and does not make any business sense therefore cannot be treated as rational.

Mr. David's credit applications would not have been accommodated if a solid organizational structure (*formal*) and professional management culture (*rational*) were in place in the NLS Bank. The Bank was run as an autocracy. The Chairperson was all-powerful and no Board member or an officer had the courage to oppose the decisions of the Chairperson (*Boss*). But the bureaucracy was also challenged by the Chairperson overruling the decisions of the credit officers of the NLS Bank in the case of David Group. Therefore, those clients who can approach him could by-pass the normal lending procedures that were rigid and not conducive for the new concepts to grow. Although the theory of Autocracy explains '*how*' the decision was made it is not helpful in explaining '*why*' the Chairperson made decisions which are *not rational* in business sense. The concept of Bureaucracy does not support the motives of the decision-maker but it is helpful to explain why the borrowers tend to resort *heuristic* methods when they seek credit from banks. Therefore, the question '*why*' remain still unanswered. However, another colleague pointed out the extent of publicity the Chairperson was aiming at and argued that, the psychological trait of ego also would have had significant influence over Mr. David, the Chairperson to make such *situational* decisions. To the extent the decisions are *heuristic*, radical and *situational* they attract more publicity. The Chairperson as an esteemed individual was leading his Group of companies, to which the NLS Bank belongs, towards the largest group of companies in Sri Lanka. Therefore, it could be argued that the Chairperson would have been influenced by such attitudes but the motives are not clear and decisions seem *not rational* but *situational*.

Case Two

Branch manager (BM)

The concepts such as ego, self-esteem, social power or even social responsibility have little relevance in the Case Two since the decision makers and credit applicant involved were not socially powerful individuals or from a social network with powerful backgrounds. They were in the middle management of the Bank. Mr. Sena of Janitorial believed that his major client (the NLS Bank) is doing a favour to him and used to reward the bank officers heuristically as Sri Lankans normally do. In the South Eastern culture traditional gifts differed from corruption and if Mr. Sena's free services at the Christmas and New Year time were treated as gifts.

Branch Manager (BM) Mr. Edwin was instrumental in building up Janitorial. Edwin, the branch manager who nurtured Mr. Sena and his business Janitorial, was promoted and transferred to head office. The *formal rational* banking practice is studying enterprising person, taking a calculated risk and deciding on nature and amount of credit entrepreneurs who in turn bring business to the bank. In this sense KBM Mr Edwin has managed Mr. Sena's account as a true banker friend (*Buddy*).

Manager-recoveries (MR); Case Two

Mr. Sena pays gratitude to the Manager-Recoveries (MR) who had re-adopted a comprehensive customer rehabilitating approach even going beyond his authority (*Heuristic*) and really reaping results of his ingenuity and *rational* approach in a rebellious manner. The MR had a practice of re-structuring and transferring such loans with new facilities beyond his authority to show very impressive performance and managed to secure raises and rapid promotions as well. Though the credit decisions made by MR in favour of Mr. Sena seem *rational* according to the research participants the decisions were made heuristically without proper authority (*Rebel*). There was no substantial evidence available for major personal gratifications offered by Mr. Sena to MR. The most probable reason for the *heuristic* decisions by MR could be that the intention of impressing the management on his performance for rewards. This shows a kind of 'petty corruption' that prevails in the system and whenever the opportunities arise, the bank officers and the customers are very well prepared to resort to heuristic methods.

The Credit Officer (CO) and the CMR; Case Two

The Credit Officer (CO), who was in charge of Mr. Sena's account after Edwin's transfer, had a personal grudge with Mr. Sena and acted *irrationally* (overcharging interest and delaying payments etc.). He pretended that he was applying *formal* banking rules and declined credit to Mr. Sena while harassing him for 'irregularities'.

Chief Manager Recoveries (CMR)

CMR was a senior banker and he too seems following banking practices to the letter (*formal*). He quickly discovered that former officer had accommodated Mr. Sena without proper authority. He was yet to impress the Management on his skills and performance. CMR did not believe Mr. Sena's appeal about his business prospects, may be due to adverse comments of the CO of the branch, declined further credits by applying credit rules (*formal* processing). Also Mr. Sena did not have qualified accountants to provide "convincing" financial projections about his business prospects. CMR was contemplating to serve a letter of demand towards legal action against Mr. Sena and his house which was mortgaged to the Bank. The Management agreed with CMR's recommendations and Mr. Sena was helpless. Both the CO and CMR had a role relationship with Mr. Sena and acted *irrationally/mechanically* (*Robot*) in evaluating Mr. Sena's credit applications.

26.5.2 Conclusion

Example 26.6 Dynamics of Credit Decision-making.

Conclusions 26.2 Dynamics of Credit Decision-making.

Conclusions

The research revealed six major dynamics under three dimensions based on which credit decisions are made; (a) the nature of decision-making (*Systematic/Formal* or *Heuristic* approach), (b) the type of relationship (*personal or role*) and, (c) manifested sense/logic or justification (*Rational* or *Irrational/Situational* judgment) and the following conclusions are drawn thereon.

Taxonomy of Influence Tactics and Personality Traits in Credit Decision-making.

The Barry's Taxonomy of Expectancy is modified with personality traits. Further the Taxonomy is replicated placing the decision-makers in appropriate quadrants as explained in the theoretical analysis sections of the two cases. Therefore, this proposed Taxonomy shows a wholistic picture of four of the six credit decision-making dynamics under two dimensions; the relationships between credit seekers and the style of decision-making of the five decision-makers who are at various levels in the bank's management hierarchy (see, Fig. 26.2: Taxonomy of Influence Tactics and Personality Traits in Credit Decision-making).

Matrix of Credit Decision-Making

Four types of decision-makers are identified in the above Discussion section (Nature of decision-making) and they are labelled as **BOSS, ROBOT, REBEL** and **BUDDY**. The *Bosses* make *situational* decisions employing *heuristic* methods. The *Bosses* (*Heuristic-Situational*) make heuristic decisions because he is powerful with legitimate authority to break the rules but he provides 'solace' to the 'needy' of course to people who are 'close' to him and therefore, more like *situational* in business sense. The *Robots* (*Formal-Irrational*) strictly follow rules, therefore, formalities are adhered but lack common sense

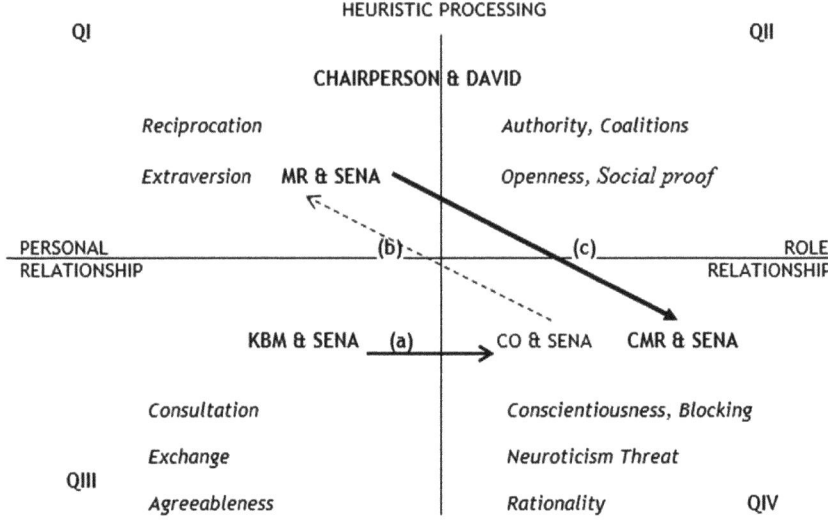

(a) Breaking the relationship with the branch staff after KBM was transferred to head office.

(b) Develop a temporary relationship with MR which did not last long and MR lost the job.

(c) Back in the due quadrant due to formalities when the applicant is from the powerless-class.

Fig. 26.2 Taxonomy of Influence Tactics and Personality Traits in Credit Decision-making

therefore, likely to be *mechanical/irrational*. The *Rebels* (Heuristic-*Rational*) break rules for the sake of just and fairness but lack enterprising logic. They make decisions which are beneficial to both the decision-maker and the credit applicant, therefore they make *rational* decisions but it is likely that *heuristic* methods are followed since formal rules forbid such decisions. *The Buddies* (*Formal-Rational*) on the other hand find ways within the formal system without breaking rules to support decision seekers as well as develop business therefore, they make *rational* decisions following *formal* procedures. These two variables and the combination of decision types can be summarized in a form of a matrix as shown in the Table 26.3.

From this research outcome, it is recommended that both the loan officers and borrowers should understand the each other's positions/expectations and employ appropriate influencing tactics to explore the maximum benefits from the credit decision such as collaterals, repayment terms, interest rates and grace periods etc. *This theorisation of research findings might be useful for future research when study similar issues in different backgrounds.*

Table 26.3 Matrix of credit decision-making

Variable		Justification	
		Rational	Situational
Evaluation procedure	Formal	BUDDY	ROBOT
	Heuristic	REBEL	BOSS

26.6 Mixed Methods Discussions and Conclusions

26.6.1 Discussion

Example 26.7 Stock Market Development and Nexus of Liquidity.

Discussion (Mixed methods) 26.5 Stock Market Development and Nexus of Liquidity.

Discussion

Market liquidity (LQDT) increases investors' confidence in the stock market promoting stock market development (Yartey, 2010). But the SPX-SMD shows no significant dependence on LQDT as per the above results, and the annual stock value traded at the SPX as a percentage of the GDP is very low compared to SEM, as shown in Fig. 26.3. The average free float of the SPX seems critically low (with only 20%) compared to 70% in the SEM, and foreign investor participation, too, shows negligible contribution to the SPX compared to that of the SEM as shown in Fig. 26.4.

Fig. 26.3 Stock value traded (% GDP). *Source* Compiled by the Author

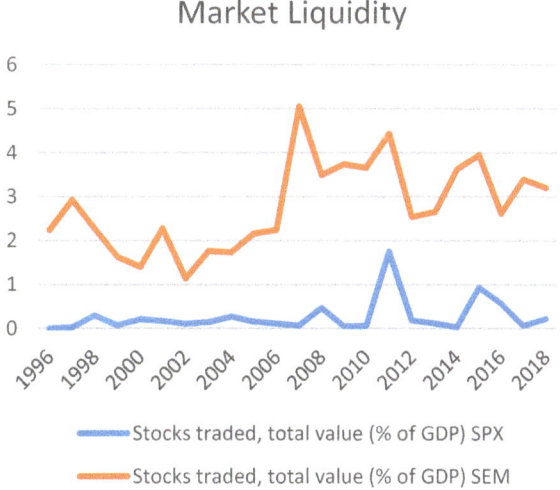

Market Liquidity

Stocks traded, total value (% of GDP) SPX

Stocks traded, total value (% of GDP) SEM

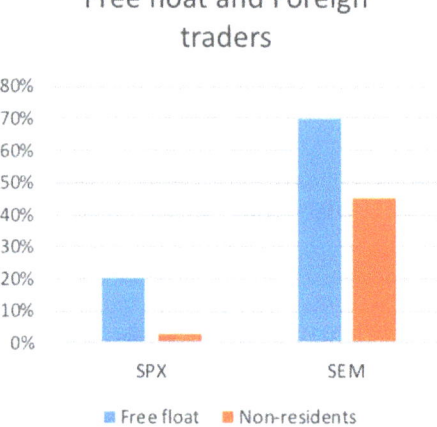

Fig. 26.4 Free float and foreign traders. *Source* Compiled by the Author

Bekaert and Harvey (2000, p. 567) further elaborated the importance of this market liquidity as follows:

...the potential of market manipulation is acute in small emerging markets and liquidity is often poor. Although there are many policy initiatives that could increase liquidity and reduce the degree of collusion among large traders, there may not be a sufficient mass of domestic speculators to ensure market liquidity and efficiency.

The regression results showed that the GDP has a significant positive influence: for every one-unit increase in GDP there was an increase in SMD growth by 1.898% suggesting that an increase in general income level can induce the performance of the SPX. This phenomenon is reflected in the annual total return generated by the SPX for past 18 years with an average of more than 10% per annum. The descriptive data in this regard are shown in Figs. 26.5 and 26.6 compared to those of Mauritius.

The correlation of FDI with SMD of SPX is also not significant despite comparatively higher FDI inflow into Fiji's economy compared to that of Mauritius where stock market development is vibrant as shown in Fig. 26.7. However, the regression results showed that the impact of FDI on SMD of SEM is estimated as significantly positive, suggesting contradictory conclusions with regards to the impact of the macroeconomic determinant FDI on the two stock markets. This is again a reflection of entry barriers to the SPX which delivered exceptionally attractive annual returns but is not openly convenient for more lucrative foreign investors.

According to the regression findings, investments, represented by capital formation, also do not show a significant impact on the change in SMD while inflation shows a significantly negative influence in line with almost all other stock markets. This phenomenon is explained theoretically when the inflation is higher, the fixed income securities can offer higher returns due to higher interest rates, persuading investors to shift their investments from shares to bonds, and pulling the performance of the stock markets down.

Financial liberalisation market integration

Fiji exchange control regulations restrict non-Fijians freely investing in Fiji. Low cost of equity capital encourages more capital formation/investments. This would induce economic growth and benefit the country to a larger extent than the financial windfall gains to a handful of domestic owner-shareholders (Henry, 2000b). However, according to the regression findings, this assertion is confirmed in the case of SPX where the regulatory quality estimate showed a negative influence on change in SMD ($\beta = -25.718$, p < 0.009) suggesting that every one-unit increase in regulatory quality has resulted in a fall in SMD growth by 25.718%.

The assertion that stock markets become more active and transparent after restrictions are removed seems more applicable for larger markets where economies of scale play a critical role in reduction of cost, and are not properly fit for smaller economies like Fiji, according to these results. However, the argument proposed by Bekaert and Havey (2000, p. 567) against this paradox is convincing;

...opening the market to foreign speculators may increase the valuation of local companies, thereby reducing the cost of equity capital. The intuition is straightforward. In segmented capital markets, the cost of equity capital is related to the local volatility of the particular market. In integrated capital markets, the cost of equity capital is related to the covariance with world market returns. Given that emerging economies have different industrial mixes and are less subject to macroeconomic shocks...[and] Since local market volatilities tend to be large, the cost of capital should decrease after capital market liberalizations.

On the other hand, when local investors are allowed to buy better foreign stocks it will affect the foreign reserves of a country adversely, especially for small economies such as Fiji's. Similarly, Kose et al. (2009) assert that there are certain prerequisites to be fulfilled for emerging economies to open up their financial markets to the world. Furthermore, financial market integration may also hinder the development of stock markets due to a loss of domestic investors (Bekaert and Harvey, 2000). This assertion is confirmed by the above results where Fiji's regulatory quality estimate showed a significant negative influence on SMD. These results can challenge the validity of the data produced by the Worldwide Governance Indicators of the World Bank on *Regulatory Quality* of countries by assessing the 'perceptions of the ability of the government to formulate and implement sound policies and regulations that permit and promote private sector development' (The World Bank, 2019). On the other hand, Bekaert and Harvey (2000) caution that "if restrictive measures are not

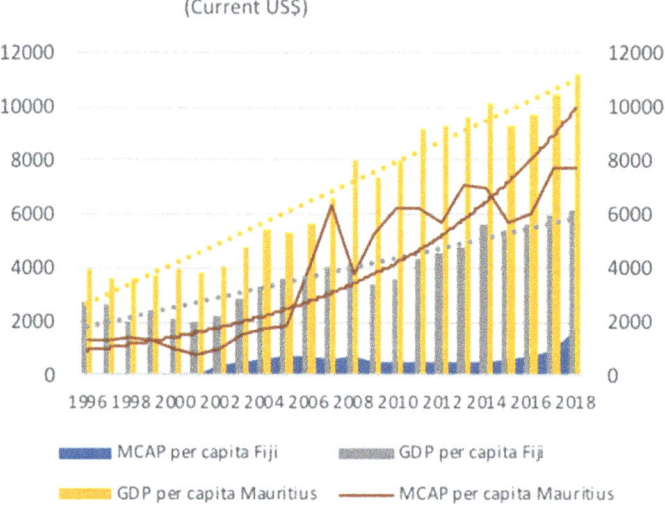

Fig. 26.5 Per capita GDP and MCAP. *Source* Compiled by the Author

conducive to international investors, one should see a decline in foreign capital flows" (p. 2).

A higher degree of integration will also act as a push factor for further development of the domestic market. For instance, investors may pressure domestic intermediaries to upgrade their trading systems or domestic regulators to modify existing legal frameworks to support a greater variety of financial instruments.

According to the messages disseminated in Annual Reports of the SPX, and the views expressed by certain officials of the SPX, brokers and other relevant institutions do not show much interest in implementing proven paths for the stock market development in many other emerging economies. Financial liberation is the initiative practiced by many stock markets to develop faster promoting stock market liquidity. Policy formulators of Fiji should seriously evaluate the costs and the benefits of relaxing the financial regulations, integrating the SPX with the global markets, and promoting foreign investments through policy reforms. None of the research participants expressed any concern on restrictions on capital flow of funds, liberalisation of financial markets and/or integration of the SPX internationally. The brokers and the SPX seem happy with the exceptional level of stock market performance as it has generated very high annual returns consistently for last five years (since 2013) as shown in Fig. 26.7 earlier, though volumes are negligible and the lion's share of the benefits are exploited by a handful of owner-shareholders.

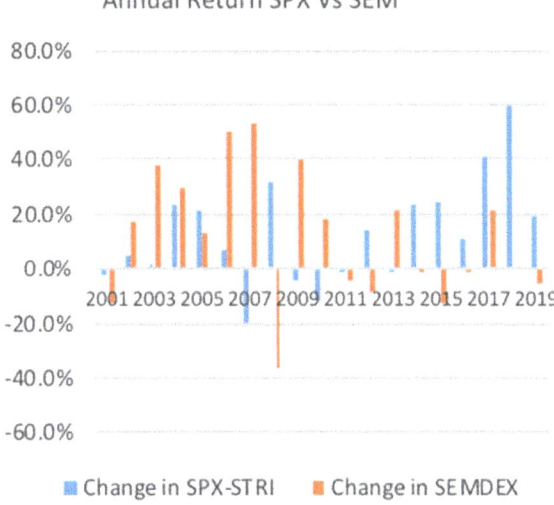

Fig. 26.6 Annual return: Change in indices. *Source* Compiled by the Author

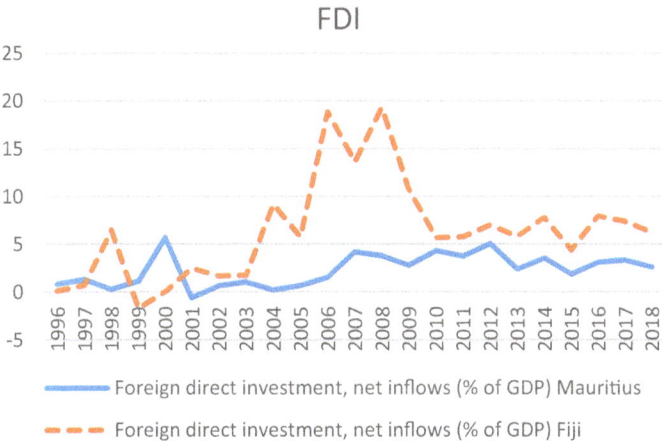

Fig. 26.7 FDI of Fiji compared with Mauritius. *Source* Compiled by the author

26.6.2   Conclusions

Example 26.8 Stock Market Development and Nexus of Liquidity.

Conclusions (Mixed methods) 26.3 Stock Market Development and Nexus of Liquidity.

Conclusions and Policy Recommendations

A growing literature on empirical studies shows the importance of efficient financial systems in reducing costs, influencing investment decisions, technological innovation, and growth rates. The theoretical literature shows that economic activity and financial liberalisation can influence financial systems, stock markets included. Only GDP shows a significant, consistent, positive impact on SMD. Surprisingly, results show that Regulatory Quality has a negative, but not significant, impact on SPX. All other macroeconomic and institutional determinants are insignificant and/or not consistent with the results of other studies in emerging markets.

Market liquidity exhibited a unique property showing no correlation to SMD and not distributed normally, contrary to the other markets. Further exploration of the nexus of the market liquidity reveals that present regulatory conditions are not conducive for the stock market to be vibrant. It is apparent that the existing operational mechanisms of the SPX create many roadblocks; paperwork, hassles, inconvenience, delays, and frustration to both the investor as well as to the broker. Eliminating these hindrances is critical, not only to attract new players to the market but also to increase trading volumes.

It is also revealed that closely held ownership is another barrier for the market to be liquid. Making more opportunities for such owners (through global integration and financial liberalisation) might encourage such behaviour as releasing more shares to the market stimulating one critical element, the supply side, of the nexus of market liquidity.

Policy initiatives could increase liquidity by relaxing entry barriers to the stock market. Allowing stock markets to integrate globally and allowing foreigners to enter would increase the efficiency of stock market operations, and would immensely contribute to the development of the SPX. More importantly, these reforms would considerably reduce the degree of collusion among the few powerful owner-shareholders, especially in the small Fijian economy. Financial liberalisation can bring a mass of domestic speculators to ensure market liquidity and efficiency.

26.6.3   Future Research

It is also important to make a note of the spaces that this research opens up for future research and present it after the conclusions.

Example 26.9 Stock Market Development and Nexus of Liquidity.

Box 26.1 Future research

This study estimated how significant the selected determinants of SPX are, and compared the results with the findings from similar studies in other emerging economies. But why are there disparities in stock market developments among countries which are in the same geographical region with similar characteristics and parallel beginnings? As the results of this study show very little significance relating to the impact of the selected macroeconomic and institutional *determinants* on the SPX, further exploration is warranted to find what factors cause these differences and why. It would also be timely to revisit the preconditions of financial liberalisation for emerging markets. Another aspect to be explored is the bi-directional causality between financial development and economic growth in emerging economies compared to advanced economies.

Taking a deeper step, future studies could aim at potential psychosocial-sociocultural dynamics such as risk appetite, social values, cultural habits and religious rituals, etc., which may be functioning as underpinning root causes influencing these determinants.

26.7 Summary

- Outcome interpretation involves moving away from detailed results and taking into account research issues, questions in a study, existing literature, and perhaps personal experience, and their greater meaning.
- The purpose of the 'Discussion' paragraph in a research paper is to define the importance of the findings in light of what is already known about the research problem under investigation.
- In interpretation one has to deal with the following four concepts. They are significance, justification, reliability and validity. Therefore, all researchers should be able to review these facts and successfully represent their findings.
- The concept of importance has a definite, statistical meaning and a general definition with a more general understanding.
- Ability to justify If other organizations, groups, or individuals have something to say about their behaviour or experience, if so, how do you know this is so? Such matters should be clarified here.
- Credibility is the ability of another researcher to provide the same results (although not necessarily the same interpretation) if they investigate these same questions in other ways under similar circumstances.
- Research methods, approaches, and techniques used to verify validity are related to whether what was actually measured was actually measured.

- A conclusion is not simply a summary of your facts or a re-statement of your research problem but a synthesis of key facts. These include: presenting the final judgment; Presenting important implications of the study; Illustrate the importance of ideas; Introducing new or expandable methods.

References/Further Reading

Bekaert, G., & Harvey, C. (2000). Foreign Speculators and Emerging Equity Markets. *The Journal of Finance, 55*, 565–613. https://doi.org/10.1111/0022-1082.00220.

Blaxter, L., Hughes, C., & Tight, M. (2008). *How to research*. Birkshire, England: Open University Press.

Bollen, K. A. (1989). Structural equations with latent variables. John Wiley & Sons. https://doi.org/10.1002/9781118619179.

Creswell, J. W. (2015). *A concise introduction to mixed methods research*. Thousand Oaks, CA: Sage.

Creswell, J. W., & Plano Clark, V. L. (2011). *Designing and conducting mixed methods research*. London: Sage.

Jarque, C. M. (2011). Jarque-Bera Test. In: Lovric, M. (eds) International Encyclopedia of Statistical Science. Springer, Berlin, Heidelberg. https://doi.org/10.1007/978-3-642-04898-2_319.

Kose, M. A., Prasad, E. S., Rogoff, K., & Wei, S. J. (2009). Financial Globalization and Economic Policies. *Handbook of Development Economics, 5*. https://doi.org/10.2139/ssrn.1392949.

Maxwell, J. A. (2013). *Qualitative research design: An interactive approach*. Thousand Oaks, CA: Sage.

O'Leary, Z. (2005). *Researching real-world problems*. London: Sage.

O'Leary, Z. (2014). *The essential guide to doing research*. London: Sage.

Saliya, C. A. (2020). Stock market development and nexus of market liquidity. *International Journal of Finance and Economics, 25*(3), 23–45.

Sacred Heart University (2020). Discussion of findings. https://library.sac142 redheart.edu/c.php?g=29803&p=185902.

Westfall, J., Judd, C. M., & Kenny, D. A. (2015). Replicating studies in which samples of participants respond to samples of stimuli. *Perspectives on Psychological Science, 10*(3), 390–399. https://doi.org/10.1177/1745691614564879.

Wolcott, H. F. (2008). *Writing up qualitative research*. USA. University of Oregon. https://library.sacredheart.edu/c.php?g=29803&p=185935.

Yartey, C. A. (2010). The institutional and macroeconomic determinants of stock market development in emerging economies. *Applied Financial Economics, 20*(21), 1615–1625. https://doi.org/10.1080/09603107.2010.522519.

Chapter 27
Writing Abstracts/Summaries

27.1 Introduction

A summary is a brief presentation of the whole. Usually a paragraph of 200 words or less, the complete tract contains a summary of the following key points in the sequence:

1. The overall purpose of the study and the research problem you investigated.
2. The basic design or methodology of the study.
3. Major findings or trends found as a result of analysis and,
4. A brief note on interpretations and conclusions.

27.1.1 What a Summary Should Not Include

- Long background information.
- References to other literature (such as "Current research suggests…" or "Studies suggest…").
- *Quotations* (ending with "…" or incomplete sentence).
- Short phrases, jargon (technical phrases) or phrases that may confuse the reader, and
- Any kind of image, illustration or table or references to them.

© The Author(s), under exclusive license to Springer Nature Singapore Pte Ltd. 2022   379
C. A. Saliya, *Doing Social Research and Publishing Results*,
https://doi.org/10.1007/978-981-19-3780-4_27

27.1.2   Purpose

Summary writing begins with a clear definition of the purpose of the research. What practical or theoretical question does research respond to, or what research questions are intended to answer? The topic may include a brief context of social or academic relevance, but should not go into detailed background information.

After identifying the problem, state the purpose of the research. Use verbs such as inquiry, test, analysis, or evaluation to describe exactly what you intend to do.

This part of the summary can be written from the present or past, but the future should never be mentioned as the research has already been completed.

27.1.3   Methods

Summary writing begins with a clear definition of the purpose of the study. What practical or theoretical question does research respond to, or what research questions are intended to answer? The topic may include a brief context of social or academic relevance, but should not go into detailed background information.

After identifying the problem, state the purpose of the research. Use verbs such as inquiry, test, analyse, or evaluate to describe exactly what you intend to do.

This part of the summary can be written from the present or past, but the future should never be mentioned as the research has already been completed.

27.1.4   Results

Thereafter, the main research results are summarized. This part of the summary may be from the present or past tense.

Depending on how long and complex the research is, not all results will be included here. Try to highlight only the most important findings that allow the reader to understand the conclusions.

27.1.5   Conclusions

Finally, state the main conclusions of the research: What is your answer to the problem or question? Your research should conclude with a clear understanding of the central point that has been proven or argued. Conclusions are usually written in the present tense.

If there are significant limitations to the research (e.g., relation to sample size or methodology), they should be briefly stated in the summary. This allows the reader to accurately assess the reliability and generalizability of the research.

If your goal is to solve a practical problem, the conclusions may include recommendations for implementation. If applicable, you can briefly suggest further research.

27.1.6 Key Words

If you submit a manuscript, you will need to add a list of keywords at the end of the abstract. These words should mention the most important elements of research to help prospective readers find your paper in their literature searches.

Some publishers have specific key words. For example, some journals require classifications such as JEL (JEL Classification System).

27.1.7 Instructions to Write an Abstract or a Summary

Summarizing your entire dissertation into hundreds of words can be a real challenge, but it is important to get it right because an abstract is the first part (and perhaps the only part) that people read. The following strategies can help you get started.

Reverse outline: It is not necessary to have the same structure in every summary. If your research has a different structure (for example, an anthropological dissertation that builds an argument through thematic chapters), you can write your summary through an inverse sketching process. That is, for each chapter or section, drafting 1–2 words and sentences that summarize the focal point or argument will provide a framework for the structure for the summary. After that, you can revise the sentence to make connections and show how the argument develops.

The abstract should tell the whole story and include only the information found in the main text. The abstract should be re-read to ensure that it gives a clear summary of the whole argument.

Read Other Summaries: The best way to learn the conventions of writing a summary of your subject is to read other summaries. You should already have read many journal article summaries while reviewing your publications. Those structures and styles can be used as a framework.

Write clearly and concisely: A good summary is short but effective, so make sure every word counts. Each sentence must clearly communicate one key point.

Avoid unnecessary filling words, and avoid jargon—the summary should be understandable to readers who are unfamiliar with your topic.

Focus on your own research: The purpose of the summary is to record the pioneering contribution of your research, so avoid discussing the work of others. You can include a sentence or two that summarizes the academic background to locate your research and demonstrate its relevance to a broader debate, but do not require specific publications. Do not include citations in a summary unless absolutely necessary (for example, if your research responds directly to another study or revolves around one leading theorist).

Formatting: If presented to a journal there are often specific forms for summaries. Author guidelines should be carefully followed and formatted accordingly. Always adhere to the word limit. If you have not been given any guidelines regarding the length of the summary, then consider 200 words as a benchmark.

27.2 Types of Abstracts

Abstracts found in recognized academic journals can be broadly classified into three categories. They are; classified abstracts, descriptive summaries, and single paragraph abstracts.

27.2.1 Classified Abstracts

Classified abstract are often found in journals published by the well-known publishing house Emerald. This classification is as follows: Purpose, Design/methodology/approach, Findings, Research limitations/implications and Originality/Value. See Example 27.1 for this type of classified abstract. The peculiarity of this illustration is that this abstract (140 Words) is only 39 words less than the research publication (179 Words). That is because this special publication is a poem.

Example 27.1 The Oldest Profession.

Abstract 27.1 The Oldest Profession.

The oldest profession

Author: C.A. Saliya

Publication: Accounting, Auditing & Accountability

Publisher: Emerald Publishing Limited

Date: 07/27/2012

The oldest profession

C.A. Saliya ▾

Accounting, Auditing & Accountability Journal

ISSN: 0951-3574

Article publication date: 27 July 2012

Abstract

Purpose

The purpose of this paper is to draw attention towards how religious beliefs are linked to today's accounting principles and legitimating exploitation.

Design/methodology/approach

The poem is a critical thought.

Findings

Religious concepts have been widely used to justify various historical social injustices. Four basic accounting principles, matching, accrual, going concern and reporting cycle perfectly correspond with life cycles, the acts and rewards of such lives, and "balancing" the pains and gains of individuals and society. These concepts, which have now become the foundation of accounting, were in existence millennia ago.

Research limitations/implications

Accounting is broader than has been widely perceived. Accounting researchers should pay more attention to how accounting helped sustain social injustices.

Originality/value

This poem provokes thoughts on the origin of basic accounting principles and their historical roles in sustaining social injustices.

While I was born rich, buoyant and preferred
You were born poor, compliant and unfavoured
Was it God? Destiny? Karma?
Or simply luck or by accident?
Was I good, were you bad in a previous life …or more recent?
I care, but please please be fully aware
When I share, I always get back my lion's share
You must never ever say that
I have not been legally fair
When accounts are destiny-balanced, then I will responsibly bear
All debits have credits and all credits have debits … an equal claim
You stupid know not the Matching principle,
while I know the worldly game
When all accounts are settled, in this fiscal Cycle
or in life's next Cycle
I follow well the accounting rules, yes, based on Going-concern
For three millennia I balanced my gain
Against your pain … ordained?
Remember you were born poor and I was born preferred … die-casted?
On Earth, my Accrual concepts always prevail,
an expertise in legal exploitation
So which profession is the oldest? You know now … just trust my reputation

27.2.2 Single Para Abstracts

Such summaries are usually no more than 150 words. Some magazines limit such summaries to 100 words, which is a very challenging task. Many journals published by well-known European publishing houses, such as Elsevier and Sage, request such summaries. Example 27.2 presents such abstracts.

Example 27.2 *Creating and Reenforcing—Accounting Forum.*

Abstract 27.2 *Creating and Reenforcing.*

Abstract

This paper examines the controversial role played by accounting within the discriminatory bank-lending practices of a privately owned bank in Sri Lanka. It reports on an analytical auto-ethnography (during the period 1994–2004) coupled with follow-up interviews and reiterated analyses. Data were analysed using Pierre Bourdieu's concepts of field, capital, habitus and symbolic violence. The empirical findings of the paper illustrate how key capitals at macro levels (social, cultural and symbolic) are mobilised in the dominance structures within the banking lending field and how individuals with given habitus behave and follow given strategies to deploy rational accounting systems at a micro level to translate discriminatory bank lending policies into practice and, as a result, create and reinforce discrimination.

27.2.3 Descriptive Abstracts

These abstracts are typically limit to 200–300 words and vary from journal to journal. Many American publishing houses such as the John Wiley, Taylor & Francis request such summaries. Example 27.3 provides such an abstract, it has 250 words.

Abstract 27.3 Stock Market Development and Nexus of Liquidity.

RESEARCH ARTICLE

Stock market development and nexus of market liquidity: The case of Fiji

Candauda Arachchige Saliya ✉

First published: 07 December 2020 | https://doi.org/10.1002/ijfe.2376

Read the full text >   🗎 PDF   🛠 TOOLS   ◁ SHARE

Abstract

The purpose of this paper is twofold: First, to explore the macroeconomic and institutional factors influencing stock market development (SMD) in Fiji during the period 1996–2018 and second, to explore the causes of the prolonged sluggish nature of the South Pacific Stock Exchange (SPX) in Fiji. A mixed-method approach was employed: autoregressive distributed lag (ARDL) testing framework was used to estimate the influence of the structural determinants in Fiji while descriptive and narrative analysis was used to explain the status of the SPX. The study findings partially confirmed the findings of previous studies. The findings that economic growth promotes SMD while inflation is negatively correlated to SMD were consistent with previous findings. However, the finding that regulatory quality and banking sector development have negative impacts on SMD was contrary to the findings of previous studies. Many determinants, which have been shown to have significant impact on SMD in other studies, did not show any significant impact on the SPX in the Fijian context, particularly the stock market liquidity showed no correlation to SMD and not distributed normally. Exploration of the nexus of the market liquidity reveals that present regulatory conditions are not conducive for the stock market to be vibrant and suggests eliminating excessive paperwork and entry restrictions are critical, not only to attract new players to the market but also to increase trading volumes. We argue that the concentration of ownership and market disintegration resulted in the current sluggish status of the SPX.

27.3   One Research Many Outputs

Eight research papers have been published based on the doctoral dissertation in Example 7.2. Six manuscripts were developed from the case studies and two were on research methods and research philosophy used there. The data from those case studies were interpreted by various theories and the manuscripts were redesigned and submitted to 25–30 international journals. Only one submission can be done for one manuscript. You have to wait for the editor's decision. If rejected you can submit it again to another journal. In the first submissions, these journals rejected those manuscripts by the editors themselves, without even considered for peer reviewing, with only editor's brief notes (Desk review and rejection).

The manuscripts were then revised and further developed and presented to other journals. Some journals take seven to eight months to provide reviews. Then some were accepted with major revisions and some were accepted with minor amendments suggested by the reviewers. Addressing these critical reviews and improving the manuscripts is challenging and time consuming, to bring the manuscripts to a level suitable for final acceptance. The data used for the case studies have been analysed and interpreted through different theories and different conclusions have been drawn. Abstracts of all eight leaflets are provided below.

27.4 Summary

- An abstract is usually a paragraph of 200 words or less,
- An abstract contains the following key items; (1) The purpose of the study and the research problem investigated. (2) The basic plan or methodology of the study. (3) Major findings (4) The interpretations and conclusions.
- Also, an abstract should indicate, what practical or theoretical question does the research respond to, or what research questions are intended to answer?
- Use verbs such as investigate, test, analyse or evaluate.
- It is usually written in the past tense.
- The research should conclude with a clear understanding of the central point that has been proven or argued. Conclusions are usually written in the present tense, but change from time to time.
- If there are significant limitations to the research, they should be briefly mentioned in the summary.
- The best way to learn the conventions of writing a summary of your subject is to read other summaries.
- A good summary is short but effective, so make sure every word counts. Each sentence must clearly communicate one key point.
- There are specific forms for summaries. Author guidelines should be carefully followed and formatted accordingly. Always adhere to the word limit.
- Many manuscripts can be crafted from a single research, employing different strategies, using the same data, case studies and different theories and drawing different conclusions.

References

https://www.scribbr.com/dissertation/abstract/.
https://library.sacredheart.edu/c.php?g=29803&p=.

Part VI
Publishing Process, Cover Letters and Addressing Reviewers' Comments

Chapter 28
The Publishing Process

28.1 Introduction

Almost all scholarly journals have a peer-review process, a quality control mechanism in which one to four scholars who are faculty experts in the author's field evaluate each article. These peer reviewers (also called referees or readers) identify inadequacies, misinterpretations, and errors; provide recommendations to the author for improvement; and aid the editor in making a decision about the value of the work (Belcher, 2019, p. 10).

A journal is a scholarly genre of publication that has standard features. It is usually five to forty pages (2500–12,000 words) and can contain up to five or fifty quotations. It discusses the publications of other scholars, and is reviewed by other scholars (peer-reviewers: peer reviewers).

It is a known fact that publishing a paper in an IPRAJ is harder than gaining a Ph.D. The main reason for this difficulty is the time it takes for peer review and the inability to submit a manuscript to several journals simultaneously.

It would seem logical that a prestigious journal with a high impact factor would have a correspondingly low acceptance rate. All researchers want their articles published in the most prestigious journal in their field, and will often submit their paper automatically, fully anticipating rejection and the need for re-formatting for submission to an alternate journal. (see below: https://www.enago.com/academy/journal-acceptance-rates/).

© The Author(s), under exclusive license to Springer Nature Singapore Pte Ltd. 2022   389
C. A. Saliya, *Doing Social Research and Publishing Results*,
https://doi.org/10.1007/978-981-19-3780-4_28

© 2020 American Psychological Association
ISSN: 0003-066X

Summary Report of Journal (

Journal	Manuscripts				
	No. received	No. accepted	No. pending[a]	Rejection rate[b]	Rejection rate (including out of scope submissions)[c]
American Psychologist[f]	234	58	58	72%	80%
Behavioral Neuroscience	176	52	17	51%	71%
Developmental Psychology	711	195	251	72%	74%
Emotion	586	141	255	73%	75%
Experimental & Clinical					

It seems logical that a journal with a high impact factor should have a high rejection rate in line with its reputation. All researchers want their papers to be published in the most prestigious journal in their field, and often their manuscript will be 'automatically' submitted to such a journal first. They hope to reject it, and the goal is to reprint it according to the reviews it receives and submit it to another journal.

28.2   Crafting Multiple Manuscripts from a Data-Set

One strategy to deal with is to submit several different manuscripts to several journals at once. That is, creating several manuscripts in a variety of ways using a set of research data. The following methods can be used for this.

- Use of different techniques for data analysis.
- Using different theories to interpret the results.
- Create manuscripts on the research methodology and methods used.
- Creating data sharing and collaborations with other researchers.

Thus, by creating several different manuscripts using the same database, on different themes, it is possible to submit several manuscripts to different journals at the same time. Having this approach in mind at the beginning of the research (data collection and survey questions) facilitates this strategy to be implemented successfully. Eight different papers were created from the case study dataset used in the Example 7.2: Ph.D., research thesis (see Fig. 28.1).

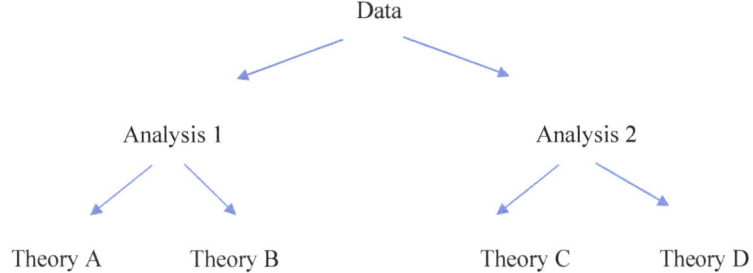

Fig. 28.1 Many output from one study

28.3 Writing a Manuscript

Writing up is not just a critical, but a continuing, part of the research process, which should start soon after the commencement of the research project, and continue to and beyond its completion. So don't be misled by this being the penultimate chapter: writing up begins as soon as you start thinking about and reading around your research (Blaxter et al., 2008).

Creative writing has no place in research...Organization is necessary for the efficient allocation of one's time and effort, and for the presentation of a paper whose internal structure is balanced and sound, and whose argument proceeds along logical lines. Conventions are vital in a context where one writes not for oneself, but for a critical public (Berry, 2004, p. 3).

28.4 Publishing Process of a Research Paper by a Journal

The publishing process of a submitted manuscript varies from publisher to publisher. The journal's objectives, its author's style, personality and vision, the editorial board's composition, its peer-review process, the knowledge and time of its support staff, and the size of the budget, and whether the article is designed for a specific area. In general, however, a journal article goes through the following stages.

28.4.1 Submission

Once a manuscript is submitted to a journal by its corresponding author, it will be sent for peer-reviewing if the editor decides the manuscript is worthy of reviewing. It is not possible to submit the same manuscript to several journals simultaneously. The author has to wait for the decision of the editor of that particular journal, before considering to submit to another journal (single submission rule). If it is rejected, it can be sent to another journal after considering and amending the manuscript accordingly for the comments provide by the rejecting-editor and the reviewers.

28.4.2   Editorial Review

The journal editor first evaluates all the manuscripts by skimming through it to ensure that it meets the basic criteria of a manuscript [e.g., adapting to the purpose and scope of the journal, scholarly quotations, grammar, and relevance to the paper in which the journal is currently published] and its absence of major deficiencies (e.g. Problematic, weak methodology or weak arguments). If the editor identifies the underlying problem, the manuscript will be rejected. This is called desk review and desk rejection.

It is becoming increasingly common for journal editors to reject manuscripts without referring to manuscripts for peer-review. Editor's Letter 28.1 to inform the author the manuscript is under scrutiny to check whether it meets the criteria of the journal, and if successful it will be sent for peer-reviewing. This editor is from the International Journal of Finance and Economics (IJFE), which is a journal published by John Wiley, a well-known American publishing house.

Editor's Letter 28.1 Letter informing the author that the manuscript has been referred to the reviewers.

IJFE-20-,0320.R1 successfully submitted

Terry Wirtel <onbehalfof@manuscriptcentral.com>   Mon, Nov 23, 2020

Dear Professor Saliya,

Your manuscript entitled "Stock Market Development and Nexus of Market Liquidity: The Case of Fiji" has been successfully submitted online. Your paper will now be sent for immediate screening. If it is successful in the screening process it will then be sent for review. If it is a resubmitted paper (R1) it will be seen by the editor and may be sent again to the referee. You may check the status of your paper at any time by going to https://mc.manuscriptcentral.com/ijfe and entering the author centre. You will be notified of the editors' decision by e-mail.

Your manuscript number is IJFE-20-0320.R1. Please mention this number in all future correspondence regarding this submission.

If you have difficulty using this site, please click the 'Get Help Now' link at the top right corner of the site.

Thank you for submitting your manuscript to International Journal of Finance and Economics.

Kind regards
Inernational Journal of Finance and Economics Editorial Office

28.4.3   Selecting Peer-Reviewers

If the author is satisfied that there are no serious problems with the manuscript, he decides to select peer-reviewers. This is not easy. The editors have to work hard to find scholars who are willing to give reviews, and according to Didham et al. (2017), 28% more invitations were sent in 2016 for review than five years ago. Peer-reviewers often consist of one member of the journal's editorial board or recently published scholars. Critic peer-reviewers rarely provide a written report quickly. Therefore, motivating and reminding reviewers is the main task of any journal editor.

Under the double-blind rule, reviewers should refrain from reviewing articles suspected of being written by their peers or students and withdraw from the review process on their own. Specifically, in such situations, it is unethical to accept for review if identities have been revealed.

28.4.4   Peer-Review Process

Peer-reviewers read the manuscript and critically review its novelty, contribution, clarity, relevance, originality, and scholarly findings, the validity of the methodology, the interesting analysis, and the strong arguments. Some journals give clear instructions to reviewers (e.g., answering specific questions, filling out a form, or ranking). Reviewers recommend whether the publication of the manuscript should be accepted or rejected. In response to reviews, the journal's editor suggests that the manuscript be rejected, accepted for publication, or that the author improve on the strengths and weaknesses of the manuscript. Editor's Letter 28.3 shows one such letter sent to an author by a journal editor.

Methods for reviewing can vary greatly. Dual anonymous reviews are common in the humanities. The purpose of adopting such procedures is to protect the authors and to motivate critical reviewers to critique the manuscript impartially, fearlessly, and honestly.

Single-blind review is sometimes used if the author is more powerful than the reviewer. Here the reviewers know the identity of the author but the author does not know the identities of the reviewers.

28.4.5   Editor's Decision

Based on the recommendations of the reviewers, the editor decides whether to accept the manuscript for publication. If all the reviewers agree that the manuscript is strong or weak, the decision is easy. The decision is challenging if one reviewer recommends the manuscript while others recommend a rejection. Then the editors will never accept the manuscript. Editor's Letter 28.2 shows a letter sent to an author by an editor informing him of such a denial. The manuscript was revised to address the issues pointed out by the reviewers and the editor and submitted to IJFE journal (see the Editor's Letter s 28.1 and 28.3).

Editor's Letter 28.2   A letter sent to an author to inform a 'rejection'

On Fri, Mar 13, 2020 at 7:00 AM Heliyon <em@editorialmanager.com> wrote:

Manuscript Number: HELIYON-D-20-00182

Title: Stock Market Development and Financial Liberalisation: The Case of Fiji

Journal: Heliyon

Dear Dr. Saliya,

Thank you for submitting your manuscript to Heliyon.

Unfortunately, the reviewers have advised against publication of your manuscript, and in agreement with this assessment we are returning the manuscript to you.

For your reference, the reviewer comments are appended at the end of this message.

Thank you for submitting your work to Heliyon and giving us the opportunity to consider your work.

Kind regards,
Ahmed Elsayed
Associate Editor—Business & Economics
Heliyon

(Following the submission of the manuscript to another journal (IJFE) with adequate answers to the shortcomings and criticisms pointed out by the reviewers provided by the previous reviewers, the editor accepted it for publication subject to minor revisions. The letter of intent to refer to the reviewer is in Editor's Letter 28.1. and the editor's letter (see Editor's Letter 28.3) informing the authors that it may be accepted with minor revisions after review).

If the paper is of interest to the editor, instead of rejecting the manuscript based on reviewers' harsh criticism, perhaps he might send the manuscript to another reviewer. Or the editor informs the author of the his/her decision ignoring negative review but representing the positive review. If the editor decides to accept the reviewer's recommendation, he or she will make recommendations to revise the manuscript to the rejected reviewer's response. That is, it provides the opportunity to improve the manuscript by providing additional evidence or further explanations for those criticisms, and recommends revising-and-resubmit notice. Otherwise, if the editor decides to accept the reviewer's recommendation for rejection, the manuscript will be rejected.

28.4.6 Author's Response

If the manuscript is rejected, the author often sends it to another journal, with or without revisions. If invited to edit and resubmit the manuscript (Editor's Letter 28.3), experienced authors will always revise the manuscript according to editorial instructions.

Editor's Letter 28.3 is a letter of recommendation sent by an editor to an author. The second paragraph states: "Reviewers suggest a few minor revisions to your manuscript. Therefore, I invite you to respond to the comments of the reviewers and to revise your manuscript."

Editor's Letter 28.3 Invitation to resubmit the manuscript with minor amendments

Terry Wirtel <onehalfof@manuscriptal.com>   Thu, Nov 12, 2020, 5:32 AM

Dear Dr. Saliya,

Manuscript ID IJFE-20-0320 entitled "Stock Market Development and Nexus of Market Liquidity: The Case of Fiji" which you submitted to International Journal of Finance and Economics, has been reviewed. The comments of the referee are included at the bottom of this letter.

The referee suggests some minor revisions to your manuscript. Therefore, I invite you to respond to the referee's comments and revise your manuscript.

You can upload your revised manuscript and submit it through your Author Center. Log into https://mc.manuscriptcentral.com/ijfe and enter your Author Center, where you will find your manuscript title listed under "Manuscripts with Decisions".

28.4.7 Second Round of Review

If the recommended revisions are minor, this second round of revised manuscripts will often be reviewed by the editor himself and the manuscript accepted. If the recommended revisions are large, the revised manuscript will be sent to the original reviewers for further review. It is important to respond to reviewers' criticisms with polite testimony. The final chapter is devoted to provide examples of such responses. Such revisions may consist of two, three or four rounds of review. Such revised manuscripts are rejected if the reviewers are not satisfied with the revisions and explanations. Chapter 30 provides examples of such lengthy communications in several rounds. The following Editor's Letter 28.4 shows such a rejection after several such rounds. This rejection has been done by this editor with a very long explanation as the manuscript has been revised and submitted several times. Therefore, the decision to reject this decision was taken with great difficulty, she said. Below are just a couple of 200-word paragraphs out of 1,000-word letter.

Editor' Letter 28.4 Article showing rejection after several rounds of reviews

Critical Perspectives on Accounting

criticalperpectives@schulich.yorku.ca

via elsevier.com

Ms. Ref. No.: YCPAC-D-10-00090R4

Title: Rational Bank Lending Faces Local Habitus and Capital: Reflections on a Closely Held Bank in Sri Lanka

Critical Perspectives on Accounting

Dear Saliya,

As you know we have received split reviews on your submission.

I have spent some time carefully reading through the paper and the reviews and I am sorry to say that I feel that the paper is not of the standard required to publish in CPA.

Nonetheless I do feel that a paper which benefits from an author who worked in the case bank for many years and on a bank which went through two restructurings should have a lot to say which would be of interest to the readership of CPA.

…Was it really down to misapplication of bank rules because of poor leadership that the bank failed?

I know that after so many rounds of reviews (and I appreciate the improvements in the paper) it is tough to receive a rejection. I hope that I have provided some helpful suggestion about how to reconfigure your research in order to publish in CPA.

For your guidance, the reviewers' comments are included below.

Thank you for giving us the opportunity to consider your work.

Yours sincerely,
Christine Cooper
Managing Editor
Critical Perspectives on Accounting

This manuscript was later divided into three papers and published in three journals. These are the Examples P, Q and R expressions summarized in Part IV.

28.4.8 Copyediting, Proofreading, and Publication

Once a revised manuscript is accepted by the editor, a copy editor edits the paper's grammar, punctuation, documentation, style, and errors. The edited manuscript is sent to the author for review, usually with the modification function enabled so that the version can be easily viewed as a tracked MS Word document Microsoft Word document (usually as a Microsoft Word document in which the Track Changes function has been turned on so that the editing is easy to see) usually giving the author a few days to answer, approve, or reject any questions the editor may have, and request that the editor make suggestions and ensure that no errors. The author may make limited changes at this point, but publishers may be angry about this and charge a fee. The Journal, for all images, is permitted to publish those images, and enters into a copyright agreement (the author grants certain rights to the paper to its publishers). The paper is then electronically compiled and published. Perhaps there is a proofreading round, in which the author has the last chance to make the article, making sure there are no errors. The author usually has about forty-eight hours to respond to this.

28.4.9 Editor's Acceptance 'As It Is'

Although it is very rare for research papers to be accepted at the sole discretion of the editor without any review, there are such instances. One such article appears in Editor's Letter 28.5.

Editor's Letter 28.5 An rare occasion when a paper was accepted by the editor for publication as it is (without any revision)

Bruce Burton (Staff) <b.m.burton@dundee.ac.uk>   Thu, Jan 30, 2020, 4:15 AM

Dear Candauda,

I am happy to do this of course. I notice that this paper was originally submitted to a special issue but this issue has been put on hold. I apologise unreservedly as this information was not conveyed to you. I have read the paper myself however, and it is a strong study so if you wanted I would be happy to accept it in its current state for publication in a regular issue. If not, I would fully understand and I will remove it from our system.

Regards,
Bruce

Bruce Burton

Professor of Finance

Editor, Qualitative Research in Financial Markets (QRFM)

Director, Centre for Qualitative Research in Finance (CQRF)

School of Business, University of Dundee

28.5   Submission Process

The process of publishing a research paper involves the following activities:

- Choosing a journal.
- Format Formatting the manuscript according to the format of the journal
- Adjust the paragraphs to match the purpose of the journal
- Focus more on the literature review section
- Preparation of cover letters and
- Respond to critical reviews.

28.5.1   *Choosing a Journal*

Today, tens of thousands of academic journals are published. The following is a ranking of active peer-reviewed numbers in online, online edition of English language journals by 2018: over 9,000 in medicine and health, over 6000 in the social sciences, 3700 in business and economics. More, more than 2000 in education, more than 1500 in arts and literature, and more than 1000 in philosophy and religion. The total number of journals in each section is over 60,000. The number of peer-reviewed, English-language journals is always increasing, and the growth rate has been around 3 percent over the past two centuries, thus doubling the number of journals every twenty years. This means that many authors are looking for material to fill their journal every year. In 2018, nearly two million articles were published (Ware & Mabe, 2015).

The cost of choosing the wrong journal is not just a denial. It was noted above that authors cannot submit their manuscript to more than one journal at a time. It is an ethical violation to present one manuscript to several journals. So if you follow the proper procedure, you should wait until the decision comes before submitting it to another journal. Choosing the wrong journal can significantly delay publication, as the review process usually takes three to four months or even more. The following screenshot shows that after sending a manuscript for reviewing, even after 4 months none of the four reviewers have responded. All five reminders, two direct to the editor using "Post Comment to Editor" facility in submission system (see the screenshot), and three inquiries to the administrator of the journal (she in turn did draw editor's attention by forwarding author's emails), are unsuccessful.

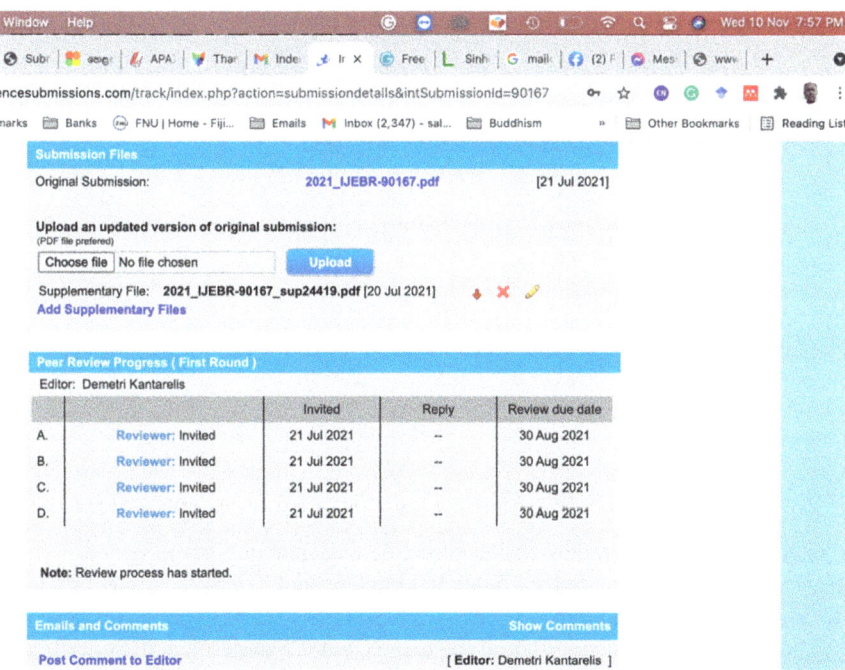

Therefore, for many other reasons, submitting a manuscript without exploring all the alternatives, not selecting the best/right journal for your manuscript, is not a good strategy.

Predatory journals: Journals that are predatory are not considered scholarly. Such journals charge a fee for processing and publishing. Such journals are not listed in databases such as Scopus, Web of Science or EBSCO. The publishing business of academic journals is now a multi-billion dollar industry. The so-called predatory journals do not even review, edit, or even read the manuscripts presented to them. Maybe they do"t even publish manuscripts. How do you identify such journals? If the journal you choose is not a publications of a reputed institution, this link is one way to find such predatory journals: https://predatoryjournals.com/journals/.

Possible journals: Be the first to check out the published journals with the sources quoted in your manuscript. Check the reputation of those selected journals in the rankings/indexing in world recognised rating agencies such as Scimago and ADBC et. Check if those journals are listed in databases such as Scopus, ProQuest or EBSCO. Then check whether the journal's objectives match the objectives of your manuscript and the content is within its scope. If the selected journals do not meet those criteria, check the reference list of the referred papers, in the list of sources you cite. Next you can inquire about the journals in the above Scopus/EBSCO and check if you can choose a journal. If none of these find a journal that fits your manucript, the next step is to find and research your title in ratings such as Scimago (the following is an example: https://www.scimagojr.com/journalrank.php).

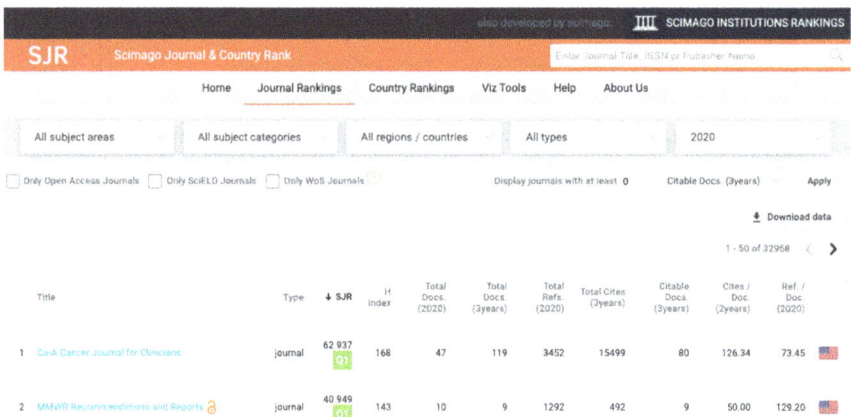

In addition, searching the websites of major academic publishers such as SAGE, Elsevier, John Wiley, Routledge (Tailor and Francis) and Inderscience can be a great way to find out about journals.

Finally, you can search for journals by using Google Scholar search. For example, when searching 'inequality' Google Scholar (left screenshot) for 'journals' published in the last three years (2019–2021) using 'Conditional' Advance search (with filters), you will find the following results (Screenshot on the right).

In addition to evaluating the quality of a journals such as reputation ranking, metric ranking and impact factor, it is helpful to consider the followings when choosing a journal.

- Editor's response to an e-mail regarding manuscript matching.
- Time taken for review.
- The number of issues published per year/volume.
- Number of articles in one issue.
- Whether there is an on-line publication.
- Percentage of rejection.
- Word limit for a mnuscript.

28.5.2 Formatting the Manuscript

Formatting a manuscript is a tedious task. This is because these formats vary not only from publisher to publisher, but also between different journals of the same publishing house. Formatting the list of references is also vary from journal to journal. Therefore, it may be helpful to keep a separate list of references for different genres (APA, Harvard and Chicago, etc.). The text formatting such as line spacing, font size and font, and positioning of the starting line in paragraphs etc. too are vary from journal to journal.

28.5.3 Re writing Certain Paras to Match the "Focus" or "Aim" of the Journal

Presenting the content of a research paper should be focused on a limited number of words, unlike an assignment, dissertation or thesis. A graduate research dissertation should often have a minimum number of words, but a research paper should often be framed within a range or maximum range of words (examples: Words in the rage of 4000–6000/6000–8000 or Up to a maximum word limit; 9000/12000 words etc.).

The most challenging part is editing a manuscript within the maximum word limit. The idea presented should be as concise and focused as possible and clearly presented with separate paragraphs.

The contents of these paragraphs vary from journal to journal. Partitioning the pamphlet Many journals state that the pamphlet should consist of parts in the following order;

1. Introduction
2. Methodology or Methods
3. Literature review
4. Results or Findings
5. Interpretations or Discussion
6. Conclusions.

Prior to the method section, the literature review section is offered in some journals. Even in the same journal, the order of presentation of these sections may vary according to the research method employed. For example, a research using grounded theory was data-based and initiated without any prior thought, therefore, the literature review comes in the discussion section in such manuscripts.

28.5.4   *Focusing the Literature Review Towards the Journal's Scope/Aim*

This section presents a brief analysis of relevant literature on the specific topic of the paper, constructs arguments, and builds hypothesis drawing from prior knowledge. Here it is important to pay more attention to the papers published in the same journal, because it will be easier to draw editors' attention. Even journal administration will encourage authors to do so as it will also improve the ranking and reputation of the journal.

28.6   Similarity Test

It may be helpful to have a similarity test before submitting a manuscript to a journal. This problem is most often encountered when creating several manuscripts from one data-set or when using the same method or model even for different research. When there are similarities like this, you have to face several issues.

28.6.1 Rejecting the Manuscript

The same method (SEM) was used for both analyse presented in two Examples 24.1 and 6.3 using same research data. Although the two research were completely different, one manuscript was rejected by the journal in the Example 24.1. The objection letter of that journal given in Editor's Letter 28.6.

Editor's Letter 28.6 Rejection letter by the journal that first presented Example 24.1 paper because the method used is similar.

Dear Dr. Candauda Arachchige Saliya,

Ref: Your submission entitled: Psychosocial factors of stock investors' inclination: The case of the South Pacific Stock Market -Fiji
Submitted to: Int. J. of Economics and Business Research
Submission code: IJEBR-80024

There is too high of a similarity index, such as from the following URL:

https://doi.org/10.1016/j.accre.2021.03.012

Kind regards,
The Inderscience Submissions Team
Inderscience Publishers Ltd.
newsubmissions@inderscience.com

28.6.2 Editing Similarities

An appeal was lodged against the above denial. The author was then instructed to edit and resubmit it. The letter is as follows Editor's Letter 28.7.

As pe their policy, they have explained that even if all the other requirements for accepting a manuscript are met, the similarity requirement must be met in order to be sent for reviewing, and they are;

- Similarities with different sources should not exceed 20%
- Similarity to the same source (including articles published by the same authors) should not exceed 5%.

The author has also been invited to edit and resubmit.

Editor's Letter 28.7 Letter informing the author to edit and resubmit the manuscript.

Article submission and peer-review system

Newsubmissions <newsubmissions@allsetbposervices.com>   Tue, Apr 6, 9:37 PM

Dear Author,

CrossCheck iThenticate, the standard Plagiarism Detection software tool used by scholarly publishers, has detected that your submission has an Index of Similarity larger than 5% with this article: https://doi.org/10.1016/j.accre.2021.03.012.

Regrettably, as long as the Index of Similarity with the above article is larger than 5% we will not be able to admit your submission to review.

The policy of Inderscience is to not admit to review, submissions with an overall Index of Similarity bigger than 20% or/and an individual Index of Similarity per source larger than 5% (including articles published by the same authors), except in the cases outlined on the Step 1 web page of the online submissions system, which you have used to submit your work.

You are welcome to resubmit a revised version of your submission, making sure to keep the total or overall Index of Similarity below 20%, and reducing the individual Index of Similarity with the above article (https://doi.org/10.1016/j.accre. 2021.03.012) below 5%.

Best regards,
Elle
newsubmissions@inderscience.com.

The presentation of the method in the above paper was rewritten in a different way and resubmitted, and the positive response receiver for that is shown in Editor's Letter 28.8.

Editor's Letter 28.8 Positive response when edited and resubmitted.

Article submission and peer-review system

Dear Dr.Candauda Arachchige Saliya,

Thank you for your recent submission, reference code IJEBR-90167, entitled,
'Psychosocial factors of stock investors' inclination: The case of the South Pacific Stock Market'

submitted to Int. J. of Economics and Business Research.

We are pleased to inform you that your article has passed the screening stage and is entering the review process.
Your article will now be checked to ensure it meets the subject scope and quality levels of the journal and will be sent
for peer-review if it is suitable.

You can track the progress of your article by logging in to the Inderscience Submissions system at https://www.
indersciencesubmissions.com/

Your username is: saliya.ca@gmail
You can get a password reminder on the log in page.

Thank you for considering this journal as a venue for your work.

28.6.3 *Providing References for Such Similarities and Avoid Plagiarism*

The methodology of the above paper was rewritten in a different way and crafted another paper (Example 28.1) using the same dataset and presented to another journal. The editor of that journal accepted the paper and recommended to provide references to the relevant diagram, which was considered similar. The letter is as follows Editor's Letter 28.9.

Editor's Letter 28.9 Letter referring to the similar paper and the advised to avoid plagiarism.

The author revised the manuscript providing proper references as advised by the editor. Editor's Letter 28.10 shows the positive response on the revision.

Editor's Letter 28.10 Positive response when edited and resubmitted.

MA3973: Notification on Submission; "Economics of Development" scientific journal.

managing_editor3@manuscript-adminsystem.com   Tue, Jun 1, 11:12 PM

Dear Candauda Arachchige Saliya,

The manuscript "Driving Forces of Individual Investors in Stock Market Participation", submitted to the "Economics of Development" journal, has been positively reviewed and is preparing for publication in the scientific journal.

Editor's comments:

Please modify the figures to show the continuation of the research that was published earlier. So that there are no replicating parts.

https://doi.org/10.1016/j.accre.2021.03.012

To make the changes visible, be sure to add a sentence in the text of the article that the results obtained are a continuation of the research carried out earlier.

In the "Conclusions" section, be sure to indicate that the results of revision and new research results allow for a new approach to solving the problem under study.

Following these instructions, the manuscript was edited for those changes.

Example 28.1 Driving Forces of Individual Investors in Stock Market Participation.

28.6.4 Websites for Similarity Checks

- https://www.turnitin.com/
- https://www.ithenticate.com/
- https://www.crossref.org/
- https://edubirdie.com/plagiarism-checker
- https://www.plagscan.com
- https://searchenginereports.net/plagiarism-checker

The following websites provide facilities to check the similarity of two paragraphs.

- https://www.youtube.com/watch?v=rpw91QhNAFA
- https://copyleaks.com/text-compare
- https://draftable.com/compare.

28.7 Popularising a Published Paper

Once a paper is published, it is also important to draw attention to it. The number of references they receive to your paper is critical to improve your academic recognition, scoring the required marks for carer advancements/promotions, and enhance the employability- more opportunities (when applying for new appointments). Here are some strategies you can follow.

28.7.1 Register in Academic Websites

The following are some of the popular websites for this purpose.

- https://www.researchgate.net
- https://www.researcherid.com (Web of Science)
- https://publons.com
- https://www.academia.edu
- https://www.mendeley.com
- https://www.emerald.com/insight/
- https://hq.ssrn.com
- https://scholar.google.com
- https://orcid.org
- https://www.scopus.com
- https://www.growkudos.com
- https://authors.repec.org
- https://ideas.repec.org
- https://gbsn.org.

28.7.2 Exhibiting or Displaying Publication

Paper display is not always possible because copyright may be violated. Then only the summary can be displayed. Publications that have open access often have the copyright with the author, so there is no issue in displaying the full paper. The above websites can be used to display leaflets in this way.

28.8 Collaboration

Developing research papers together with peers or colleagues is a very effective strategy and it is called developing articles "in collaboration". There, the division of labour allows more publications to published under their own name in a shorter period of time. Examples 8.1, 12.1, 27.1 and 30.1 are such papers developed in collaboration.

28.9 Summary

- Writing is an end-to-end task, but you should start writing as soon as you start thinking and reading about your research.
- There is no recognition of creative writing in research.
- Proper planning and organising enable to balance and strengthen the internal structure and present manuscripts that moves the argument along logical lines.
- A manuscript cannot be submitted several journals simultaneously.
- The review process is time consuming.
- Creating several different leaflets using the same database allows multiple manuscripts to be submitted to different journals at once.
- A manuscript goes through the following stages

  - Submission
  - Editorial review
  - Selecting peer-reviewers
  - Peer-review process
  - Editor's decision
  - Author's response
  - Copyediting, proofreading, and publication.

- Although it is very rare for research papers to be accepted at the sole discretion of the editor without any review, there are such instances.
- Predatory journals are not considered scholarly. Such journals are not listed in databases such as Scopus, Web of Science or EBSCO.
- Factors to be considered, when choosing a journal. Paper processing fees, editors' response to e-mails regarding suitability/matching with the scope/aim, review time, number of issues/volume. number of papers published per issue, whether there is an on-line publication.
- Rate/percentage of rejection of journals and word limit etc. are also critical when choosing a journal.

References

Belcher, W. L. (2019). *Writing Your journal article in twelve weeks, a guide to Academic Publishing Success*. Chicago: Chicago Press.

Berry, R. (2004). *The research project: How to write It*. London: Routledge.

Blaxter, L., Hughes, C., & Tight, M. (2008). How to Research. Berkshire, England: Open University Press.

Didham, R. K., Leather, S. R., & Basset, Y. (2017). Don't be a zero-sum reviewer. *Insect Conservation and Diversity, 10*(1), 1–4. https://doi.org/10.1111/icad.12208.

Holliday, A. (2002). *Doing and writing qualitative research*. London: Sage.

Murray, R. (2005). *Writing for academic journals*. Maidenhead: Open University Press.

Pears, R., & Shield, G. (2005). *Cite them right: The essential guide to referencing and plagiarism*. Newcastle upon Tyne: Pear Tree Books.

Seely, J. (2004). *Oxford A-Z of grammar and punctuation*. Oxford: Oxford University Press.

*The Chicago manual of styl*e, 17th ed. Chicago: University of Chicago Press (2017d). https://doi.org/10.7208/cmos17.

Ware, M., & Mabe, M. (2015). International Association of Scientifific. Technical and Medical Publishers. The STM Report: An overview of scientific and scholarly journal publishing.

Chapter 29
Cover Letters

A cover letter highlights the novelty of the research and provides an insight into the key points of your study. It is wise to include a cover letter even when it is not necessary, as it may be useful to bring it to the attention of the editor even before reviewing the manuscript. Sometimes, accepting or rejecting your manuscript may depend on the attractiveness of your cover letter. This chapter attempts to explain and facilitate how to write an effective cover letter to successfully present your manuscript to a reputable magazine.

29.1 Essential Features of a Cover Letter

You should first check the points/suggestions mentioned in the journal when drafting a cover letter for a particular selected journal. Cover letter allows you to include all the requirements. It is sometimes recommended to use specific expressions or words and tone that are unique to the journal when writing a cover letter. It will be useful for you to be aware of some of the key findings of your research paper. Here are some important points to be included:

- Use the name and position to address the editor.
- The title of the script should be mentioned as the headline.
- Use the name of the journal specifically when necessary.
- Explain why your study fits the scope and/or aims of the journal.
- A brief note on the research you have done, it should explain its importance.
- Highlight key points such as innovations and solutions in your research.
- Highlight the uniqueness of your article.
- Use a header to show your address and date.

29.2 Format of a Cover Letter

The recommended size for a cover letter is a single page. However, it is not important to count the words in a cover letter if they are effective within the range given in the journal. Highlight the importance and uniqueness of your research and try to link it with the purpose of the target journal. Below are some of the author's letters attached to copies of the illustrated publications discussed above that have been submitted to ten different journals.

29.2.1 Cover Letters to Be Submitted with Submisions

Examine how the Author's letter 29.1 and 29.2 highlight the content of the presentation match with the objectives/aim of the journal. Also pay attention how your research and manuscript contribute to the field of that particular journal (Bold italics).

Author's letter 29.2: A short cover letter to a Marxist magazine

Dear Editor,

Submission of a manuscript: Bank Credit System as an Exploitation Mechanism in Aggravating Income/Wealth Inequality

This paper contributes to fill a gap in the literature pertaining to the role of credit weapon and its co-integration with world inequality. This paper is a first of its kind to analyse critically certain banking practices in Sri Lanka which promote exploitation of public bank-deposits and capital accumulation to the private affluent class in the society from Marxist point of view. Therefore, **this paper addresses an area which was not adequately addressed by the Capital and Class before**.

I very much appreciate if you could please consider this paper for reviewing and to publish in your prestigious journal.

Best Regards,

29.3 Drafting Cover Letters After Checking the Suitability

Preparing a research manuscript for a journal format is a difficult task. Therefore, before submitting the manuscript, you may send an e-mail directly to the editor's email account to see if your paper fits the purpose of that journal. Depending on

Author's letter 29.1 The cover letter sent to International Journal of Finance and Economics

Edited By: Mark P. Taylor
Impact factor: 0.943
2019 Journal Citation Reports (Clarivate Analytics) B7/109 (Business, Finance)
Online ISSN: 1099-1158
© John Wiley & Sons Ltd

SCOPE

The International Journal of Finance and Economics aims to publish articles of **high quality** dealing with issues in international finance which impact on national and global economies. While maintaining the **high standards** of a fully refereed academic journal, **policy-makers** and practitioners …written at a non-technical, but academically **rigorous,** level. IJFE will concern itself with issues such as … and portfolio management, **financial market regulation, Third World** debt… …the financial aspects of transition economies, financial instruments and international financial policy co-ordination. IJFE is aimed at practitioners, researchers and graduate students in:-international economics-international finance-**financial economics**-international political economy -financial analysis and treasury management-**policy making**

Professor Mark P. Taylor
International Journal of Finance and Economics (IJFE)
Dear Editor,
Manuscript title: Stock Market Development and Nexus of Market Liquidity: The Case of Fiji
I am sending this paper to your prestigious journal **IJFE** because this research paper is a study which used a mixed-method approach; *a rigorous econometric analytics* were used to estimate the influence of the structural determinants while *in-depth personal communications and analysis …*
The study findings are important and useful to the **policy makers and financial market regulators** in Fiji. However…
Many determinants which have shown significant impact on SMD in other studies did not show any significant impact on **South Pacific Stock Exchange (SPX) in the Fijian context**
This paper contributes to fill a gap in the literature pertaining to the economic activities in **Fiji, a leading small-island nation in the pacific region.** This paper might be a first of its kind to **analyse critically the influential factors** of SMD in the SPX
I very much appreciate if you could please consider this paper for reviewing and to publish in your prestigious journal
Best Regards, 13 April 2020

the response of the editor, you can decide how much time and commitment you will spend on it. Below are two cover letters prepared after two such short email communications.

29.3.1 Negative Responses and Sample Cover Letters

The following is a detailed cover letter to the International Journal of Qualitative Method (IJQM), which specializes in qualitative research methods. This cover article

Author's letter 29.2   The cover letter submitter with Driving Forces of Individual Investors in Stock Market Participation

Aims and Scope	Editor,
The journal is aimed at … deal with problems of economics of development—…The realities of global economic changes prove the relevance of studying the economics of development from the standpoint of studying the uneven processes, …general tendencies of functioning of the world countries' economies (including with the use of quantitative methods)	Review of Economics and Finance (REF)
	Dear Editor,
	Manuscript title: Driving Forces of Individual Investors in Stock Market Participation
	I am sending this paper to your prestigious journal because the REF is a forum for debate and discussion on the theory and practice of finance in emerging markets. Also because of its emphasis on articles that are of practical significance, this study too is an empirical work, the results are significant both statistically and economically, and provide important information to Fijian policy formulators
Key topics:	
microeconomics;	
macroeconomics and monetary economics	
…formation of macroeconomic policy;	
financial economics …financial institutions and services;	
business economics;	
economic development, human development, financial markets, shadow economy, planning models and policy, innovations and inventions, processes and stimuli…	The main purpose of this research paper to introduce a rigorous quantitative design and a model to estimate the influence of the socioeconomic and psychic-cultural drivers by constructing a mediator variable 'investors' willingness', using AMOS. We use data from the South Pacific Stock Market (SPX) and Fijian university-educated people (employed) for this estimation
	This paper might be a first of its kind to estimate the influence of psychosocial factors on stock investors' behaviour
	I very much appreciate if you could please consider this paper for reviewing and to publish in your prestigious journal

was prepared following a brief text exchange with Professor Norman K. Denzin, a distinguished scholar who is the Editor-in-Chief of Qualitative Inquiry (QI), another prestigious quality journal. In this communication, Professor Norman has said that this paper is more in line with IJQM journal than his QI journal. This response was very friendly but negative. But on that advice the following cover article was prepared.

Email 29.1: A message sent to another magazine editor on the advice of one editor

Professor Linda Liebenberg

Dalhousie University, Canada

Editor-in-Chief: International Journal of Qualitative Methods (IJQM)

Dear Editor-in-Chief,

Doing Case-study Research: A Concise Practical Framework for Research Students

I am Dr. ….. attached to ……. University, New Zealand. Professor Norman K. Denzin, Editor: Qualitative Inquiry, suggested that IJQM is an appropriate journal for my article; "Doing Case-study Research: A Concise Practical Framework for Research Students". After referring to a few relevant articles in IJQM (for example; https://journals.sagepub.com/doi/full/10.1177/160940691 9862424 and https://journals.sagepub.com/doi/full/10.1177/1609406918801730), I communicated with your Associate Publishing Editor, Ms. Kristina Moulton and her response to the Abstract of my article; "This paper sounds like a great fit for the journal", encouraged me to format it to IJQM and submit.

This paper is not a highly technical article, and therefore easily accessible to an audience with a general methodological interest. The key methodological contributions are (a) attempts to encapsulate the basic components of social research into seven pillars (7Ps) namely; Paradigm, Perspective, Purpose, Plot, Practice, Procedures, and Persuasion.; (b) demonstrates, with real world examples, the holistic approach to case study research as a concise practical guide for business research students.

Irrespective of the methods followed and the types of data, the credibility of research outcomes largely depends on auditability of all research engagements and verifications. Therefore, a cohesive arrangement of the entire research exercise in the most transparent manner is essential for thorough scrutiny. The suggested 7Ps framework serves that purpose.

I draw case study examples from my own Ph.D. thesis available online to show how to align the research context with philosophical issues (ontological and epistemological); articulate research problems and choose a methodology; theorise research findings using classical reasoning methods of abduction, deduction, and induction and; expand the theorisation beyond the original research problem.

The structure and discussion of this paper are presented in a very simple format and language so that it would especially appeal to budding academic researchers.

Best Regards,

Sincerely,

29.3.2   Positive Responses and Sample Cover Letters

Editor's Email 29.1: Positive responses

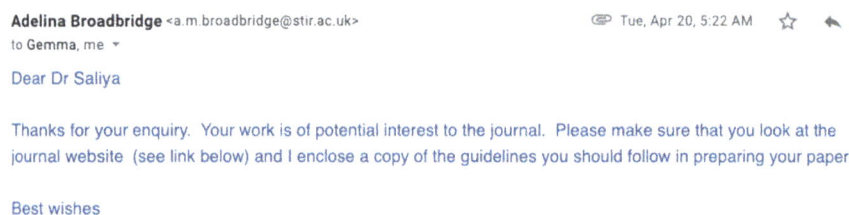

Adelina Broadbridge <a.m.broadbridge@stir.ac.uk>   ☞ Tue, Apr 20, 5:22 AM   ☆   ↰
to Gemma, me ▾

Dear Dr Saliya

Thanks for your enquiry. Your work is of potential interest to the journal. Please make sure that you look at the
journal website (see link below) and I enclose a copy of the guidelines you should follow in preparing your paper

Best wishes

Adelina

Editor, *Gender in Management: An International Journal*
https://www.emeraldgrouppublishing.com/journal/gm

Adelina Broadbridge
MWO, Stirling Management School
University of Stirling
STIRLING
FK9 4LA

29.4   Summary

- The cover letter is useful to get the editor's attention before reviewing the manuscript.
- Check the points/suggestions contained in the cover letter provided by the selected journal.
- Highlight the uniqueness of this study in a cover article, how it fits into the scope and/or aims of the journal, the novelty of the research, the objectives and the targeted solution.
- Whenever possible it is helpful to consult the editor before submitting the manuscript.
- The recommended size for a cover letter is a single page.
- ☐ The cover article highlights the novelty of the research and gives an insight into the important highlights of your study.
- Acceptance or rejection of your manuscript may depend on the attractiveness of your cover letter.
- It is useful to highlight important key findings in your cover article and innovations in your research paper.

References/Further Reading

Murray, R. (2005). *Writing for academic journals*. Maidenhead: Open University Press.

Seely, J. (2004). *Oxford A-Z of grammar and punctuation*. Oxford: Oxford University Press.

Chapter 30
Addressing Reviewers' Critics

30.1 Introduction

Receiving reviewer comments can be hard to swallow, but ultimately the constructive criticism will make for much stronger written work. Responding to reviewer comments should be complete, polite and supported with evidence (Aurini et al. 2016, p. 217).

Reviewers' critics can often embarrass you, but at the end creative criticism will lead to a stronger paper. Responses to reviews should be complete, polite, and evidence-based.

You can gain an in-depth understanding of some real-world examples discussed above and also how to present complete, polite, and evidence-based response to reviewers' critics. These reviews and responses can be categorized into four categories.

A. Accepted with minor amendments: Example 28.1 (Driving Forces in SMP) and 6.1 (Conducting Case-study Research).
B. Invited to resubmit with major amendments: Example 12.1 (FBACC), 13.1 (Risk Preparedness) and 27.2 (Creating and Reinforcing Discrimination).
C. Invited to resubmit with major revisions: Example 28.4 (The Role of Credit Weapon) and 30.5 (Cultural Politics of Bank Lending).
D. Rejected with serious lapses. Taking such reviews into account, the pamphlet can be redesigned and submitted to another journal: Example 6.4 (SMD in Fiji), 12.1 (FBACC) and 25.2 (Credit and Employment).

30.2 Criticisms and Feedback Leading to Minor Edits

For such reviews, editors expect minor edits and explanations. Authors' responses to such reviews are not sent back to reviewers for their approval. It is enough for the editor to be satisfied with the authors' responses. One such instance is the Editor's letter 28.9, received for Example 28.1, with a version of paragraph 28.5.3 above.

Criticisms and feedback leading to significant edits

Examples for a Quality Research Review

Reviewers' comments and authors' responses 30.1 Provides an examples of reviews received by the editor and forwarded to the authors for Example 12.1 (FBACC), and how the feedback from the authors was crafted. The specialty of these reviews is that the two reviewers have conflicting opinions. The first reviewer states that the methodology is not explicitly reported (Methodology: The methodology is not clear and appropriate) and the second reviewer states that the methodology is clearly presented. (The methodology considered in the study is comprehendable [sic] and all the requirements has been properly fulfilled as the researcher used all the relevant points in a comprehensive manner).

Reviewers' comments and authors' responses 30.1: Reviews by reviewers and authors.

Dear Prof. Bruce Burton.

Manuscript ID QRFM-05-2020-0087: Financial Battle Against Climate Change—Assessing Effectiveness using a Scorecard.

We are indeed very thankful to you for giving us the opportunity to address the comments made by the reviewers and to make amendments for you to consider in making your decision.

We also wish to thank the reviewers for their constructive feedback, and for their appreciation of our hard work on this important study, especially for emerging small economies disproportionately exposed to climate change.

We have made relevant amendments to the paper, indicated in **bold**, and provided below supported by our explanations in *italics* wherever necessary.

We also note that concerns/issues raised by one reviewer are perceived positively by the other reviewer; providing explanations, we have underlined them.

Comment; reviewer 1	Response/amendments/revision
1.1. As a whole the authors should have taken more care while writing sampling techniques, results and discussions more elaborately	*We added the following para to Sub-Sect. 2.1: Procedures in the Methodology section*: **The survey participants were selected using our professional networks of accountants and bankers, academic colleagues and postgraduate students, therefore the sampling techniques we use are convenient and purposive ones** Results and discussion: *This is a good suggestion. We added the following analysis to Sub-Sect. 4.3: Survey results which provide a few graphic presentations with very brief descriptions, because this section already has over 1600 words (17% of 9400 words).* **Figure 3, which consists of three charts, shows the relative contributions of CFs and perspectives in constructing the overall score**

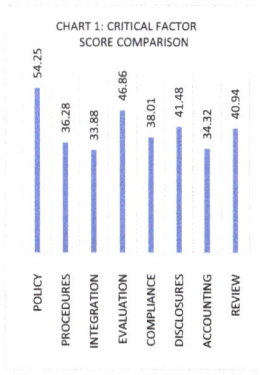

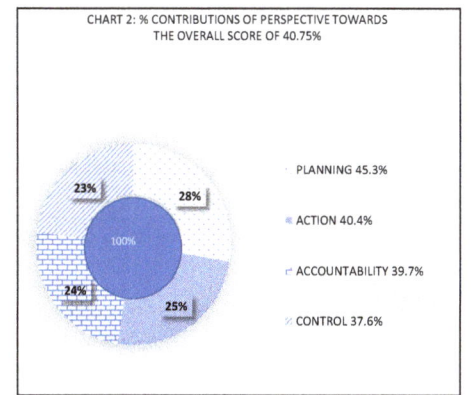

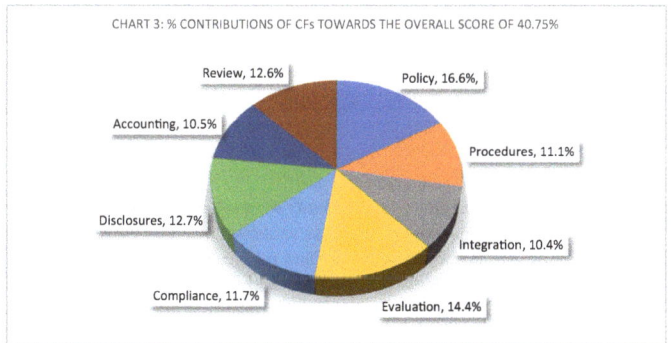

Figure 3: Relative contributions and importance of CFs and Perspectives in the final score
Chart 1 portrays the relative importance of the CFs while Chart 3 graphically depicts the relative contributions of the CFs; Chart 2 shows the relative contributions of the Perspectives towards the overall assessment of FFBACC

(continued)

(continued)

Comment; reviewer 1	Response/amendments/revision
1.2. Abstract missing some important elements like methodology, important findings and one or two recommendations	*We added the following sentence to the Abstract* **"This study uses a mixed-method methodology"** *We further added the following to the findings;* **The relative contributions of each perspective in constructing the overall score are distributed as 28%, 25%, 24% and 23% respectively between Planning, Action, Accountability and Control** *For recommendations, we reworded the following;* **Research limitations/implications**—These results …confirmed that the existing political initiatives **need to be** effectively communicated **and/or** implemented in the financial system by the regulatory agencies. *Also under practical implications we have elaborated the usefulness of our research findings*
1.3. Justifications of the study is not clearly explained and needs to focus on the research gap	*We added the following sentence to the Introduction* **With regard to financial systems, many researchers have examined how climate change and mitigating/preventive/resilient policies could affect a central bank's ability to meet its economic and financial stability objectives (see Batten, Sowerbutts, & Tanaka, 2016). There are also a few theoretical frameworks, such as the Sustainable Banking Assessment (SUSBA) and the Climate Change Governance Index (CCGI), that have been introduced to assess the preparedness of financial institutions for climate-related risks. However, little research has been carried out to assess the preparedness of a whole financial system of a country, especially in the Pacific region where climate-related risks are pertinent. This study attempts to fill this knowledge gap by introducing a Scorecard to assess climate-related financial risks in Fiji**
1.4. There is a need provide more clarity on methodology	*We added more detail, however, the other reviewer says*: "The methodology considered in the study is comprehendable [sic]…"
1.5. How did you select high ranked officials to conduct interviews? How many high ranked officials selected for the conducting interviews?	*We have provided these details under Sect. 2.1: Procedures in the methodology section* **The survey participants were selected using our professional networks of accountants and bankers, academic colleagues and postgraduate students; therefore the sampling techniques we used were convenient and purposive ones** *The first few high ranked officials were contacted thanks to the fellow Chartered Accountants, Bankers and Postgraduate student. Then, we asked each interviewee to put us in touch with other similar officials known to them*

(continued)

Comment; reviewer 1	Response/amendments/revision
1.6. How did you select auditors and academicians? How many auditors and how many academicians?	*We contacted Auditors through our personal connections, we are both chartered accountants. Academicians are our colleagues* *On page 6, under procedures we have provided these numbers as; "…23 accountants/auditors, 107 related academics/students, and 82 bankers/financiers/stock brokers."*
1.7. What type of sampling techniques did you use for selecting six commercial banks and other financial institutions?	*Sampling techniques were explained above in 1.3* *We have contacted all six commercial banks*
Further questions	
1. Originality: The paper contain the original content but there is no clarity on the purpose and justification of the paper	*Please refer to 1.3 above*
2. Methodology: The methodology is not clear and appropriate	*We attempted to refine this section further by obtaining an independent fellow researcher, and we added the following paragraph:* *Reviewer 2 contradicts this comment since he commented:* "The methodology considered in the study is comprehendable [sic] and all the requirements has been properly fulfilled as the researcher used all the relevant points in a comprehensive manner" *in his comment under Methodology*
3. Results: The data analysed by using BSC and presented clearly	
4. Implications: clearly explained the implication of the research findings	
5. Quality of Communication: well-structured and well written	
Comment; reviewer 2	**Response/amendments/revision**
2.1 Yes, the paper has certain amount of originality value but doesn't contain significant information to justify the publication as all the information provided is not in synchronized manner and has many quotations but no proper citation And the linkage between the variables has not been explained properly with citations	*We attempted to address this issue and provided all citations promptly* *We added the following para to address this issue, please also refer explanation given for Comment 2.2 below* **We followed a data-driven, qualitative, inductive approach where linkages between variables are not established in advance (as in the theory-driven deductive tradition) (see Chang & Chen, 2018; Niemeijer, 2002) but later in the results and discussion while assessing the impact of them on latent variables such as Critical Factor and Perspectives and finally, the overall score (see Kaplan & Norton, 1996; Saliya, 1999)** *We added the following references;* Niemeijer, D. (2002). Developing indicators for environmental policy: Data-driven and theory-driven approaches examined by example. *Environmental Science & Policy, 5*(2), 91–103. DOI: 10.1016/S1462-9011(02)00026-6 Chang, W., & Chen, X. (2018). Monthly Rainfall-Runoff Modeling at Watershed Scale: A Comparative Study of Data-Driven and Theory-Driven Approaches. *Water, 10*(9), 2–21. DOI: 10.3390/w10091116

(continued)

(continued)

Comment; reviewer 1	Response/amendments/revision
2.2 The literature is inadequate and doesn't give apt information about the variables considered in the study as no past study related with these variables has been explained properly The relationship between the variables that shows the effect of independent variable on dependent variable has not been explained in a systematic manner. The variables are explained individually but not in terms of relationship they need to show	*The latent Critical Factors (CFs) are described, under each Perspective, in detail in Table 2 with the measured criteria (independent variables) used to assess the scores of those CFs, then the Perspectives* *Unlike the theory-driven deductive quantitative, approach, in this study, we followed the data-driven qualitative, inductive approach, hence the establishing of links/relationships are discussed in the results presentation and discussion stage. Therefore, we did not refer to literature in order to construct hypotheses as we do not intend to run any regression but rather, to analyse the survey data using describing techniques, backed by thematic analysis of the narratives provided by the interviewees. The results and analysis are positively acknowledged by both the reviewers elsewhere*
3. Methodology: The methodology considered in the study is comprehendable and all the requirements has been properly fulfilled as the researcher used all the relevant points in a comprehensive manner	*We attempted to address this issue further as explained in 2.1 above as follows (repeat):* **We followed the data-driven, qualitative, inductive, approach where linkages between variables are not established in advance (as is done in the theory-driven deductive tradition) (see Chang & Chen, 2018; Niemeijer, 2002) but later in the results and discussion while assessing the impact of these on latent variables such as Critical**
4. Results: Yes, the results is presented clearly and analysed properly. As the result are below the benchmark therefore, the conclusion inadequately tie other elements all together	**Factor and Perspectives and finally the overall score (see Kaplan & Norton, 1996; Saliya, 1999)** *We added the following references;* Niemeijer, D. (2002). Developing indicators for environmental policy: Data-driven and theory-driven approaches examined by example. *Environmental Science & Policy, 5*(2), 91–103. DOI: 10.1016/S1462-9011(02)00026-6 Chang, W., & Chen, X. (2018). Monthly Rainfall-Runoff Modeling at Watershed Scale: A Comparative Study of Data-Driven and Theory-Driven Approaches. *Water, 10*(9), 2–21. DOI: 10.3390/w10091116

(continued)

(continued)

Comment; reviewer 1	Response/amendments/revision
5. Implications for research, practice and/or society: the paper doesn't clearly identifies any implications as the the researcher only suggests the broad aspect related with climate change and financial battle Moreover, the paper doesn't bridge the gap between theory and practice as the results of the survey shows all the perspectives below the benchmark. Hence, the results are underprivileged to show the relation between theory and practice The implication of the research shows the overall effectiveness of climate change on financial battle but doesn't explain the adequacy in an detailed manner The researcher compiled the results for the implication in a broad and general manner and not specified the role of the research for the betterment to combat financial battle	*We have added the following paras to the Conclusions section* **The theoretical, managerial and practical implications of this type of research include: (a) the findings could help financial institutions in the development of internal policy in relation to the effectiveness of climate change impacts on financial products, including financial lending; (b) these results, which were complemented by the information shared during the interviews, confirmed that the existing political initiatives need to be effectively communicated and/or implemented in the financial system by the regulatory agencies Further, this BACC scorecard can be applied to other under-developed financial systems in emerging countries to assess the effectiveness of sustainable banking regulations and/or guidelines in those countries in relation to the FBACC. It can also be applied to individual firms to assess their contribution to the FBACC** *We have also added the following to the Introduction the newest literature applicable to this study which might shed light on these concerns as well, in general* **28 Recently, the New Zealand Government confirmed it is seeking to make climate-related financial disclosures mandatory for all publicly listed companies and large financial services organisations. While this amendment to the Financial Markets Conduct Act (2013) still needs to be approved by Parliament, impacted organisations could be required to make disclosures as early as 2023 (PWC NZ, 2020)** *We have also added the following to the Introduction appropriately.* **According to the proposed amendment to the Financial Markets Conduct Act (2013) of New Zealand, the impacted organisations will be required to comply with the disclosure reporting requirements, and to assess the risks and opportunities of climate change to their business across four thematic areas: Governance, strategy, risk management, metrics and targets (PWC NZ, 2020). This study also encompasses these four areas by identifying related critical factors and the set of measuring criteria** PWC NZ (2020). Climate-related financial disclosure to be mandatory for NZX and financial services sector. https://www.pwc.co.nz/services/consulting/sustainability/insights/climate-related-financial-disclosure-to-be-mandatory-for-nzx-and-financial-services-sector.html

(continued)

(continued)

Comment; reviewer 1	Response/amendments/revision
6. Quality of Communication: The language used in the paper seems more technical rather than simple and precise as if non finance background individual would not be able to understand it The terminologies are also not explained properly by the author. whatever the paper is trying to comprehend is not coming through effectively and clearly The communication of the paper is bit more confusing due to jargon	*We quote the following comments of both the reviewers which say the paper is well written, however, we have attempted to remove jargon as much as possible*
	Reviewer I: 4. Results: The data analysed by using BSC and **presented clearly** Reviewer II: 4. Results: Yes, the results is **presented clearly and analysed properly** Reviewer I: 5. Implications for research, practice and/or society: clearly explained the implication of the research findings Reviewer I: 6. Quality of Communication: well-structured and well written

We hope that our responses are sufficient and fully address, to your satisfaction, the issues raised by the reviewers. We would very much appreciate it if you could let us know if further refinement is necessary for publication.

Yours faithfully,

Please note the politeness, even if you do not agree, you can agree to disagree, and highlighting the points you want to stress using different colours etc. in the examples provided here.

Examples for quantitative reviews

Reviewers' comments and authors' responses 30.2 provides the reviews forwarded to the authors by the editor of Advances in Climate Change Research (ACCR) journal for the manuscript titled "Risk Preparedness….(Example 13.1) and the responses of the authors. The speciality about these reviews is that even the issues pointed out by the reviewers were not critical, the responses are very long and technical.

Reviewers' comments and authors' responses 30.2.

The Editor,

Advances in Climate Change Research Journal.

Here we indicate our responses to the reviewers. We also wish to thank the reviewers for their constructive feedback, and for their appreciation of our work focusing on emerging small economies which are struggling to manage their financial risks associated with climate change.

We have addressed all reviewers' concerns and our responses are presented below in **bold**. The added texts are underlined, and the relevant line numbers of the manuscript are provided in *italics*.

Reviewer 1

(1) The opening paragraph should establish a clearer message and stronger connection between the sentences. For instance, while the emphasis of the first sentence appears to be on the vulnerability of the financial system to climate change, the second sentence has jumped right into the topic of sustainable finance.

This is a very good suggestion. We have added the following sentence to connect the two premises.

It is noticeable that financial sectors have been late to respond to issues related to environmental sustainability and climate change (Carbon Disclosure Project, 2013).

Line numbers: 29–30.

(2) Messages appear to be unclear in the final sentence of the second paragraph on P.3.

We have edited the sentence as follows;

According to Australian Accounting Standards Board/Auditing and Assurance Standards Board [AASB] (2018), information (omitted or misstated) becomes material if it could influence decisions that users make on the basis of financial information about a specific reporting entity.

Line numbers: 59–62.

(3) There appears to be a repeated sentence "There were no significant cross-factor loadings" toward the end of P.10/start of P.11.

Thank you for pointing out this mistake, we deleted the repeated sentence.

(4) A minor typo on line 6/P.10, should be suggest instead of suggests.

We have corrected this typo.

(5) In the conclusion section, the first paragraph on P.15 may need a bit more elaboration.

Thanks, the following sentence was added to elaborate this paragraph.

These results confirmed that the existing political initiatives need to be effectively communicated and/or implemented in the financial system by the regulatory agencies.

Line numbers: 308–309.

Reviewer 2

1. The first problem is that there are endogenous problems in the questionnaire and principal component analysis. The author used sixteen questions to do principal component analysis. However, the author seems to have formulated four factors first, and did not let the correlation of variables form factors like principal component analysis, and then name the factors. So the author may have problems setting the four factors first, and I have not seen the correlation coefficient matrix of the factors and the correlation coefficient matrix of the variables.

Thank you for this comment.

We took the magnitudes and significance of correlation coefficients of study variables into account in forming the four perspectives. Correlation coefficient matrix are provided in the Annexure 1.

2. This article does not discuss financial risks and climate risks in empirical studies. How to confirm that these four factors are important factors affecting climate risk? It also does not measure the transition risks and physical risks in climate risk, so there is no direct evidence to directly identify these four factors **as important factors affecting climate risk.**

We thank the reviewer for raising this important point. Our focus is only on financial risks as of perceived climate change. Therefore, we assess the impact of the perceived climate change on the stability of the financial system using identified variables as proxies. The four factors are hypothesized based on extant literature and the two frameworks; SUSBA and CCGI (Cogan, 2006, RBA, 2019, RBF, 2018, WWF, 2019a & 2019b). SUSBA used six pillars (Purpose, Policies, Processes, People, Products and Portfolio) based on 11 indicators, and CCGI used 14 indicators to evaluate corporate climate change activities in five governance areas; Board Oversight, Management Execution, Public disclosure, Emission Accounting and Strategic Planning.

After reviewing all these pillars, governance areas and indicators, we initially formulated 30 questions and conducted Principal Components Analysis (PCA) which is a variable-reduction technique in SPSS. PCA results produced nine 'principal components' covering 69.42% of cumulative variance of the variance in the original variables (Annexure 2—Table A and Graph 1. Based on SUSBA and CCGI theories, knowledge on the Fiji context, together with the level of contributions of each factor (e.g., Eigenvalues of more than 2, please see graph 1 below) and correlations and component matrices among all items we identified substantively important and empirically powerful four components/factors and constituent items (Annexure 2—Table B) which explained 50% of variance. The following sentences are added appropriately;

We really blended our experience and background knowledge of Fijian context in this exercise. As shown in Fig. 1, this was a process combining theories, local knowledge and, empirical findings (data).

Line numbers: 136, 137 & 140.

3. How did the sixteen questionnaire questions of this paper come out?

Thank you for this comment. We also added the following paragraph to clarify this issue on page 7 under methodology.

We considered questions having more than 0.45 of loadings under the four most prominent of components with the total Eigenvalues of higher than two. We really blended our experience and background knowledge of Fijian context in this exercise.

Line numbers: 138–139.

4. How to deal with neutrality in the questionnaire?

Weights of 100, 0 and 50 were attached for the responses Yes, No and Neutral respectively for statistical analysis (page 8, under 2. Methodology).

5. How to identify or check "yes" and "no" in the questionnaire? Questionnaires often ask if you know? Is this far from climate risk? Knowing is only the first step in perceiving risk. If the questionnaire asks if you know, why is it neutral?

The questions are very straight forward prompting the answers 'Yes', 'No' or 'Neutral' in three columns/boxes which the participants check the relevant box. None of the questions starts with" If you know…". Therefore, one can choose Neutral if they are not aware of the issue.

Thanking you,

30.2.1 Authors' response for a reviews to an abstract

Reviewers' comments and authors' responses 30.3 provides the reviews on the abstract of Example 26.5 received by the editor from the reviewers, and the authors' responses thereof.

Reviewers' comments and authors' responses 30.3: Authors' responses to reviews.

Responses to the reviewer's comments.

Thank you for the constructive comments and I would like to respond to those comments as follows in ***bold italics***.

Responses to the reviewer's comments.

Thank you for the constructive comments and I would like to respond to those comments as follows in *italics*.

Abstract Number: 17–113

Abstract Title: revised to—**Dynamics of credit decision-making: a taxonomy and a typological matrix**, *Review of Behavioral Finance*

Tract Acceptance:

Please indicate if the abstract is acceptable or needs further development

If further development is needed, please provide feedback for the author(s)

Acceptable

Needs Further Development

Definitely needs more refinement/and stating of outcomes; the results don't explain how credit decision making is going to be better understood or improved

The results and conclusions revised as follows;

Results: The research revealed two major traits/characteristics, based on which credit decisions are made; (a) the nature of decision-making (**F**ormal or **I**nformal approach) and, (b) manifested sense/logic (**R**ational or **I**rrational judgment)

Conclusions: Based on the above results, four types of decision-makers are identified; **BOSS, ROBOT, REBEL** and **BUDDY** as illustrated in the table below. ...these findings provide guidance for better understanding of how and why credit decisions are made and could empower decision-makers and decision-seekers to employ appropriate techniques to make effective credit decisions

Topic	Objectives	evidence	Conclusions
Topic Clear communication of ideas	Objectives and Key points are soundly presented	Authors discuss findings and/or supportive evidence Key points and analysis are viewed as credible	The conclusions/implications are sound and justified

(continued)

(continued)

Unclear objectives		
The title needs to be more descriptive of the research problem to be solved	Methods are stated, but findings not summarized	No conclusions given; I was unsure what research problem was being solved
The introduction is not very clear to me	Now the results section read as; "The research revealed two major traits/characteristics, based on which credit decisions are made; (a) the nature of decision-making (**F**ormal or **I**nformal approach) and, (b) manifested sense/logic (**R**ational or **I**rrational judgment)."	Research problem of "how and why certain credit decisions are made" is solved by revealing some influencing factors of credit decision-making processes and identification of four types of decision-makers as described above. The research results also could be helpful in selecting appropriate influencing tools/methods/techniques for better credit decision-making
The title changed to "Influencing factors of credit decision-making beyond key 5Cs of credit evaluation"		
Introduction section attempts to explain **a** few of existing traits/characteristics or influencing factors of credit decision-making beyond the key 5Cs; Capital, Character, Condition, Collateral and Capacity. This paper is challenging those factors and tries to introduce other influencing factors as explained in the results section		
Reciprocity means "a relationship between people involving the exchange of goods, services, favours, or obligations, especially a mutual exchange of privileges between trading nations" (Encarta Dictionary English UK)		

Any other comments/feedback

30.3   Serious Reviews Expecting Major Revisions

Examples for reviews of qualitative manuscripts

Reviewers' comments and authors' responses 30.4 shows the criticisms provided by the reviewers on the manuscript (**Example Q**) submitted to a reputed international journal "Critical Perspectives on Accounting" (CPA). Author provided in-dept responses to these critiques, yet the it was rejected and the disclaimer is provided as shown in Editor's letter 28.4. This manuscript was later published by Journal of Applied Economics.

These reviewers have focused heavily on the theory used to interpret research results. It was a Marxist point of view. You can get some idea of how to respond to such a critique by paying attention to how the responded were crafted.

Reviewers' comments and authors' responses 30.4: Reviews and authors' responses.

*Thank you very much for your very constructive comments. Please see my efforts to address the issues (given below in **Bold italics**) giving references to the relevant pages of the article.*

Discriminatory lending: evidence from Sri Lanka

Review

This is an interpretation of two specific "cases" concerning credit lending by Sri Lankan commercial banks that reflect somewhat lenders' discriminatory behaviour. The paper is theorized from a broadly Marxian perspective. While the paper has some potential, as it stands now, it needs major revisions. I have several major concerns.

Firstly, on the use of Marx: I am convinced that Marx was skeptical of credit system as it represents power of a social class. However, I am not convinced why Marx came to be central in the paper. While I agree that frameworks drawn from the reading of Marx is pertinent to a CPA publication, I cannot see any discussion approaching the author's reading of Marx. I thus suggest that paper must need a thorough discussion of alternative approaches (e.g. functional analysis versus critical analysis) to theorizing lending while making the case for the use of Marx. Even though there is a case for the use of Marx, I cannot see a coherent framework in the paper. Rather, there are some ad hoc references to secondary materials dealing with reading Marx.

Finally, as a paper to be published in CPA, there is no discussion showing how this paper makes a contribution to accounting research, despite there are references to accounting papers published in CPA. Areas such as development finance or micro fiancé (there is a reference to Yunus which is in the micro finance literature) would be accounting (broadly defined), but I cannot see that this paper has been located even in that literature.

Hopefully, the above comments give a good indication of my concerns. If the paper seeks to comment about process within a particular context, then it needs to provide information about the process and the context.

****

Marcia Annisette
Co-Editor, Critical Perspectives in Accounting
Schulich School of Business
York University
4700 Keele Street, Toronto M3J 1P3
Canada

These responses were very lengthy. However, it exemplifies the fact that this paper was eventually rejected with a very empathetic note by the editor.

Examples for quantitative reviews

Reviewers' comments and authors' responses 30.5 shows examples of serious criticism on the manuscript C, when it was first presented to the prestigious journal HELIYON (6.4). The authors responded at length, in-dept, but they were not sufficient to convince either the editor nor the reviewers. However, taking those critiques into account the manuscript was revised and submitted to the International Journal of Finance and Economics published by renowned a US based John Wiley publishers (Scimago ranked Q.2), they accepted with minor revisions for publication (Editor's letter 28.3).

It is worth examining carefully how critical these reviewers are. You can get some idea of how to respond to such a critique by also paying attention to these authors' responses carefully.

Reviewers' comments and authors' responses 30.5: Reviewers' critiques and authors' responses

Manuscript Number: HELIYON-D-20-00182

Title: Stock Market Development and Financial Liberalisation: The Case of Fiji

Journal: Heliyon.

Dear Editor,

Thank you for all your prompt feedback and giving me the time and opportunity to respond to the reviewers.

Please note that my paper is an analysis from the perspective of critical finance, supplemented with more descriptive and narrative analysis.

I also take this opportunity to thank both reviewers for their valuable comments.

Editor and reviewer comments	Author response revision/amendment [page no.]
Reviewer #1: **Methods**: Why are you using panel data? You have one country and Fiji's stock market so why panel data?	Thanks. This is an error. I meant longitudinal data, now changed. The term 'panel data' is replaced with the term 'time series' in two places. **Revision/amendment**: The present study uses annual time-series data ranging from 1996 to 2018. [**Page 10**]. Therefore, the time-series regression…[**Page 11**]
You need to connect your regression framework to the literature. For example there are studies on Fiji's stock market that you simply are unaware of. See Narayan and Smyth Journal of Asian Economics. You can use this paper as a starting point to extend the work	I am aware of these articles but they provided only limited help as they are not on stock market development (SMD) but are rather on other economic issues—whereas mine is more financial analysis specifically on SMD of SPX with a blend of critical theories. **Revision/amendment**: Apart from an econometric study on the determinants of price clustering in the SPX (Narayan and Smyth, 2013), to our knowledge, there are hardly any publication on SMD, especially from the perspective of critical finance. [**Page 3**]
I would recommend using an ARDL framework–see the Fiji literature in journals such as Economic Modelling, Applied Economics, Journal of Developing Areas etc.	Thank you for this suggestion. The ARDL procedure followed and amended/revised the whole section accordingly. The interpretation & discussion updated with new statistics (coefficients and p-values) in the text. I also added the normality tests, the unit root tests for non-stationarity and Dickey-Fuller tests for co-integration; long-run relationships (as suggested by Reviewer 2). **Revision/amendment**: Tables 4, 5, 6 & 7 [**Pages 12–17**]

(continued)

(continued)

Results: I am not sure what to make of your results because I do not understand your econometric approach. There is a large literature on Fiji–see the Narayan papers	*Sorry for this. This paper goes beyond econometric analysis into deeper "how" and "why" reasoning. My attempt is to analyse the SMD and SPX from a critical financial point of view. However, the title "Financial Liberalisation" is now replaced with "Nexus of Market Liquidity".* **Revision/amendment**: The title: Stock Market Development and **Nexus of Market Liquidity**: The Case of Fiji
Interpretation: Neaten up your tables. That's not the way to make tables. Similarly, equations are incorrectly written, they are missing the subscripts and parameters. Follow some papers and do a professional job of it that makes the reader and reviewer in this case interested in reading your work	*Thanks for this comment and I have done the required work.* *Formatting is also revised.* [**Pages 11–18**]
Your presentation is too boring an unattractive. I suggest after writing a draft send it to some of your department colleagues (or collaborators) to read and give you some feedback because your paper will need a complete overhaul before it can be assessed for publication	*Thanks. Many of my colleagues confirm that the paper is interesting and conclusions are plausible.* *More importantly, the senior editor of the IJOEM Martins, Henrique Castro asserts*: "**The topic is clearly interesting and relevant, and the manuscript is well written**."
It is clear that you have not read enough and understood what you are doing	*I read all the articles suggested. However, this assertion contradicts the comment from* **Reviewer #2**. "**The article presents a solid survey of literature**."
Other comments: Overall, the idea s good but the implementation is poor and un-motivated. You need to also re-write the introduction telling the reader why you doing this study (what is the motivation), what is new in your work (obviously here need to explain how your work is different from others), discuss your contributions	*I have further clarified the purpose of the study by amending as follows*: **Revision/amendment**: …to explore the causes of the prolonged sluggish status of the South Pacific Stock Exchange (SPX) in Fiji in the context of market liquidity, financial liberalisation and stock market integration. [**Page 2**]
…think of some robustness tests and conduct it	*Thanks, the unit root stationarity tests, cointegrated lon-run relationship tests, and correlation tests are added.* **Revision/amendment**: Tables 5, 6, 7, 8 & 9 [**Pages 13–17**]

(continued)

(continued)

Comment	*Author response* revision/amendment [page no.]
Reviewer #2: Methods: The article presents a solid survey of literature	*Thank you, much appreciated*
It seems, however, that an important contribution neglected in the text is Kose et al. 2009. M Ayhan Kose & Eswar Prasad & Kenneth Rogoff & Shang-Jin Wei, 2009…It contains an important IMF's idea that financial globalization and openness is good only if some pre-conditions are met (which might not be the case for Fiji and should be discussed)	*I really appreciate this suggestion, thank you. I have added a para on this. I also feel that now those pre-conditions may have changed since then, and that discussion, I believe, is beyond the scope of this article. Opening up of the stock market is only a part of total financial liberalisation.* **Revision/amendment**: After an extensive theoretical and empirical study, Kose et al. (2009) conclude that there are many unanswered questions about how a country could open up financially to the world in the midst of rapidly growing empirical evidence supporting a significant role for financial globalization. [**Page 9**] *and* Similarly, Kose et al. (2009) assert that there are certain prerequisites to be fulfilled for emerging economies to open up their financial markets to the world. [**Page 23**] *Under future research* It would also be timely to revisit the preconditions of financial liberalisation for emerging markets. [**Page 25**]
The Author remarks (p. 8, lines 8–12, after Ho and Lyke, 2017) that the relationship between SMD and growth could be bi-directional (endogeneous). This should call for some more adapted econometric techniques	*Yes, I agree. However, this paper is now too long at over 9000 words. I will certainly take all of these suggestions seriously in my future research.* **Revision/amendment**: Another aspect to be explored is the bi-directional causality between financial development and economic growth in emerging economies compared to advanced economies. [**Page 25**]

(continued)

(continued)

The quantitative empirical approach is limited by the very small size of the sample. Still, some possible endogeneity issues should be at least discussed, possibly some co-integration techniques e.g. ARDL or Granger causality tests could be run. I would also recommend adding unit root tests to the table 4 as well as a correlation table between the variables included in the analysis	*Thank you for this suggestion. The ARDL procedure followed and amended/revised the whole section accordingly. The interpretation & discussion updated with new statistics (coefficients and p-values) in the text.* *I also added the normality tests, the unit root tests for non-stationarity, Dickey-Fuller tests for co-integration; long-run relationships, and correlation tests as well.* **Revision/amendment**: Tables 4, 5, 6, 7, 8 & 9 [**Pages 12–17**]
The remarks to Eqs. 2 and 3 ("The present study uses annual panel data ranging from 1996 to 2018." [p. 11, line 26] and "This regression model can be estimated using the panel data analysis methods." [p. 12, line 12]) are inaccurate. The study is just a time-series analysis (it concerns only Fiji)	*Thank you, now corrected (pages 10 and 11)*
The qualitative part of the research (interviews) is not presented in a sufficiently detailed manner. Additionally, especially given these limitations	*I feel I have discussed necessary aspects in a very concise way. The length of the paper is also a limitation for further details*
I would recommend adding some more details about SPX (I suppose that via interviews at least the missing details of Table 1 could be completed)	*Thank you, yes now the question marks in this area are removed with estimates and perceived knowledge*
Results: The empirical part is based on a very limited sample and thus the results are not fully convincing	*Thank you, Yes the sample size is small due to availability of information especially because the SPX started operations only in 1996. Now robustness is added with the unit root tests, cointegration tests and ARDL procedure, as suggested*
I do not think that Eqs. 1 and 2 should be as explicit. Moreover, in Eq. 4 the variable SAV disappears (as compared to Eq. 3)	*Yes, I agree. Equations 1 and 2 removed, and explanation incorporated with previous Eq. 3.* (Now Eq. 1. **Pages 10 & 11**)
Table 4 could also (apart from unit root tests mentioned above) contain some explicit normality tests rather than a mere remark about skewness	*Thank you. Done.* Table 4 [**Page 13**]
At page 17 (lines 38–42) the Author provides some details on the participants of the SPX. Maybe such data could be available with a higher frequency that it could constitute a better proxy for SMD than MCAP over GDP? Or at least it could be presented in a way to complete the picture presented in Figs. 2–5?	*Actually, data availability also is limited. Also MCAP over GDP is better for the purpose of comparison with other studies.* *Thank you, yes, I love to have a detailed picture on the level of participation but brokers are not willing to provide the data or they have not tabulated them. This data were extracted from the annual reports of 2018. Previous reports do not provide these data*

(continued)

(continued)

Interpretation: I am generally skeptical about the validity of results from such a small sample, but maybe after separating the determinants into macroeconomic and institutional determinants the Author could run a model with all the variables which were significant in these separate estimates? This would also constitute some kind of a robustness check (which are generally missing)	*Thank you. Running all significant variables together did not help. To increase the robustness, I added the normality tests, the unit root tests for non-stationarity, Dickey-Fuller tests for co-integration; long-run relationships, and correlation tests as well.* **Revision/amendment**: Tables 4, 5, 6, 7, 8 & 9 [**Pages 12–17**]
Interpretation is generally appropriate, I would just recommend rewording of a "one-unit increase in inflation"	*Thank you.* **Revision/amendment**: …every one percent increase in inflation percentage has resulted in a fall in SMD growth [**Page 16**]
In the result tables the last column of tables 5 and 6 should be p-value rather than "Sig". In table 5, last line, "Gross capi"… appears unexpectedly. Other comments	*Thank you. Done.* Tables 8 & 9 [**Pages 16 & 17**]
There is a repeated problem with Ho and Lyke (2017) or Ho and Iyke (2017). The spelling of the second author should be checked and corrected throughout the text (especially pages 3–5, but also afterwards)	*Thank you. Corrected.* *They are* Ho and Iyke (2017)
Again, the Author compiled a number of studies on one topic. As a further research I would recommend some kind of a meta-analysis	*Thank you. I will certainly look into this*
There is a number of typos, e.g.: Electronic Trading Flat form (instead of Platform, in Table 1, p. 5, line 25) Jordon instead of Jordan (in Table 2, p. 6, line 46 and lines 53–54)	*Thank you. Corrected*
p. 11, line 49 (in table 3): PS/AV: there is something missing after and/or p. 11, line 53, (in table 3): DMCY: there is something missing after Voice and… Fuji instead of Fiji (p. 14, line 32)	*Thank you. Corrected* **Revision/amendment**: PS/AV *Political Stability and Absence of Violence* Perceptions of the likelihood of political instability and/or Politically motivated violence
The sentence "The three brokers revealed that they deal with only a few hundred clients totalling fewer than 2,000 of which the active number of clients are also numbered." (p. 16, lines 35–39) is unclear to me and should be reformulated	*Thank you. Reformulated.* **Revision/amendment**: The three brokers revealed that they deal with only a few hundred clients totaling fewer than 2,000 but again only a very few are active, because they are mostly retail customers. [**Page 18**]

(continued)

(continued)

p. 19, line 8: "of Mauritius. 6" (is it a former footnote?). I would also recommend adding sources to Figs. 2–7 (and where is Fig. 6?)	*Thank you. Corrected. I am so sorry about these typos. The Fig. 6 had erroneously been labelled as 7. The sources were added to all the figures.* **Revision/amendment:** Source: Compiled by the author
p. 20, lines 54–57: There are two sentences starting with "On the other hand"	*Thank you. Now the second sentence starts as;* 'Furthermore…' [**Page 23**]
p. 7, line 12, there should be: Philippines (Ho and Odhiambo, 2018), Malaysia (Ho, 2019b). p. 7, line 13: there should be: (Jun et al., 2015) [Hongzhong et al. should be deleted] p. 7, line 22: there is a reference missing after 6 GCC countries	*Thank you. Corrected. Now there is only one Ho (2019). Corrected. Corrected.* (Al Samman and Jamil, 2018) *added, thank you*
p. 8, line 21–22, instead of Nyasha and Odhiambo, 2020 and Nyasha and Odhiambo, 2017, there should be Nyasha and Odhiambo 2017 and 2020	*Thank you. Corrected.* Nyasha and Odhiambo 2017 and 2020
Additionally, most of referencing styles recommends sorting references first alphabetically and then chronologically	*Thank you. Corrected*
p. 12, line 34: there should be Grimm et al., 2016 instead of Grimm, Ram, & Estabrook (or all the other references should at the first usage include all the authors instead of et al.)	*Thank you. Corrected.* *First appearance,* …determinants (Grimm, Ram, & Estabrook, 2016). *Subsequently,* (Grimm et al., 2016) [**Page 11**]
A number of entries included in the list of references are not cited in the text. This includes: Ahmed (2017), Alam and Uddin (2009) Huybens and Smith (1998 and 1999) Tsetsenzaya et al. (2019)	*All deleted, thanks*
Additionally, it seems that Ho (2019a) is just a published version of Ho (2017)	*Yes, thank you. Corrected. Now only one: Ho (2017)*
Moreover, for the articles published in journals, probably adding "available at:" is not necessary (on the other hand, the use of doi numbers is recommended)	*Yes, thank you. Corrected*
	I am sincerely grateful to you for your valuable comments and they helped me to improve this paper tremendously. I hope now the paper is in a publishable state

30.4   Reviews Leading for Rejection of a Manuscript

Reviewers' comments and authors' responses 30.6 provides comments by the editor of another prestigious magazine, AAAJ, rejecting the manuscript, which was critically reviewed by the reviewers. In light of these criticisms, this paper was revised and submitted to the Emerald Publishers' own journal, the Review of Behavioral Finance, as shown in above **Reviewers' comments and authors' responses 30.1**. Financial Battle Against Climate Change (Example 12.1: **FBACC**).

Reviewers' comments and authors' responses 30.6: Reviewers critiques and authors' responses (AAAJ).

Accounting, Auditing & Accountability Journal <onbehalfof@manuscri...   ⊕ Aug 4, 2020, 6:38 PM   ☆   ↰   ⋮
to me, ca_saliya ▾

04-Aug-2020

Dear Dr. Saliya,

I am writing regarding manuscript # **AAAJ**-05-2020-4576 entitled "Financial Battle against Climate Change – Assessing Effectiveness using a Scorecard" which you submitted to the Accounting, Auditing & Accountability Journal.

In view of the referees' reports included and attached, and having assessed the paper's prospects of ultimate acceptance for publication, I have decided against proceeding further with consideration of your paper for publication in the Accounting, Auditing & Accountability Journal.

Thank you for considering the Accounting, Auditing & Accountability Journal for the publication of your research. I hope the outcome of this specific submission will not discourage you from the submission of future manuscripts.

Best wishes,

RMIT Distinguished Professor Lee Parker
Joint Editor
Accounting, Auditing and Accountability Journal (ISI Listed)
aaaj@rmit.edu.au | tel: +61 8 8223 1092
http://www.emeraldinsight.com/**aaaj**.htm

Those reviews are given below. In these reviews, serious critiques are **bold**. The most serious criticisms that can be considered the most likely reason for rejection are marked as underlined. Also, both of these reviewers have provided conflicting reviews, with one is commenting the paper was very well written while the other reviewer was in the view that it was difficult to read. (Reviewer 2: Following the flow of arguments is hard for the reader. *Reviewer 1 contradicts this assertion by saying* '*While the paper is relatively well written*'). The positive comments are highlighted in ***bold italics***.

Review of: Financial battle against climate change—Assessing the effectiveness using a scorecard

Review conducted for: Accounting, Auditing and Accountability Journal

Reviewer 1

The stated focus of the paper is to investigate how and to what extent the Fijian sustainable banking regulations and guidelines and designed, communicated and implemented. A mixed method approach is used wherein desktop research is combined with interview and survey data, the analysis of which is performed based on a Balanced Scorecard-type framework designed by the author(s) for this purpose. Given the focus and setup of the current version of the paper, **it appears of marginal potential interest to the readership of AAAJ. The paper also needs significant revision** in several respects and some editorial attention. I outline some suggestions below that I hope will help the author(s) in developing the paper further.

Motivation/contribution

At present, **the motivation for the paper is not at all clear**. The introduction (page 2) includes discussion of what the study sets out to do, including claims that this study is the first of its kind focussing on Fiji. OK, but the fact that no other study has done what is proposed here does not constitute a strong/tight motivation for this paper. Missing for me is a clear narrative about why the paper has been written and <u>why I simply must keep reading.</u> There is the start of some comments in this regard on around line 41 of page 2, but I would encourage the author(s) to develop this further. What role could the banking sector in Fiji truly have, and why do we need to know this? Put another way, given the focus of the study, the conclusion that I would have predicted ex ante, is that the banking sector is doing some good things with respect to climate change but it is also deficient on a number of aspects.

This is entirely consistent with the experience/activity of banking sectors around the world and is also entirely consistent with what the study finds. As such, <u>what is new here?</u>

In short, much more careful consideration on the actual motivation for the paper is needed here, and I would urge the author(s) to consider whether they can build any tension in this story.

Related, **the contribution of the paper is not clear either**. OK, no known study of this type and focus has been conducted on Fiji, but why is it needed and what can we learn about Fiji and other settings? I suspect the core issue at play here is that the author(s) is/are not entirely clear on what the real contribution is … whether it is shedding light on what is happening in Fiji, or whether it is the development of a climate change based framework (FBACCS). I am not sure which of these is the best avenue to pursue. There is a mention (page 15) of other assessment frameworks that have been used prior. The framework developed here claims to build on these frameworks, so perhaps this is an opportunity to develop a clearer sense of the contribution. What are the gaps in these frameworks and why is this study needed to fill those gaps?

The role of the Balanced Scorecard (BSC) is not clear. Why is this a good framework to use in this sense? The essence of this tool is that it is 'balanced' in various ways and can enable effective performance management in complex settings where goals are multi-faceted. Why is such a tool effective here? Prior knowledge of the BSC is clearly assumed on page 5, in a discussion that is presently far too loose. For example, the BSC as originally proposed has four dimensions, but this is contingent on the setting. What is it about this setting that suggests 4 perspectives to be appropriate? Where do'prevent', 'mitigate' and 'resilience' come from? Perhaps I have missed this and if so, I apologise.

Research approach (model, execution etc.)

I am concerned with several aspects of the research approach. First, on a superficial level, I did not expect the methodology section to come straight after the introduction. Perhaps as a result, it includes some information that is probably not needed and/or which requires sharper focus (the discussion in the BSC page 5 is an example). Also, regarding the 'procedures', more clarity is required about how the different activities fit together. Take the interviews for example, were these needed to supplement the desktop research and provide the basis for generating the survey? Alternatively, there may have been another objective. Did the interviews give rise to any additional questions or perspectives relevant to the subsequent survey?

How was the survey developed? Was it piloted? If so, with whom and did this result in any refinements of questions? Further, who were the survey respondents and how were they selected? There is a couple of lines at the end of page 6, but more please. What was their sector/position and experience? Also the content of questions is not immediately clear and in this instance, I would certainly have benefited from seeing a copy of the survey. I note that various procedures were performed (page 9) to provide the reader with comfort as to reliability and internal consistency of the responses, but **I am more concerned with its broader validity**.

The presentation of the FBACCS (p. 17) is not clear to me. For greater clarity, inclusion of specific measures/targets would help and this combined with having access to the actual questions asked would enable the reader to gain greater confidence about what is being asked and measured.

I would also urge the tightening of the discussion of results. For example on page 22, why should the theoretical benchmark be 50% and even if so, why is 40% unsatisfactory? I also note that the author(s) claim the paper tells us about the effectiveness of Fiji's fight against climate change. In its current form, I do not have any degree of confidence that the paper actually, delivers on this promise/claim.

Narrative/editorial

Several aspects of the narrative require attention. In addition to the point already raised regarding the placement of the methodology discussion, **the literature review also needs work**. At present it seems that this discussion comprises several relatively disconnected sections, some of which certainly don't need their own sub-heading. A good, powerful literature review highlights only those areas relevant to the study and on which it seeks to build. I would also consider moving the literature review earlier in the study.

Related—there are a great deal of acronyms in this study. Consider whether they are all really needed and/or whether a list somewhere would help the reader.

A general edit of the document would help. *While the paper is relatively well written, there are several instances of clumsy phrases and typos that need to be addressed.*

I wish the authors well in developing this paper. I hope these comments are helpful and apologise, sincerely, if they come across as unnecessarily negative.

Reviewer 2

Manuscript ID: AAAJ-05-2020-4576

Title: Financial Battle against Climate Change-Assessing Effectiveness using a Scorecard

Overview

The paper examines how and to what extent the Fijian sustainable banking regulations and guidelines are designed, communicated, implemented, and monitored within the financial system in Fiji.

Overall, I find the research topic interesting. The background information, related conferences and news regarding the climate change issue in Fiji are all given good coverage at the introductory parts of the paper. However, all the background information is **not presented cohesively**. It is hard to connect this background information with the main research questions. While the reader is expected to know the main research question and the research contributions to the climate change issues in Fiji, the lack of cohesivity and some disconnects between the background information and research question makes it hard to understand the connections among the sections.

There are multiple acronyms in the text which in some cases are not well defined. Following the flow of arguments is **hard for the reader**. (*Reviewer 1 contradicts this assertion by saying* 'While the paper is relatively well written') Specially after the first page the reader is confused how the sections are connected. Preparing a glossary for the terms may make it easier for the reader to follow the arguments. For example, some acronyms are not even defined before being first used (e.g. FBACC in the second paragraph). Overall, given the issues I raised above, **it is hard to comprehend the key research question and Contributions**.

Here are my comments on each section of the paper:

1. Introduction

Relying on the background information on the climate change conferences and guidelines, the author(s) claim that they provide evidence on how the Fijian sustainable banking regulations are implemented. However, the limited information on Fiji banking system leaves the question of "**how climate change issues are addressed?**" **not sufficiently responded**.

By explaining the Fijian financial sector, the authors highlight the measurement of the factors that are important in the financial sector, but it is not clear why these factors are important and how they affect the financial sector in Fiji.

This section is disjointed, especially when comparison is made between the financial sector indicators and the CO2 emission. It is not clear why Table 1 is demonstrated at the end of the introduction, why the information provided is important and how they are connected to this study.

2. Methodology

Immediately after the introduction, the reader is directed to the methodology while research question has not yet well articulated. Although one of the main data collection methods is survey, **the questionnaire items are not provided in the paper**. No statistical analysis relevant to the reliability of the research methods is presented. Moreover, a detailed diagram of the balanced scorecard is illustrated in the method section drawing on Kaplan & Nolan. The author(s) could have explained their own framework and how they have adapted Kaplan and Nolan's framework in this paper, instead of discussing a general balanced scorecard framework.

2.1. Procedures

Again, in the procedure section there are details about the data collection while how the data collection procedures relate to the research question is **not thoroughly explained**.

2.2. Scoring System and 2.3. Benchmarking

The scoring system and benchmarking are defined in detail which could have been useful only if the reader knew what perspectives, critical factors, and criteria were used and how they were defined in this study.

3. Literature Review

While the reader is still struggling with the basic definitions, the literature review starts after method section. However even by reading the background **it is hard to understand what the main goal of the paper is**.

Another major disconnection is in the assessment framework, the author(s) mention(s) three frameworks are used. While, the third assessment is after the first one. It is not clear what the second assessment is.

3.6. Assessment Frameworks; Scorecards, and Indices

The assessments are not well defined in the presented assessment frameworks, scorecards, and indices. At the end of this section, the question arises as **why FBACCS is important** and how the perspectives are relevant to this study.

4. Conclusion

When the reader gets to the results and conclusion, what the key research question or contributions are still remain unclear. Furthermore, **the results are not conclusive**. Specially the survey results are all insignificant and below the acceptable thresholds

(Fig. 4). Thus, the question is how the conclusion can be a strong support for the research hypotheses (which is not even clearly mentioned in the paper) while **all the results are insignificant**.

On the one hand, in the conclusion, the author(s) conclude that Fiji banking regulations and guidelines are not properly implemented, monitored and complied with the related standards. On the other hand, in the limitation and future research section, one of the main limitations highlighted by the author(s) is lack of access to data. Thus, it questions the contributions of the paper. How the vague results can help the authorities to design and communicate the guidelines to the financial sector?

Finally, there are numerous typos and grammatical errors that require correction.

Responses to editors' comments and reviewers' critiques should be complete, polite, and with sufficient evidence.

Reference

Aurini, J. D., Heath, M., & Howells, S. (2016). *The How To of Qualitative Research*. London: Sage